Convergent Issues
in Genetics and Demography

Convergent Issues in Genetics and Demography

Editors

JULIAN ADAMS

DAVID A. LAM

ALBERT I. HERMALIN

PETER E. SMOUSE

New York   Oxford
OXFORD UNIVERSITY PRESS
1990

Oxford University Press

Oxford   New York   Toronto
Delhi   Bombay   Calcutta   Madras   Karachi
Petaling Jaya   Singapore   Hong Kong   Tokyo
Nairobi   Dar es Salaam   Cape Town
Melbourne   Auckland

and associated companies in
Berlin   Ibadan

Published by Oxford University Press, Inc.
200 Madison Avenue, New York, New York 10016

Oxford is a registered trademark of Oxford University Press

Library of Congress Cataloging-in-Publication Data
Convergent issues in genetics and demography / editors, Julian
Adams . . . [et al.].
p. cm.   Based on the conference held at the University of Michigan,
in Oct. 1988.   Includes bibliographical references.
ISBN 0-19-506287-6
1. Human population genetics—Congresses.   2. Demography—Congresses.
3. Genetic epidemiology—Congresses.   I. Adams, Julian, 1944–   .
[DNLM: 1. Genetics, Population—congresses. QH 455 C766 1988]
QH431.C765   1990   573.2′15—dc20
DNLM/DLC   for Library of Congress   89-72196

1 2 3 4 5 6 7 8 9

Printed in the United States of America
on acid-free paper

Preface

Traditionally, population research has been based in two areas; in the biological and in the social sciences. Research activities in Biology departments have been concentrated in the areas of population genetics and population ecology, while Sociology and Economics departments are the traditional homes of demographers. Population geneticists and population ecologists have long recognized the importance of demographic theories and research, and demographers are similarly well aware of the importance of biological parameters in determining and predicting population change. In spite of this manifest common interest, the fields of population biology and demography travel along different paths. As an example, demographers are currently interested in the concept of unobserved heterogeneity; namely, that unidentified or unidentifiable variables may contribute to different vital rates or other outcomes among different sectors of a population. On the other hand the essence of research in population genetics is the understanding of the nature and maintenance of variation in populations. Thus, both disciplines have a fundamental interest in the same phenomenon, but because they exist independently from one another, their approaches can be quite different.

In 1987, Al Hermalin, at that time director of the Population Studies Center at the University of Michigan, organized a series of meetings to encourage interactions between population biologists, demographers, and other interested groups. The stimulus for these meetings was a grant awarded to the University of Michigan, by Michael Teitelbaum of the Alfred P. Sloan Foundation, to foster cross-disciplinary connections between workers involved in population-based research. During those meetings, the seeds of a conference, devoted to a discussion of the issues and questions common to the disciplines of population genetics and demography, were planted and allowed to germinate. This volume represents the fruits of those earlier meetings. The chapters were derived from a series of papers presented at an international conference, held at the University of Michigan, Ann Arbor, on October 7-8 1988, entitled "Convergent Questions in Genetics and Demography". The conference brought together geneticists and demographers from the U.S.A., Europe and Latin America, who have contributed significantly to their respective fields in recent years. This conference in many ways continued the dialogue which began at a series of four conferences held during the sixties in Princeton and at Asilomar, California, as well as a more recent conference held in 1981 in Boston.

As is usually the case, the conference and this volume would not have been possible without the help of a number of different individuals and financial assistance from a number of different sources. Lora Myers, with the assistance of Ruth Crankshaw and other members of the staff at the Population Studies Center at the University of Michigan, were outstanding in coordinating and organizing the myriad tasks which made this conference a success, not the least of which included keeping the four of us out of debtors prison. Judy Mullin of the Population Studies Center and Sandie Schaerer of the Biology department were coolly efficient in typing the manuscripts and converting them into $\TeX$. Carol Crawford, of the Population Studies Center was an invaluable source of assistance in deciphering some of the arcane details and idiosyncrasies of $\TeX$. Thanks are also due to Bill Curtis and Susan Hannan of Oxford University Press, who responded patiently and creatively to our requests, and to our sometimes naive questions about the publishing process.

In addition to the invited speakers, approximately 80 individuals attended the conference, and we are grateful to all the participants for the stimulating and spirited discussions which followed almost every talk. We are especially appreciative of the authors of the chapters of this volume who modified their contributions (sometimes substantially) in the light of these discussions. We also extend our thanks to our colleagues throughout the world, unfortunately too numerous to list here, who critically reviewed the manuscripts, and played a significant role in strengthening the contribution of this book to the fields of genetics and demography.

Major financial support for this conference was provided by a grant from the Alfred P. Sloan Foundation, with help also from the Departments of Biology and Economics, the Population Studies Center, the College of Literature, Science and the Arts, the Office of the Vice-President for Research, the H.H. Rackham Graduate School and the N.I.H. Graduate Training Program in Genetics (S-T32-GM07544) of the University of Michigan.

The volume is organized into four sections, each representing major points of congruence between the disciplines of demography and genetics. The first chapter of each section provides an overview of the main issues discussed in the subsequent chapters of the section. The first section is concerned with the use of historical information, specifically pedigree and genealogical data in genetics and demography, while the second section deals with the treatment and analysis of variation in the two fields. In the third section a third discipline, epidemiology provides the common ground where the disciplines of genetics and demography converge. Finally, the fourth section considers some issues in genetics and demography which have long attracted the attention of workers in both fields.

<table>
<tr><td>*Ann Arbor, MI*</td><td>J.A.</td></tr>
<tr><td>*Rio de Janeiro, Brazil*</td><td>D.L</td></tr>
<tr><td>*Palo Alto, CA*</td><td>A.I.H.</td></tr>
<tr><td>*New Brunswick, NJ*</td><td>P.E.S.</td></tr>
<tr><td>*April 1990*</td><td></td></tr>
</table>

Contents

II. Heterogeneity, Phenotypic Variation, and Frailty

III. Genetics, Demography, and Epidemiology

IV. Persistent Issues in Genetics and Demography

Contributors

Julian Adams   Department of Biology, University of Michigan, Ann Arbor, MI   48109-1048

Lee L. Bean   The Middle East Center, University of Utah, Salt Lake City, UT   84112

Eduardo E. Castilla   Departamento de Genética, Fundaçao Oswaldo Cruz, Caixa Postal 926, Avenida Brasil, Rio de Janeiro   20.000, Brazil

Ansley J. Coale   Office of Population Research, Princeton University, 21 Prospect Avenue, Princeton, NJ   08540

Freddy Bugge Christiansen   Department of Ecology and Genetics, University of Åarhus, Ny Munkegade, DK-8000, Åarhus C., Denmark

Warren J. Ewens   Department of Biology, University of Pennsylvania, Philadelphia, PA   19104, and Department of Mathematics, Monash University, Clayton, Victoria 3186, Australia

Albert I. Hermalin   Population Studies Center, University of Michigan, 1225 South University Avenue, Ann Arbor, MI   48104-2590

Jan M. Hoem   Demography Section, Stockholm University, S-106 91 Stockholm, Sweden, and Center for Demography and Ecology, University of Wisconsin, Social Science Building, Madison, WI   53706

David A. Lam   Population Studies Center, University of Michigan, 1225 South University Avenue, Ann Arbor, MI   48104-2590

Paul W. Leslie   Department of Anthropology, SUNY-Binghamton, Binghamton, NY   13901

Massimo Livi-Bacci   Dipartimento Statistico, Università di Firenze, 18 Via Baldesi, 50131   Firenze, Italy

Jurg Ott   Department of Psychiatry, Columbia University, 722 W. 168th St., New York, NY   10032

Alberto Piazza   Dipartimento di Genetica, Biologia, e Chimica Medica, Università di Torino, Via Santena 19, Torino 10126, Italy

Robert A. Pollak   Department of Economics, University of Pennsylvania, 3718 Locust Walk, Philadelphia, PA   19174, and Department of Economics DK-30, University of Washington, Seattle, WA   98195

Samuel H. Preston   Population Studies Center, University of Pennsylvania, 3718 Locust Walk, Philadelphia, PA   19174

Germán Rodríguez   Office of Population Research, Princeton University, 21 Prospect Avenue, Princeton, NJ   08540

William J. Schull   Center for Demographic and Population Genetics, Graduate School of Biomedical Sciences, University of Texas Health Science Center, P.O. Box 20334, Houston, TX   77225

Burton Singer   Department of Biostatistics, Yale University, 60 College Street, New Haven, CT   06510

Peter E. Smouse   Department of Human Genetics, University of Michigan, Ann Arbor, and Center for Theoretical and Applied Genetics, Rutgers University, New Brunswick, NJ   08903

Michael S. Teitelbaum   Alfred P. Sloan Foundation, 630 Fifth Avenue, New York, NY   10111

Elizabeth A. Thompson   Department of Statistics, GN-22, University of Washington, Seattle, WA   98195

James Trussell   Office of Population Research, Princeton University, 21 Prospect Avenue, Princeton, NJ   08540

James W. Vaupel   Center for Population Analysis and Policy, Humphrey Institute, 301 19th Avenue South, Minneapolis, MN   55455

Maxine Weinstein   Department of Demography, Georgetown University, Washington, DC   20057

Kenneth M. Weiss   Department of Anthropology, 409 Carpenter Building, Pennsylvania State University, University Park, PA   16802

James W. Wood   Department of Anthropology, 409 Carpenter Building, Pennsylvania State University, University Park, PA   16802

I

GENETICS, DEMOGRAPHY, AND HISTORICAL INFORMATION

1

Introduction:
Genetics and Demography
and Historical Information

JULIAN ADAMS

In the opening sentences of Chapter 2, Livi-Bacci asserts that the most important question in demography is, "Why does a population change in size over time?" Yet it would be just as reasonable to suggest that this is one of the most important questions in population genetics. Biological fitness, the basic parameter of population genetics, which describes the relative ability of a group of individuals with a common characteristic to survive and transmit its genes to the next generation, is probably best measured by the intrinsic rate of natural increase (Fisher, 1958, Cavalli-Sforza and Bodmer, 1971, and see Chapter 19 by Christiansen). Even though the foundations of the two fields are different there are some noteworthy similarities. To introduce the relationship between the two disciplines it is instructive to consider these similarities and differences in some detail.

The discipline of population genetics is founded on the two cornerstones of Darwinian theory and Mendelian genetics. The first is best stated in Darwin's own words (Darwin, 1859):

> As many more individuals of each species are born that can survive, and as consequently there is a frequently recurring struggle for existence, it follows that any being, if it vary somewhat slightly in any manner profitable to itself, under the complex and sometimes varying conditions of life, will have a better chance of surviving, and thus be *naturally selected*. From the strong principle of inheritance, any selected variety will tend to propagate its new and modified form.

The focus of contemporary population genetics on the differences between individuals in their ability to survive and reproduce reflects Darwin's overwhelming influence.

The axioms of Mendelian inheritance define the second cornerstone, the most important of which, for our purposes are, (1) that inheritance is particulate (that is determined by discrete units or *genes*), and (2) that male and female parents contribute equal amounts of genetic information to their offspring.

In formulating his theory of evolution, Darwin was influenced strongly by the writings of Malthus, a founder of the field of demography. Malthus, whose

3

contribution to Darwin's theory is that populations grow exponentially, but that resources do not, was more concerned with the relationship between social factors and the rate of natural increase of populations. For Malthus, the intrinsic rate of increase of a population was a variable that was dependent on the ability by which the population modified its environment. Such a concept is certainly not alien to the field of population genetics, where it is usually defined as (emphasizing naturally the *genetic* differences among individuals) a genotype/environment interaction. However, Malthus' belief that there were important feedback effects of population growth rate and environmental modification (expressed in economic terms as subsistence-employment) is quintessentially a demographic concept, though it does appear (albeit in disguised form) in population genetic models of density-dependent selection.

The emphasis on socioeconomic factors stemming from Malthus' original proposition that there is a fundamental relationship between population growth and subsistence or employment, has naturally led to the traditional home for demographers being Economics and Sociology departments. On the other hand, Darwinian theory emphasizes genetic differences between individuals and populations, which has led to the traditional home for population geneticists being biology departments. These differing emphases on socioeconomic (presumably mostly nongenetic) factors on the one hand and genetic factors on the other is probably responsible more than anything else for the independent development of the two fields.

POPULATION HISTORY AND GENEALOGIES

Although much is made of the predictive nature of population sciences, a central goal of both population genetics and demography is the explanation of past events in terms of either social or genetic laws governing population change. In this respect, a common denominator of both fields is the use of pedigree and genealogical information. In the first section of this volume, concerned with the uses of genealogies in population genetics and demography, a common theme emerges from most of the papers; that is, an interest in the roles of migration and inbreeding in promoting population change. Although genetic differences between individuals and their effect on the growth and survival of populations represent the focus of population genetic theory and experiment, from the viewpoint of population genetics these two forces (migration and inbreeding) are important, as they can modulate the rate of genetic change due to selection. From a demographic viewpoint, interest in migration stems from its direct effect on population size. In addition, both migration and inbreeding can indirectly affect the fertility and mortality of a population, and thus the overall rate of growth. Demographers, much more than geneticists are also concerned with the *causes* of migration, generally socioeconomic in origin.

MIGRATION AS A DEMOGRAPHIC AND GENETIC FORCE

Immigration has the primary and simple effect of increasing population size, though the effect on the population growth rate cannot always be easily predicted, as migrants may not retain the same characteristics or cultural practises in their new environment (Bogin, 1988; Roberts, 1988). For some groups, such as the founders of Québec (see Chapter 2 by Livi-Bacci), or the Amish (Cross and McKusick, 1970) increased fertility may follow migration, whereas others may experience decreased fertility (Lee and Faber, 1984; Pool *et al.*, 1987). On the other hand, emigration may have a negative effect on population growth rate, beyond a simple reduction in numbers, due to separation of husbands and wives during their childbearing years, creation of an imbalance in the sex ratio, and other factors (Van de Walle, 1975; Leslie, 1985). The interest of geneticists in migration stems from its role in promoting or reducing *genetic* differentiation between populations. By breaking down isolation between populations, migration will result in a reduction in the frequency of consanguineous matings, and consequently of inbreeding (see below), as shown by Leslie in Chapter 5. Thus, in general, migration and inbreeding can be considered to be countervailing population pressures. Although predisposition to migrate may not be genetically determined, if migrants do not form a random sample of the population, then the role of migration in population change may become amplified. Thus, a tendency for less inbred individuals to emigrate noted by Castilla and Adams (Chapter 4) resulted in a unexpectedly large increase in inbreeding levels in the population of origin.

It is generally known that surnames can convey historical and demographic information. Chatwin's (1977) observation that, "The history of Buenos Aires is written in its telephone directory," expresses this succinctly. Two chapters in the first section of this volume, by Castilla and Adams (Chapter 4) and by Piazza (Chapter 6), make use of surname data to infer migration rates and population histories. Geneticists have known for some time that the inheritance of surnames in most societies mimics a particular category of genes (those located on the Y chromosome), and have taken advantage of this to generate estimators of migration rates and inbreeding (see also Chapter 4 by Castilla and Adams) based on surname data (Gottlieb, 1983). Chapter 6, by Piazza, deals with immigration to provinces of Italy whereas the data of Castilla and Adams (Chapter 4) analyzes migration in Argentina on a much finer scale, between a cluster of five rural Argentinean villages. In both cases inferences concerning migration from genetic data correlate well with those gained from demographic data. A similar correlation between genetic and demographic data is seen for inter-island migration in the Faroe Islands (see Chapter 3 by Thompson).

The use of surnames and genetic data is particularly intriguing, as such data are freely available for many societies and populations and may substitute for demographic data, which is much more costly and at times impossible to obtain. Genetic data become particularly important in the analysis of large-scale migration events occurring on longer time scales and in eras for which demographic data are virtually nonexistent. The importance of a Celtic contribution to the colonization of Iceland, previously thought to be colonized by the Vikings with Irish slaves, was elucidated with genetic data (see Chapter 3 by Thompson). Genetic data have also been invaluable in tracing the origins of the population of Italy (see Chapter 6 by Piazza).

POPULATION SIZE, MATE CHOICE AND INBREEDING

Population size is an issue that is common to both population genetics and demography, and presents both fields with the same dilemma. Much of human society is organized into small groups or isolates, whether defined by geographical, religious or social barriers. Large, mobile, urban populations have arisen only recently, most since the industrial revolution. Thus, in understanding the demography and genetics of historical populations, populations of small size are of particular interest. On the other hand, both geneticists and demographers are well aware that to achieve statistical power large sample sizes are necessary, and these will only be attainable with large populations. For some questions, particularly in the field of demography, the structuring of a population into isolates may not pose a problem, as statistics such as fertility rates may be calculated without regard to population structure. There are, however, some special characteristics of small populations relating to an altered mating structure that cannot be inferred from a consideration of larger populations, and that may profoundly affect the demography and genetics of such populations.

The number of mates available in small populations necessarily becomes restricted. For example, Dyke (1971) estimates that the number of potential mates available to single males of mating age in the Northside French community of St. Thomas, U.S. Virgin Islands, was only 51, whereas the population size at that time was 661 individuals (see Leslie, 1985 for other examples). This, coupled with incest taboos and proscriptions against consanguineous marriages, will contribute to a reduction in fertility (Leslie, 1985) and in extreme cases will result in population extinction (Hammel, McDaniel, and Wachter, 1979). Of course it is also likely that individuals faced with a restricted choice of mates will try to increase the pool by emigration or by ignoring previous norms concerning eligible mates (Leslie, 1985). Frequently in small populations, social and religious conventions which discourage marriages between relatives, will be relaxed or ignored, and there will be an increase in the frequency of consanguineous marriages. In this volume, Leslie (Chapter 5) examines a number of factors that may influence the frequency of consanguineous marriages, including population growth, migration, and the age correlation between spouses (first shown by Hajnal, 1963). An extension of Hajnal's model allows assessment of the impact of age correlation between spouses on the probabilities of various kinds of consanguineous matings (Leslie, 1983). Leslie shows that the variance as well as the mean of the distribution of the age difference between mates will significantly affect inbreeding in a population. The age correlation between mates will also render certain categories of consanguineous matings more important in the determination of the total level of inbreeding, as seen in the data of Castilla and Adams (Chapter 4).

The interest of population geneticists in small populations, however, goes beyond a consideration of the relative frequencies of consanguineous marriages. Consanguinity that results in inbreeding has two important consequences in population genetics. Firstly, it can effectively further subdivide a population into a series of smaller subpopulations (though as mentioned above, migration can reverse that process). For example, a population of n individuals which exclusively practices mating between full-sibs (brother and sister) effectively consists of $n/2$ subpopulations of size 2. Similarly, a population which exclusively practices octuple third cousin mating becomes sub-divided into $n/16$ populations of size 16 (Li, 1955). In practice, of course, such regular patterns of mating never occur, and it is impossible to tease out and define the subpopulations in a real population where a variety of degrees of consanguineous matings occur irregularly. It is nevertheless true that all such populations behave *as if they were* divided into a series of small populations. Thus, the *effective size* of the population may be considerably smaller than the *actual* size. The concept of effective size of a population, introduced originally by Wright (1931) is a concept which has considerable heuristic value, as well as utility, in population genetics theory. It can be defined simply as the size of an ideal simple random mating population with no structure, which is equivalent to the actual "nonideal" population. Chapter 20, by Ewens, describes this metric in more detail and shows how this value may be calculated under different assumptions. The difference between the actual population size and the effective population size can be quite substantial. For example, Emigh and Pollak (1979), using life table data from the 1970 U.S. census, estimated the effective population size to be 41% of the actual population size (see also Pollak, 1980). Although the concept of effective population size is not well known in demography, it has the potential to be a useful construct in understanding the behavior of populations over time.

Small population size is also of interest as an important feature of models of evolutionary change. Random sampling events that occur during the transitions from one generation to the next in small populations can result in genetic change, or *genetic drift*. Population differentiation occurring by genetic drift is the basis of the so-called neutral theory of evolution (Kimura, 1983). In addition, the shifting balance theory of evolution, advanced by Sewall Wright (Wright, 1982) postulates that maximum rates of evolutionary change occur in a series of small partially isolated subpopulations within which genetic drift and selective change operate in consort.

Finally, inbreeding is considered important, as it increases the frequency of rare recessive phenotypes (see Chapters 3 and 4 by Thompson and Castilla and Adams) that are often deleterious (see Chapter 16 by Ott). To the extent that fertility is genetically controlled, inbreeding may have demographic consequences in reducing fertility levels. There is an abundant literature concerning the relation between inbreeding and morbidity and mortality (Freire-Maia and Elisbão, 1984) as well as fertility (Schull and Neel, 1965; Schull and Neel, 1972; Hann, 1985). An extension of this reasoning has resulted in the concept of a "minimum viable population size" in animal populations (see Chapter 20 by Ewens), a concept important in conservation programs for endangered species.

It can therefore be seen that inbreeding, although ostensibly a uniquely genetic factor affecting population change, also can be considered to influence the demographic structure of a population (see Bonné, 1963; Bonné-Tamir, 1980; Brennan, 1981; Hajnal, 1953, for some examples).

DIFFERING PERSPECTIVES OF GENETICS AND DEMOGRAPHY

In spite of the fact that many of the basic questions and issues are common to both disciplines, it is unfortunately true that the two fields have developed along parallel but independent paths. It is reasonable to ask why this has happened. In order to examine this question, and to propose ways in which the fields may become more closely allied, it is perhaps instructive to consider the ways in which the fields differ.

The concept of the population

The concept of the population lies at the heart of both fields, and the ways in which the two fields define a population differ significantly. For demographers, a population can be defined simply as an aggregation of organisms occupying a given area, while the population genetic definition of a population, a community of *randomly mating* individuals, is much narrower. In a few cases, such as small island populations, the "demographic" and the "genetic" population may be identical or insignificantly different. However, typically the two definitions result in widely divergent specifications of a population. Thus, the population of the U.S. may be considered to be a unit demographically, but it clearly cannot be construed as a single random-mating genetic population, as important determinants of mating probability will be spatial isolation as well socioeconomic status and racial identification. Where race is an issue, a single demographic (albeit stratified) population may be regarded as two separate genetic populations even though they may be located in the same area (Adams and Ward, 1973). The definition of a genetic population for aggregations of individuals that are distributed continuously over a large area is more problematic, and in such situations a statistical construct, the *neighborhood* (Wright, 1969), has been of value. The preoccupation of geneticists with mating and the differential transmission of genes, and of demographers with socioeconomic factors is reflected in these two differing concepts of a population.

Family reconstitution and pedigrees

Although both demographers and population geneticists are concerned with temporal changes in populations, for geneticists the most important temporal variable is generations, and the most important dependent variable, the differential transmission of genes. For this reason, the value of multi-generational pedigrees becomes particularly high. If, however, the focus is on socioeconomic factors, many questions can be answered using parent-offspring data (single generation data) or by cross-sectional analysis of populations at one time-point. Thus, family reconstitution, an important methodology in historical demography (see Chapter 2 by Livi-Bacci; Henry, 1980; Willigan and Lynch, 1982) typically only covers single-generation transitions. On the other hand, pedigrees reconstructed by geneticists (see Chapters 3 and 4 by Thompson, and Castilla and Adams) have a depth of multiple generations. The difference reflects the nature of the fields; the informational value of multigenerational pedigrees is much greater in genetics than in demography. It is important to point out that the collection of multigenerational data involves extensive record linking and is often very labor intensive, even for pre-Industrial (Knodel and Shorter, 1976) or rural populations (see Chapter 3 by Thompson, and Chapter 4 by Castilla and Adams). The decision by demographers to focus on single generational transitions has no doubt been made in full awareness of the difficulty and time involved to construct multi-generational genealogies. Recent endeavors by demographers, however, point to a greater emphasis being placed on multi-generational family reconstitutions. In Chapter 2, Livi-Bacci describes the initiative to reconstitute the entire population of Québec, and in Chapter 15, Bean describes the recent construction of multi-generational genealogies in the Utah Population Database

(UPDB). As Bean points out, the UPDB has been particularly important to both genetic and demographic research.

Fertility

Both demographers and geneticists are fundamentally concerned with fertility levels in populations, and variation in those levels. Yet the perspectives of the two disciplines have generated differing approaches. From the point of view of the geneticist, evolutionary change in the frequency or expression of traits that are under some degree of genetic control will be determined equally by male fertility levels as well as female fertility levels. In contrast, demographers, concerned principally with population growth, consider primarily female-dominant fertility schedules. As Pollak points out in Chapter 21, "Males play virtually no role in classical population theory." Consequently it is not surprising that Keyfitz and Flieger (1968) were able to find male-dominant fertility figures for only seven countries, whereas equivalent female-dominant figures were available for over 400 countries. Pollak's model (see Chapter 21), which generalizes classical stable population theory to incorporate both female and male fertility rates, is an important step towards the convergence of genetic and demographic perspectives on fertility.

Furthermore, demographers, concerned with the *potential* for population growth, are often interested in the deviations from a theoretical maximum fertility level. The concept of "natural fertility," developed initially by Henry (1972), which is often defined as the fertility of couples that do not practise any form of contraception and make no parity-dependent effort to limit births, is basic to much of demography (see Knodel, 1983 and Chapter 18, by Hermalin and Coale, for a more extensive discussion). Consequently, an understanding of the determinants of natural fertility, such as the risk of fetal loss, coital frequency, etc., are important areas of research in demography. In this volume, Wood and Weinstein (Chapter 11), discuss one aspect of this topic, the effect of fetal loss, and suggest that variation among women in the risk of fetal loss is important in explaining heterogeneity in fecundability (the monthly probability of a fertile conception) rates.

On the other hand, geneticists are primarily interested in the transmission of genetic information from one generation to the next, and so in the offspring that survive and reproduce. The number of suriviving and reproducing offspring may be termed the *realized* fertility. To choose an example from the animal world, a female codfish spawns on the average one million eggs, yet in a stable population, only one of these million will hatch and survive to maturity. We may say that a demographer will be interested in the ability of the codfish to be so fecund, whereas the geneticist is much more interested in the one that survives to maturity, whether it is a random sample of the one million, or whether it possesses some particular characteristics that distinguishes it from the rest (Fisher, 1958). Knowledge of the population composition at different stages, e.g., birth, maturity, and adulthood, are naturally important to determine when natural selection may be acting. Christiansen (Chapter 19) shows how this may be accomplished in two marine species. In many cases such detailed information will be lacking, and only the data on the segment of the population that is reproducing will usually be available. Although much of population genetic theory and experiment is built around the assumption that these are the only data available, historical-demographic data are also frequently restricted to the reproductive sector of the population only. Parish records often fail to record stillbirths and neonatal deaths, and immigrants into the population will only be recorded should they marry and/or produce issue which are registered within the community (see Chapter 2 by Livi-Bacci). For these historical demographic investigations, the analytical tools of population genetics, which assume complete information is unavailable, may prove to be valuable.

Phenotypic variation and heterogeneity

Perhaps the most striking difference in approach and paradoxically the best hope for a convergence of the two fields can be seen in the way in which the fields view within-populational variation. For demographers, within-populational variation, or heterogeneity, has in the past been considered a "nuisance" variable, to be factored out and forgotten. On the other hand, intrapopulational variation, or *phenotypic variation*, is the essence of population genetics and the Darwinian theory of evolution. Much of population genetics is concerned with the detection, measurement, and partitioning of this variation among various genetic and nongenetic etiologies. Although the thrust and emphasis of the field has been towards the genetic components, many of the questions statistical tools are relevant, and transferable across disciplines. This theme is discussed in more detail in chapter 7 by Lam and Smouse and in the subsequent chapters in Part II.

GENETIC AND DEMOGRAPHIC CONVERGENCE

That the two sciences of genetics and demography have different philosophical orientations is inescapably true, and one that is unlikely to be modified in time. Future progress in population genetics will be governed by the biological and genetic laws governing the transmission of hereditary information from one generation to another. Similarly, it is unlikely that the socio-economic orientation of demography will change in any radical way in the foreseeable future.

However, convergence of the two disciplines can prove fruitful in three important and common spheres of interest.

Determination of fertility and mortality

Understanding of the factors governing human fertility and mortality is the central goal of demography, and as our understanding of the significant variables grows, socioeconomic factors will continue to be important. Yet few would argue that we can ignore biological factors. The field of reproductive endocrinology is clearly central to our (biological) understanding of fertility; it is also true that fertility is a variable which is at least under partial genetic control. Similarly, mortality rates can also be influenced by biological and genetic factors, either through the biology and genetics of the human population, or through the population of a infectious disease agent or its vectors. The discipline of epidemiology lies at the intersection of these issues, and Chapter 13 by Smouse and Teitelbaum, and the subsequent chapters in Part III, explore the various ramifications of the interaction of genetics and demography with epidemiology.

Methodology

Whatever the philosophical orientation of the two disciplines may be, their focus of interest is the same; the population. It is thus inevitable that the statistical, observational, and theoretical tools of the two disciplines will be similar, even though the way in which they may be used is different. In the past, path analysis (Li, 1975), developed by Wright (1921, 1934) to determine dependencies amongst genetic variables, has proved to be useful in the analysis of interactions between socioeconomic variables (Blau and Duncan, 1967; Li, 1975; Hauser and Featherman, 1977). Similarly, theories developed for the analysis of genetic evolutionary change have proved to be useful in understanding cultural evolution (Cavalli-Sforza and Feldman, 1981).

Data

In a similar fashion, data gathered by one field may be used by another to answer different questions. Integration of a set of possibly disparate specific aims, resulting in a unified program of data collection, not only makes economic sense, but can also bring about the synergistic development of the two disciplines. The Utah (see Bean, Chapter 15) and French Canadian (see Chapter 2 by Livi-Bacci) population databases serve as excellent examples.

It is the hope that within the context of the three broad foci listed above, this volume will illustrate the intersections of the two fields rather than highlight their differences.

ACKNOWLEDGMENTS

I thank Al Hermalin, Paul Leslie, Frank Rosenzweig, and Peter Smouse for helpful criticism. This work was supported in part by grants from the H.H. Rackham School of Graduate Studies, University of Michigan.

REFERENCES

Adams, J., and R.H. Ward, 1973   Admixture studies and the detection of selection. Science 180: 1137-1143.

Blau, P.M., and O.D. Duncan, 1967   *The American Occupational Structure.* John Wiley, New York.

Bogin, B., 1988   Rural-to-urban migration, pp. 90-129 in *Biological Aspects of Human Migration*, edited by C. G. N. Mascie-Taylor and G. W. Lasker. Cambridge University Press.

Bonné, B., 1963   The Samaritans: A demographic study. Hum. Biol. 35: 61-89.

Bonné-Tamir, B., 1980.   The Samaritans. A living ancient isolate, pp 27-41 in *Population Structure and Genetic Disorders*, edited by A. W. Eriksson, H. R. Forsius, H. R. Nevanlinna, P. L. Workman and R. K. Norio, Academic Press, New York.

Brennan, E. R., 1981   Kinship, demographic, social and geographic characteristics of mate choice in Sanday, Orkney Islands. Am. J. phys. Anthropol. 55: 129-138.

Cavalli-Sforza, L. L., and W. F. Bodmer, 1971   *The Genetics of Human Populations.* W. H. Freeman and Co., San Francisco.

Cavalli-Sforza, L. L. and M. W. Feldman, 1981   *Cultural Transmission and Evolution: A Quantitative Approach.* Princeton University Press, Princeton, NJ.

Chatwin, B., 1977   *In Patagonia.* Summit Books, New York.

Cross, H. E., and V. A. McKusick, 1970   Amish demography. Soc. Biol. 17: 83-101.

Darwin, C., 1859   *On the Origin of Species by Means of Natural Selection.* John Murray, Ltd., London.

Dyke, B., 1971   Potential mates in a small human population. Soc. Biol. 18: 28-39.

Emigh, T. H., and E. Pollak, 1979   Fixation probabilities and effective population numbers in diploid populations with overlapping generations. Theoret. Pop. Biol. 15: 86-107.

Fisher, R. A., 1958   *The Genetical Theory of Natural Selection*, 2nd. revised Edition. Dover, New York.

Freire-Maia, N., and T. Elisbão, 1984.   Inbreeding effect on morbidity. Am. J. Med. Genet. 18: 381-400.

Hammel, E. A., C. K. McDaniel, and K. W. Wachter, 1979   Demographic consequences of incest tabus: a microsimulation analysis. Science 205: 972-977.

Hajnal, J., 1953   Ages at marriage and proportions marrying. Pop. Studies 7: 111-136.

Hajnal, J., 1963   Concepts of random mating and the frequency of consanguineous marriages. Proc. R. Soc. London B 159: 125-177.

Hann, K. L., 1985   Inbreeding and fertility in a south Indian population. Ann. Hum. Biol. 12: 267-274.

Hauser, R. M. and D. L. Featherman, 1977   *The Process of Stratification: Trends and Analyses.* Academic Press, New York.

Henry, L., 1972   *On the Measurement of Human Fertility: Selected Writings.* Elsevier, Amsterdam.

Henry, L., 1980   *Techniques d'Analyse en Démographie Historique.* Institut National d'Études Démographiqes.

Keyfitz, N., and W. Flieger, 1968   *Population: Facts and Methods of Demography.* W. H. Freeman and Co., San Francisco.

Kimura, M., 1983   *The Neutral Theory of Molecular Evolution.* Cambridge University Press, Cambridge.

Knodel, J., 1983   Natural fertility: age patterns, levels, and trends, pp. 61-102 in *Determinants of Fertility in Developing Countries*, edited by R. A. Bulatao and R. D. Lee. Academic Press, New York.

Knodel, J., and R. Shorter, 1976   The reliability of family reconstruction data in a German village. Annales de Démographie Historique 1976: 115-154.

Lee, B. S., and S.C. Faber, 1984   Fertility adaptation by rural-urban migrants in developing countries. Pop. Studies 38: 141-155.

Leslie, P. W., 1983   Age correlation between mates and average consanguinity in age-structured human populations. Am. J. Hum. Genet. 35: 962-977.

Leslie, P. W., 1985   Potential mates analysis and the study of human population structure. Yearbook Phys. Anthropol. 28: 53-78.

Li, C. C., 1955   *Population Genetics.* University of Chicago Press.

Li, C. C., 1975   *Path Analysis - A Primer.* Boxwood Press, Pacific Grove, California.

Pollak, E., 1980   Effective population numbers and mean times to extinction in dioecious populations with overlapping generations. Math. Biosci. 52:1-25.

Pool, I., J. E. Sceats, A. Hooper, J. Huntsman, E. Plummer, and I. Prior, 1987   Social change, migration and pregnancy intervals. J. Biosoc. Sci. 19: 1-15.

Roberts, D.F., 1988   Migration in the recent past: societies with records, pp. 41-69 in *Biological Aspects of Human Migration*, edited by C. G. N. Mascie-Taylor and G. W. Lasker. Cambridge University Press, Cambridge.

Schull, W. J., and J. V. Neel, 1965   *The Effects of Inbreeding on Japanese Children.* Harper and Row, New York.

Schull, W. J., and J. V. Neel, 1972   The effects of parental consanguinity and inbreeding in Hirado, Japan. V. Summary and interpretation. Am. J. Hum. Genet. 24: 425-453.

Van de Walle, F., 1975   Migration and fertility in Ticino. Pop. Studies 29: 447-462.

Willigan, J. D. and K. A. Lynch, 1982   *Sources and Methods of Historical Demography.* Academic Press, New York.

Wright, S., 1921   Correlation and causation. J. Agr. Res. 20: 557-585.

Wright, S., 1931   Evolution in Mendelian populations. Genetics 16: 97-159.

Wright, S., 1934   The method of path coefficients. Ann. Math. Stat. 5: 161-215.

Wright, S., 1969   *Evolution and the Genetics of Populations. Volume II. The Theory of Gene Frequencies.* University of Chicago Press.

Wright, S., 1982   Character change, speciation and the higher taxa. Evolution 36: 427-443.

2

Macro versus Micro

MASSIMO LIVI-BACCI

One of the main questions, it is tempting to say the main question, that demography tries to answer is an extremely simple one. Why does a population change in size over time? Why does mankind number five billion people, and not five million, as in the late Paleolithic, or five trillion, a figure that would be achieved in less than four centuries if the current rate of increase were to remain unchecked? And with this basic question comes another one, as a necessary corollary: why have some populations been more successful than others, multiplying and peopling vast territories? Why have the aborigines of Tasmania disappeared, while a few thousand French men and women are the founders of most of the six million French-speaking Canadians of today? (Roth, 1899; Charbonneau *et al.*, 1987). In the answer to these questions is, in a nutshell, the essence of the discipline of demography and the contribution that historical demography gives to population studies.

The use of historical material has advanced a great deal during the last decades, in terms of the analytical and methodological skills employed, of the mass of data and information treated, and of the output of results. In order not to lose track of the essential issues, some simplification is necessary here. Roughly until the middle of this century, investigators interested in the history of population were divided principally into two categories of scholars. The first category was composed mainly of those who felt it necessary to build a quantitative basis for the comprehension of past history. Geography, climate, natural resources, agriculture, population, etc. were considered important chapters of the ideal book describing the frame in which historical events took place. Building the frame was as useful, although certainly not as essential, for grasping history as the frame built by the artisan for the artist's painting. Evaluating the sources and the documents, assembling facts and data, and giving them some simple quantitative order were the essential tasks of this abstract type of scholar. Julius Beloch's painstaking research on ancient populations or on medieval and modern Italy is a distinguished case (Beloch, 1937, 1939, 1961). In the second category fell another type of scholar more interesting for those who saw population as an essential component of society. These scholars had a concept of society, or of some important aspect of society, of which population is an essential part. Malthus is the obvious leader of this category of scholars, who generated ideas and hypotheses concerning the functional interrelations between population and society.

15

Although these categories may be somewhat contrived, as many scholars spanned the two approaches or went beyond them, neither group was very much interested in analyzing the "demographic system" *per se*. It is only during the last four or five decades that a distinct approach to historical demography has rapidly developed. Not only it is important to measure population change (in size, structure, distribution), but it is also important to know how change comes about; this implies not only the reconstruction of the components of change, but also the decomposition of those components (birth, death, migration) into their "proximate" or "intermediate" determinants and beyond (Bongaarts, 1983). The proximate or intermediate determinants are the delimited set of social and biological factors through which any other variable must act to affect fertility. Broadly stated these include: factors affecting intercourse (age at marriage, rates of marital dissolution, coital frequency); factors affecting exposure to conception (fecundity, use of contraception) and factors affecting gestation (fetal mortality, voluntary and involuntary). This set of determinants also implies the grasp of the feedback between the various components: in one word, the knowledge of the functioning of the demographic system. Since the system possesses a great deal of inertia, the integration of contemporary with historical materials provides longer series and better views of population problems.

Most of the efforts of scholars during the last four decades or so have been devoted to the task of grasping the functioning of the demographic system. These scholars were generally demographers (or historical demographers) because, independent of their origins or training, they were mainly interested in the functioning of the demographic system. In doing so, they have used a variety of methods and approaches, both macro and micro (through the nominative linkages (that is by surnames) leading to the reconstitution of families and genealogies). Erroneously, modern historical demography has often been solely equated with the micro approach.

THE MICRO APPROACH

The micro approach, used here as synonym for family reconstitution, has attracted most of the efforts and resources in the field of historical analysis. The publication of the manual for parish register analysis by Henry and Fleury (1965) and of the first complete parish monograph, the Norman village of Crulai, by Henry and Gautier exactly thirty years ago (Gautier and Henry, 1958) marks the beginning of a new era in historical studies. It is true that many scholars had used genealogies for demographic analysis; that Ortssipenbücher had been established in Germany in the thirties, although not exactly for pure scientific purposes; that Hannes Hyrenius had pioneered family reconstitution for the study of the Swedish minority in Estonia in the nineteenth century (Dupaquier, 1984). However, in initiating this new era, the respective merits of Louis Henry and those of his predecessors could be equated to the credit given to Columbus for the discovery of America and that given to the first Viking landing a few centuries before.

The invention of family reconstitution has produced many fruits that are well known. The most obvious advances relate to fertility, its measurement and the mechanisms of proximate determinants. The very concept of natural fertility, proposed by Henry, has proved to be a fecund idea. The proof that even in the absence of voluntary control, fertility varies greatly, has stimulated research on the factors determining this variation, independently of the obvious effects of age and of marriage duration. The role of fecundability, a concept already developed in the early 1920s by Gini (Gini, 1924), and the presumable influence of frequency of intercourse as well as the evident heterogeneity of its distribution have been investigated. Central to this research has also been the impact of the varying length of the nonsusceptible period and its link with breastfeeding, and the impact of early infant mortality on breastfeeding, birth intervals, and early fetal loss (see Chapter 11 by Wood and Weinstein). When relatively large bodies of data were available, the level of permanent sterility has been also investigated, as well as subfertility after puberty. Larger bodies of data have also enabled the study of the complex interactions between age of the woman (and in some cases also of the age of the man), marriage duration, and parity. The accumulation of data concerning puberty and menopause, fecundability, nonsusceptible period, intrauterine mortality, and infertility by age, have permitted the reconstruction of models of reproduction and birth, useful tools for the description of a generation during its reproductive period (Leridon, 1984).

Although born to describe fertility, family reconstitution also has yielded results on nuptiality and mortality, but of much inferior depth and quality. Marriage records enable the analysis of age at first marriage and of remarriage, according to age and duration of widowhood. Less frequently available is information on the proportion remaining single (unless family reconstitution can be coupled with counts or censuses) since mortality and migration make uncertain the estimate of survivors that enter the marriage market. This is a major drawback, since exclusion from marriage may be one of the most powerful mechanisms of selection in human populations and it would be important to know what proportion remains unmarried (and therefore usually fails to reproduce), and who are those who remain unmarried (Livi-Bacci, 1984). The demographer is particularly interested in understanding what kind of characteristics impede or promote the chances of marriage. Belonging to a given social class, professing a given creed or religion, the exercise of a certain profession, may all increase or decrease access to marriage and therefore to reproduction. Geneticists are well acquainted with similar problems; demographers also use this information for gaining insight into the functioning of the social system. There are also many reasons to believe that selectivity of marriage determines the fertility performance and our ignorance of marriage reflects negatively on our grasp of fertility patterns. Much more satisfactory is the use of information on remarriage which, used in connection with age, duration of widowhood, number of surviving children, and age and marital status of the partner, permits a thorough investigation of this component of the marriage market (Dupaquier *et al.*, 1981).

Family reconstitution has been much less relevant for the study of mortality. What happens outside stable families is a matter of guesswork, since mobility makes it difficult to allocate deaths to the population at risk. Infant mortality can be studied, but registration of infant deaths in the parish books is plagued by omissions, and corrections are difficult. Mobility physically separates children from the family, and measurement of mortality becomes increasingly difficult with age. Estimates of adult mortality for the married population is possible but its significance limited.

RECENT ADVANCES IN FAMILY RECONSTITUTION

Family reconstitution using surnames, however, has not remained unchanged since the publication of Crulai 30 years ago. Three main areas of advance can be identified. The first, with the development of computer capabilities and skills, regards the enormous advance in the ability to handle increasing amount of information. Data banks can be created and managed in order to yield the most varied kinds of analysis. A second, decisive, advance, is conceptual more than technical, and consists in the attempt of linking successive generations belonging to the same family in order to compare behaviors. The third is the attempt to link information which is nondemographic in nature to the individual or the family, primary objects of the observation. Such information includes data concerning, for instance, landownership and property, social characteristics, and the like. Advances in automatic linkage are at the base of progress in this area, but it is surprising how much has been done with simple manual techniques. The sophistication of the field is increasing; examples of the profit that can be drawn from these new approaches are the Utah Population Database (UPDB) where the genealogies are linked to the "tumor registry and other medical records for the purpose of identification of high risk pedigrees" (Bean, Chapter 15). Perhaps the best known example is the Research Program in Historical Demography (PRDH), of the University of Montreal, which has created a comprehensive register of the French Canadian population of Québec, now completed for the period starting with the foundation of the colony at the beginning of the seventeenth century to 1730, but which is expected to continue to mid nineteenth century (Legaré, 1981). The project integrates parish registers, censuses and many other information; besides classic family history, it permits multigenerational analysis of the entire population. The population of the Saguenay, a region of Québec, is the object of a sophisticated work of reconstitution from 1842 to 1971. This project, managed by SOREP (Centre Interuniversitaire de Recherches sur les Populations), has recently tried to incorporate biological and genetic variables in the reconstitution (Bouchard, Roy, and Casgrain 1985, Bouchard 1988).

The main object of the project is to develop an approach to genetic epidemiology which combines several fields, such as history, demography, population genetics, molecular biology and epidemiology. It is hoped that this approach will take advantage of the merging of the recombinant DNA technology with the population register or genealogical database. The main disorders specific to the population are tyrosemia, rickets, spastic ataxia, agenesis of the corpus callosum, hemocheromatosis, cystic fibrosis, and Sipple's syndrome. Finally, the ambitious project of the so called "survey of the 3,000 families", should be mentioned (Dupaquier, 1984). This is a sample of 3,000 marriages of the first Empire from different areas of France whose genealogies will be followed until the present day, incorporating geographical mobility as well as an ambitious set of socioeconomic data. The purpose of this survey is to integrate demographic and social analysis, to link successive generations, and to observe territorial and social mobility during the formation of modern France.

The list could probably be longer, because other initiatives of varying scale and importance are being made in several places, in Europe and in North America. However the common characteristics of the projects briefly referred to here, are the large body of data; the development of procedures of automatic linkage (except for the "survey of the 3,000 families") and the integration of nondemographic data. They represent a powerful extension of the reconstitution of families, Crulai-style, although, as discussed in the following section, they also have characteristics that limit their application.

LIMITATIONS OF THE MICRO APPROACH

Nominative reconstitution, both in its pioneering form and in the most sophisticated extensions, has limitations, some of which are obvious while others are not. It is obvious, for instance, that reconstitutions require written records, with a fair degree of completeness, thus restraining their application to the last three or four centuries in a limited part of the world. It is also obvious that reconstitutions are very costly and require great financial and human resources, although technical advances are reducing some of these costs. It is evident, as already mentioned, that the success of fertility analysis is not matched by the results on nuptiality and mortality.

All this is well known and need not be stressed further. Another limitation, often echoed, is that family reconstitution limits itself to stable families that do not move around, and this selects against families with particular characteristics, for instance families with few or no children, landless families, or families of lower socioeconomic status. In some cases when comparisons of different types of families have been possible, differential behavior has appeared to be less important than one would have expected; however this is an important drawback in most instances. Only in some fortunate cases, where the population of an entire region is being reconstituted (as in the case of Québec), is it possible to overcome this criticism.

Family reconstitution being costly, the number of events recorded cannot be very high; therefore it is necessary to assemble events belonging to long periods in order to have significant results. Cohorts are grouped by periods of 50, 25 or, more rarely, 10 years. Impacts of fluctuations caused by particular events (famine, epidemic, war) are ironed out over long periods and cannot be studied. A similar restraint applies to differential analysis which needs large bodies of data in order to yield significant comparisons.

However, it is unreasonable to reproach a methodology that has innovated so much and in so many ways for the failure to yield results that obviously are out of reach. If elephants do not lay eggs, it's not their fault. Let us take note that they cannot do it and stop pressing them. Indeed, the main criticism is not for the elephants, but for those who continue to watch them, waiting for the eggs.

As mentioned earlier, in the last four decades or so, historical material has been used in order to reconstruct the demographic system in its complexity. This was an innovative departure from the traditional pattern, where historical material was used in order to build a frame for historical events or to provide information to fit in some theory of society. It is proper, therefore, to discuss briefly the significance of micro analysis for the reconstruction of the demographic system. Going back to Crulai, the reconstitution says almost everything that an eager demographer would have wanted to know about fertility, as well some valuable information about nuptiality and, less so, about mortality. On the other hand, the authors confess that they know very little about population change, although ''we are inclined to consider as constant the population of Crulai between 1675 and 1750.'' At the macro level, the size of the French population before the eighteenth century is still very much of a mystery, notwithstanding the flourishing of historical demography. We have advanced, of course, from the times of Montesquieu, the Marechal de Saxe, or Mirabeau who, in the eighteenth century, still believed France's population to be declining. We have advanced, but not too much. In other words, the answer to the simple but vital question concerning the sign and intensity of population change is frequently eluded. At another level, family reconstitutions often say nothing about reproduction of the population—about its fitness, as geneticists would say. Building a fertility function is difficult, because little is known about fertility of the unmarried population and because not enough is known about the mortality function. Therefore net maternity functions and net reproduction rates are ignored. Still some effort could be made. By linking one generation with the next, the number of children of each newly married couple reaching marriage can be determined (which, in many societies, is equivalent to access to reproduction). The ratio between married children and married couples in the parents cohort is an estimate of the net reproduction rate at the age at marriage (under the conditions of constant levels of celibacy and lack of mortality differentials by marital status), an estimate which requires no knowledge of the fertility and mortality functions. Other shortcuts could be imagined. In Canada, Charbonneau and coauthors have demonstrated that among the first thousand or so founders, each married couple had, on average, 4.2 children reaching marriage, a net reproduction rate of 2.1, implying a rate of growth of some 2.5%, which explains the success of the founders (Charbonneau *et al.*, 1987).

If aggregate growth or reproduction rates are often difficult to obtain, much less is known at the level of subpopulations defined by social or economic characteristics. But one of the main curiosities that demographers ought to have concerns differential growth, a main concern also of geneticists. In other words, who, in one generation, contributes more and who contributes less to the next generation? And are these differences due to random or to systematic factors? Systematic factors include permanent (or semipermanent) social or economic ''markers,'' inherited from one generation to the next: being a landowner or not, belonging to the category of servants or slaves, or to the bourgeoisie or the aristocracy, and professing a given religion. Each group may obey particular rules concerning marriage and celibacy, sexual behavior and breastfeeding, abortion, infanticide, and child care, and these rules may influence its reproductive patterns (or ''fitness''). Now, when family reconstitution is carried out on a large scale (as is sometimes possible) and individual data are linked to nondemographic characteristics, it is theoretically possible to proceed to differential analysis. In other words, it is possible to subdivide the population into meaningful groups such as those described above and study their "differential" patterns of growth and reproduction. But in so doing, quantitative information becomes rather thin, and in order to accrue a sufficiently large body of data it is necessary to group together many birth cohorts, thus adding new difficulties to the interpretation of the results.

Family reconstitution has been a powerful tool in helping to reconstruct the mechanisms of the demographic system, but, with some exceptions, it tends to be a peculiar system: the system of those who marry and move little, of those who lead a stable existence, of those whose roots are deeper and stronger. Migrants, paupers, vagrants; urban and institutional populations; the unmarried, widows and widowers, are often left out of the field of investigation. A parallel exists in the study of economic systems: what is studied is the official labor market, the registered firms that declare profits and pay taxes, savings invested in the banking systems, and whatever flows through the channels of the official market. Often left out is the hidden economy; illegal labor, black market transactions, firms operating out of the official system, savings kept under the mattress, consumption of goods produced in the backyard or grown in the vegetable garden. Sophisticated models can be built on the official data, and economists are often surprised that the economy has not yet collapsed.

Accordingly, it is necessary to explore the "hidden" demography to understand the real functioning of the system.

THE MACRO APPROACH

In order to reconstruct the demographic system in its totality, the micro approach needs to be integrated with the macro which, in opposition to the definition of micro given earlier, is identified with aggregative, nonnominative methods. Demographers have felt this need in the last decade or so, dedicating more attention and more skills to the analysis of aggregate series and data. Not that aggregative studies had been abandoned. On the contrary, several examples immediately come to mind: The Princeton European fertility project and the large body of information on fertility produced (Coale and Watkins, 1986), the analysis of mortality crises, and models for the analysis of historical series. An example of how fruitful the aggregative analysis can be is provided by the reconstruction of the English population by the Cambridge Group for the period 1541-1871 (Wrigley and Schofield, 1981). This study, which provides a comprehensive account of the functioning of the English demographic system, hidden demography included, shows the declining role of mortality in guiding population trends and the increasing role of fertility and nuptiality in determining demographic change. It solves the "conundrum" (to use Wrigley's term) of the acceleration of the English population in the eighteenth century (Wrigley, 1983). The English example is being followed in other areas of Europe, and with profit.

Macro analysis can increase our knowledge in the area of mortality, still very unsatisfactorily covered by historical studies. Knowledge of age patterns of mortality is rudimentary, while no satisfactory attempt has been made to conceptualize its proximate or primary determinants. But mortality of the past cannot. be understood without a clear grasp of causes of death and, particularly, of the factors generating infectious diseases. This means that the study of mortality must incorporate epidemiological models designed to explain the incidence of infectious diseases and the pattern of their epidemic outbreaks (Singer, Chapter 17). Virgin soil effects are probably the only explanation of the rapid decline and sometimes of the extinction of populations after contact with the Europeans: in Australia, in New Zealand and in many Pacific islands, in Patagonia, in Peru and Mesoamerica, in the Caribbean, and in North America. Since 1348 there have been frequent, chronologically coincident and severe outbreaks of plague, but starting in the following fifteenth century, their frequency and intensity tended to decline until the complete disappearance of the disease from Europe. Sweating sickness, certainly a new disease, comes and goes leaving no traces; typhus appears to be a new disease as is syphilis, but the latter loses its virulence while the former remains unchanged. Many of these diseases, although probably density dependent in some way, are largely exogenous to the system.

Does a given disease immunize the individual or not? How lasting is the immunization? Which are the factors that determine the transmission of the disease? Does a disease select for more resistant individuals and can the decreasing gravity of a disease be explained by this process? Is it also true that less lethal pathogenic agents tend to be selected because "infection is a balanced conflict between parasite and host in which both species survive," to use Burnet's words? (Burnet, 1972; Singer, Chapter 17). Without a grasp of these evolutionary genetic issues it seems difficult for demographers to progress very much beyond the mere description of mortality trends—which, however, is itself a challenging task. The analysis of fluctuations of deaths (by age and sex, when possible); the determination of the disease causing the crisis through the study of contemporary literature or other information; the ecological analysis of the fluctuations; the determination of their frequency and gravity, are all elements that can greatly improve our knowledge of past mortality.

CONCLUSIONS

Micro and macro analysis must be conceptually reconciled to reconstruct the various components of the demographic system of the past. Micro analysis provides knowledge of the inner mechanisms of change which has to be tested with macro approaches that incorporate "hidden" demography.

The demographic system is not a closed one. To understand the reasons for its modifications, it is necessary to make connections with other systems, in particular the socioeconomic one. Otherwise, demographers will be confined to description and classification, and population science will not advance but will remain at a stage similar to that of biology before Darwin and Wallace, or of astronomy before Copernicus and Galileo. Populations operate in a set of constraints—climate, epidemic aggression, space, land, food, energy—and since Malthus the search for theories explaining the interrelation between these constraints and populations has been a central topic for research. All the factors named above are in some measure density-dependent, and they certainly affect demographic behavior. But density depends on population change, on macro, aggregate trends: if these are neglected (and this sometime happens when research focuses exclusively on micro analysis) the opportunity for explaining behavior at the micro level is lost. The argument, of course, runs in both directions. Two books, among many, could be cited as representative of attempts to relate the demographic system to the socioeconomic system: the one by Connell on Ireland (Connell, 1950) and the one by Wrigley and Schofield (1981) on England, published 30 years later. The second was able to avail itself of the detailed reconstruction of the demographic system, while the other worked with raw demographic material. But in both books population change is functionally integrated with a set of socioeconomic variables. In Ireland, rapid population growth was a consequence of increased land availability and productivity, leading to a high rate of household formation until the system collapsed with the Great Famine of 1845-46. In England, population growth responded to price increases and declining real wages by moderating nuptiality and fertility and *vice versa*, through a process of adaptation or homeostasis. Protestant moderation and Catholic intemperance, one would think.

The geographically close but socially distant cases of Ireland and England are cited not only as exemplary studies relating the demographic and the social systems, but also because, in both cases, mortality seems unrelated to population growth, contrary to one of the postulates of the Malthusian model. In England real wages, in the long run, are positively related to mortality; in Ireland, during the century before the Famine, mortality (in Connell's interpretation) did not deteriorate in spite of demographic explosion. Now, one of the major puzzles of demographic history during the 500 years before the industrial revolution is the interpretation of mortality trends and levels. Indeed, from time immemorial scholars have accepted the notion that mortality depends on food availability, and that food depends on land, and land is short when population grows and abundant when population declines. Changing mortality is, therefore, one of the mechanisms through which populations respond to the scarcity or abundance of resources.

Does the historical material support this notion? It does as far as catastrophic mortality is concerned, since years of famine are often also years of epidemic and therefore of high mortality (although many mortality crises, produced by plague, smallpox or cholera are almost totally unrelated to nutritional levels). It does not, if mortality in ''normal'' years is considered. There are many reasons for this assertion and these are described elsewhere in more detail (Livi-Bacci, 1987). For illustrative purposes some of these are briefly stated here. The immune systems deteriorate when nutrition is deficient, but only in presence of serious or severe malnutrition, do the risks and gravity of infection increase. In normal years the nutritional level of European populations was probably sufficient, as the usual average per-capita rations of 500-1,000 grams of bread seem to demonstrate. Quality of diet, as measured by consumption of meat, was good in the late medieval period and at its worst, in most of Europe, at the end of the

18th century. Access to food was not the main factor of mortality differential, since the mortality levels of the aristocracy, bourgeoisie, and well-to-do urban classes, all usually very well fed, were similar to levels of the less fortunate people. Real wages, inasmuch as they can be taken as an indicator of the tendency of the standard of living, are inversely related, in the long run, to levels of mortality, in England and in other parts of Europe. The first transition of mortality in England, France and Scandinavian countries, from mid-eighteenth century to the first decades of the nineteenth, takes place during a period of declining real wages, declining or stationary stature of conscripts and children, and deterioration, or stability, of other indicators of the standard of living. Infant and child mortality variation (which could account for differences in expectation of life of many years) was not related to food availability (the majority of infants being breastfed) but to social habits. If the European historical material is properly organized, it appears that nutrition has to be acquitted of the charge of being the main factor of variation of past mortality patterns (except, as already pointed out, for part of the mortality crisis specific to the "ancien régime"). Epidemiological, ecological, and social factors, largely exogenous to the economic system, are probably responsible of mortality variation, while the adaptive ability of humans to withstand scarcity and to survive hardship accounts for its resilience in difficult times (Livi-Bacci, 1987; Preston, Chapter 22).

In European populations, one of the mechanisms of the Malthusian system seems not to work properly according to theory. On the other hand, availability of land because of new techniques such as the introduction of new crops, reclamation of wasteland, (such as bogs, fens and marshes), subdivision of latifundia etc., was probably an important mechanism of change, not because of improved nutrition but because of the increasing opportunities for marriage and, therefore, for fertility.

A reading of historical material at the macro level is, therefore, extremely important. Micro studies may tell us whether families with no land had different behavior patterns from those with land, but macro analysis may tell us whether land was available at all for further expansion of the population.

The fundamental questions raised at the beginning of this chapter have to be answered. Demographers have a dangerous tendency to shy away from them, and leave others to provide the answers. But without a theory of society and of its functioning, our explanation of the demographic system remain incomplete. That is why it is necessary to reconcile micro and macro approaches and attempt to integrate the demographic system in a wider view of society.

REFERENCES

Beloch, K. J., 1937, 1939, 1961   *Bevolkerungsgeschichte Italiens*, 3 Volumes. De Gruyter, Berlin.

Bongaarts, J., 1983   The proximate determinants of natural marital fertility, pp. 103-138 in *Determinants of Fertility in Developing Countries*, Vol. I, edited by R. A. Bulatao and R. D. Lee. Academic Press, New York.

Bouchard, G., R. Roy, and B. Casgrain, 1985   Reconstitution automatique des famillies. Le systeme SOREP, Centre Interuniversitaire de Recherches sur les Populations,

Université du Québec à Chicoutimi, Chicoutimi, Canada.

Bouchard, G., 1988   The Population History and the Gene Pool of the North-Eastern Populations of the Province of Québec, Proceedings of the SOREP International Symposium, September 1987, Centre Interuniversitaire de Recherches sur les Populations, Chicoutimi, Canada.

Burnet, Sir Frank MacFarlane, 1972   *Natural History of Infectious Disease*. Cambridge University Press, Cambridge, England..

Charbonneau, H., B. Desjardins, A. Guillemette, Y. Landry, J. Legaré, and F. Nault, 1987   *Naissance d'une population*. Presses Universitaires de France and Presses de l'Université de Montreal, Paris-Montreal.

Coale, A. J., and S. C. Watkins, Editors, 1986   *The Decline of Fertility in Europe*. Princeton University Press, Princeton, NJ.

Connell, K. H., 1950   *The Population of Ireland, 1750-1845*. Clarendon Press, Oxford.

Dupaquier, J., 1984   *Pour la demographie historique*. Presses Universitaires de France, Paris.

Dupaquier, J., E. Helin, P. Laslett, M. Livi-Bacci, and S. Sogner, Editors, 1981   *Marriage and Remarriage in Populations of the Past*. Academic Press, London..

Gautier, E., and L. Henry, 1958   *La Population de Crulai, Paroisse Normande*. Presses Universitaires de France.

Gini, C., 1924   Premieres recherches sur la fecondabilité de la femme. Proc. Int. Math. Congress 2: 889-892.

Henry, L., and M. Fleury, 1965   *Nouveau manuel de Dépouillement et d'Exploitation de l'État Civil Ancien*. Institut National d'Études Démographiques, Paris.

Legaré, J., 1981   Le programme de recherche en demographie historique de l'Université de Montreal: fondements, methodes, moyens et resultats, Études Canadiennes/Canadian Studies 10: 149-182.

Leridon, H., 1984   Selective effects of sterility and fertility, pp. 83-97 in *Population and Biology*, edited by N. Keyfitz, Ordina, Liège.

Livi-Bacci, M., 1987   *Popolazione e Alimentazione. Saggio Sulla Storia Demografica Europea*. Bologna, Il Mulino.

Livi-Bacci, M., 1984   Selectivity of marriage and mortality: Notes for future research, pp. 98-108 in *Population and Biology*, edited by N. Keyfitz, Ordina, Liège.

Preston, S., 1989   Sources of variation in vital rates: an overview. This volume.

Roth, H. L., 1899   *The Aborigines of Tasmania*. Halifax, England.

Wrigley, E. A., 1983   The Growth of Population in Eighteenth-Century England: A Conundrum Resolved. Past & Present 98:121-150.

Wrigley, E. A., and R. S. Schofield, 1981   *The Population History of England, 1541-1871*. Arnold, London.

3

From History to Genes:
From Genes to History

ELIZABETH THOMPSON

The theme of this chapter is centered on the importance of demographic information in providing, together with models for population genetic processes, explanations of genetic data. As a general premise this is rather obvious; all population genetics presupposes some level of demographic framework. Even if the assumption is of no population structure, the model of a random mating population of N individuals, or $2N$ genes, is a demographic assumption. However, in the analysis of human population data, we shall normally wish to impose some knowledge of a more structured population. The second level of demographic assumption would include features such as external migration and population origins—the relation of the population studied to other populations. The third level of structure would concern population subdivision and intrapopulation migration, and fourthly, the most detailed level would specify mating patterns, family size distributions, and other elements of the familial structure within the population. Finally, the demographic processes combine to produce the genealogy of individuals in the population: the complete population genealogy is the demographic realization. There are thus many different levels, time scales, and population sizes for demographic analysis and genetic inference. Rather than focus on any particular problem, this paper reviews a broad range of studies in which both genetics and demography play a role.

GENES AND HISTORY

The genetic make-up of a population is the result of its genealogical history; its detailed genealogy is the result of demographic history. Past demography affects the distribution of genes and alleles; the current observed distributions of traits are a reflection of history. Only with more detailed demographic information can detailed inferences be made from genetic analysis, but also the genetic data must have the power to differentiate the alternatives of interest. The geneticist's question is: To what extent do details of history affect current genetic make-up? At least a partial answer is provided by the extent to which history can be inferred from genetic data. If genetic data do not distinguish alternative historical

Table 3-1 Examples of pedigrees of populations

	Generations Depth	Numbers of Founders	Approx. Final Size	Approx. Total Size
(a) Geographic Isolates				
Tristan da Cunha, 1816-1961	8	13	268	350
West-coast Newfoundland communities, 1820-1975	8	2 + ~400	1728	6000
Utah population data base	6 to 8	over 50,000	500,000	1.2 million
(b) Pedigrees				
Iceland/Scandinavia	~17 or more	NA	NA	NA
The British Peerage	~17 or more	NA	NA	NA
(c) Religious Isolates				
Samaritans from c.1720	12	50	500	800
Aicuña from c.1700	~15	~10 + later immigrants	2000	8500
Hutterites from 1698	12	77	28,000	50,000

hypotheses, then, conversely, either of these alternative histories could have lead to the genetic data. From a purely genetic viewpoint, the extent to which history can be inferred is the extent to which history matters.

Population genetics provides mutation, selection, and drift as models for genetic change in populations. However, it is the drift process, or more precisely the Mendelian segregation of genes, which is of primary importance in small populations and on the short time scale of interest to historical demographers. The combination of Mendelian segregation with demographic processes produces random genetic drift, which leads to models for the distribution of allele frequencies and differentiation between populations under demographic models of population isolation, admixture and migration. Conditional on the population genealogy (the realized demography), Mendelian segregation provides models for the distribution of genetic traits amongst related individuals in the pedigree. Whether with reference to individual data or population data, the demography is a part of the underlying process that determines the distributions of genetic traits.

THE AVAILABILITY OF PEDIGREE DATA

The most detailed form of historical data is a detailed genealogy of the population involved, and genealogical information has been pieced together by demographers, historians and geneticists in surprising amounts (see Table 3-1). Many populations have genealogies that have been traced back eight or so generations; this seems to be the sort of depth where information is available, albeit requiring much effort to compile.

Some isolated populations have been studied by population geneticists for largely fortuitous reasons. A prime example is the population of Tristan da Cunha, the small island in the middle of the South Atlantic. This population was studied after being evacuated due to eruption of the volcano in 1961. The complete eight-generation genealogy (1815 to 1961) of this very small population (13 founders, 268 individuals in 1961) could be reconstructed "by hand" from local family records. Although by current standards this genealogy is very small (about 350 individuals living and dead), it is larger than any other for which joint genotype probabilities had been computed, and it is very complex. There is one quite early family whose marriages involve 10 interlocking loops, bringing us to the limit of computational feasibility for that time (see Figure 3-1). More specifically, there is a woman born before 1870

> whose first sister's husband is her father's half-sister's first son's son, and her father's mother's sister's daughter's son;
>
> whose first brother's wife is her father's half-sister's other son's daughter and her husband's sister's daughter;
>
> whose second sister's husband is her husband's mother's brother's son, and her father's first sister's son;
>
> whose third sister's husband is her father's second sister's son, and her husband's mother's sister's son's son;
>
> and whose second brother's wife is her father's second sister's daughter, and her husband's mother's sister's son's daughter.

Thus, even a small pedigree can lead to complex interrelationships between descendants of a small set of founders.

In other cases the impetus for study of a particular genetic isolate has come from the medical community. One example here is a group of West Coast Newfoundland communities with an exceptional level of lympho-reticular malignancies (Salmon *et al.*, 1980). This current population of about 1,750 individuals has over 400 immigrant founders, but in terms of genes, an overwhelming contribution has been made by the original couple, who with their children and grandchildren were the first permanent settlers in 1820. A more extensive example, with greater time depth, is that of the Aicuña population (discussed by Castilla and Adams, Chapter 4); a population genealogy stretching back 300 years has been reconstructed. This population, like the West Coast Newfoundland population, is not closed to immigration, but the major genetic contribution comes from the few early founders. The study is also similar to the Newfoundland case, in that the impetus for the study was the high frequency of a genetic trait (albinism) with potentially deleterious effects. The Utah Population Database (Bean, Chapter 15) is, by and large, shallower, but is on a far more massive scale, both in terms of numbers of founders, and size of current population. Another larger population whose genealogy is being reconstructed and analyzed is that of the French-Canadian population of northeastern Quebec (Bouchard, 1988). These larger populations contribute valuable demographic information, and scope for genetic epidemiological analyses of a variety of familial traits, but it is not practicable to obtain the complete genealogy of the entire population, specifying the complete lines of ascent of every individual.

In many populations it is possible to trace particular genealogies much further back. For example, some genealogies of rare genetic traits can be traced back to medieval ancestors, particularly in Iceland and in Scandinavia where the records are exceptionally good. However, these particular genealogies have less to say to the interaction between demography and population genetics. There are also

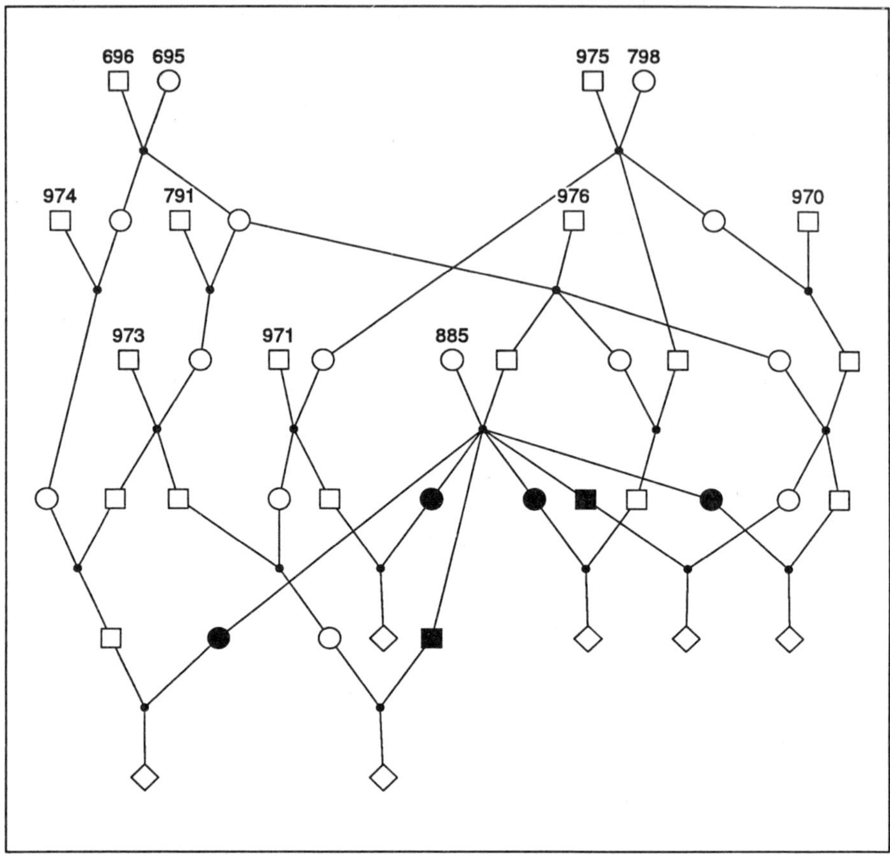

Figure 3-1. Complexities of a pedigree; pedigree of part of the early Tristan da Cunha population, showing the eleven early founders, and the complex interlocking loops generated by marriages of a family of six sibs to other families of the population. The members of the sibship are shown as solid symbols (males as squares, females as circles). The offspring sibships of the six individuals, through whom there is descent to the current population, are shown as diamonds. There are also other lines of descent from the eleven early founders to the current population, which do not pass through the six individuals, and are therefore not shown. (This pedigree was drawn using the software of Geyer, 1988.)

a few deeper genealogies of entire populations, usually of religious isolates. For example, the Samaritans, a very small and ancient genetic isolate now number-ing about 500 individuals, have a genealogy which has been traced back 12 generations, although many of the lines of ascent are incomplete (Bonné-Tamir, 1980). Even more impressive genealogies, leading to wonderful genealogical and genetic complexities, are those of the Hutterites (Hostetler, 1974). These are also of about 12 generations (300 years) depth, but are of a population closed to immigration now numbering 28,000. The genealogies are complete in that every individual has two parents specified, and each individual's ancestry traces back via both parents to a subset of just 77 founder individuals.

Accuracy of genealogical information is always a question, particularly in a large population, but here often the ancestral group is very small. Moreover, from a geneticist's viewpoint, the relevant genealogy is often limited to the ancestry of current sampled individuals. This genealogy determines the currently observable distributions for genetic traits. For genetic analysis of data observed on a current population, individuals who have left no descendants are irrelevant. This more limited view of genealogy, consisting of ancestors of sampled individuals, is often easier to research accurately. In this sense it is the "ascending genealogies" from current individuals that are the primary focus of genetic analysis. Note that, although available genetic data are often powerful enough to validate or even correct a current genealogy, they seldom provide reason to doubt accuracy of ancestral links. While more powerful DNA data, now becoming available, may make it possible to validate ancestral relationships, this will seldom be worth while, since the important question here is how far errors in the genealogy will affect genetic inferences. Generally, for problems of tracing the ancestry of rare genes or inferring ancestral genotypes, the answer is often insensitive to low levels of error, although for any particular genetic trait there are often a few key links that do affect conclusions.

GENETIC INFERENCES FROM THE PEDIGREES OF ISOLATES

The interesting genetic questions arising in the complex genealogies of small genetic isolates fall into two broad classes, exemplified here by two studies. The first is that of inferences of allelic origins, as illustrated by analysis of the Tristan da Cunha study. The second relates to a range of questions in the analysis of the genetic epidemiology of a trait, such as arise in analysis of the West Coast Newfoundland study.

Allelic origins on Tristan da Cunha

Allelic origins are again of interest for two broad reasons. The first is as a measure of information; the extent to which history can be inferred is the extent to which history is important. The second, perhaps more practical, is in inferring the origins of disease alleles, which again is only possible where that origin has affected the distribution of observable phenotypes in the population. The Tristan da Cunha genealogy is a small one, by current standards. The population was founded in 1816, and descends entirely from 13 founders of diverse origin, nine of whom were present on the island by 1835 (Roberts, 1971). In 1961 the population numbered 268, 243 of whom were sampled for a variety of genetic traits following the temporary evacuation of the population after eruption of the volcano.

Statistical methodology will not be given here. However, given a complex, but not too large genealogy, such as this one, there are methods for computing the likelihood of founder genotypes given the current genetic information (Thompson, 1986). That is, given the current data, it is possible to compute, simultaneously for every combination of founder genotypes, the likelihood of that combination—the probability that the data on the current population would be as in fact observed if that were the actual founder genotype combination. In the case of 13 founders, for a simple genetic system with 3 genotypes, this is a list of 3^{13} numbers. For example, in the case of the MN blood types, one of the simplest genetic systems, there are two codominant alleles M and N, and thus 3 phenotypically distinguishable genotypes MM, MN and NN. Then the list of 3^{13} numbers may be denoted by

$$L(G_1, \cdots, G_{13}) \tag{3-1}$$

where $G_j = MM, MN$ or NN is the genotype of founder j. Some method of examining this very high dimensional likelihood "surface" is desirable! A useful approach is to sum out founders, weighting with respect to some prior probabilities for their genotypes. In the extreme, inferences about single founders may be considered. That is, all other founders are summed out of the joint likelihood:

$$L^*(G_j) = \sum_{G_l, l=1,\ldots,13; l \neq j} \pi(G_l, l = 1,\ldots,13; l \neq j) L(G_1,\ldots,G_{13}) \tag{3-2}$$

where $\pi(G_l, l = 1,\ldots,13; l \neq j)$ denotes the prior (joint) genotype probability for all founders other than j. If, for simplicity, it is assumed that founder genotypes are independent, and that each is of genotype MM, MN or NN with probabilities p^2, $2p(1-p)$ and $(1-p)^2$, the result is a set of three functions for each founder,

$$L^*(G_j); \quad G_j = MM, MN \quad \text{and} \quad NN, \tag{3-3}$$

each a function of the single nuisance parameter p. Finally, since only relative likelihoods are important, a sufficient summary is provided by the ratios

$$\frac{L^*(MM)}{L^*(MN)} \quad \text{and} \quad \frac{L^*(NN)}{L^*(MN)} \tag{3-4}$$

for the two homozygote genotypes, relative to the heterozygote.

Figure 3-2 shows some results of this process, using as an example the MN blood types, which gives clear results for the Tristan da Cunha population. There are eleven early founders of the Tristan da Cunha population, who divide conveniently into two groups [Figure 3-1]. Group 1 consists of founders 885, 971, 970, 975, 798 and 976: Group 2 consists of founders 791 695, 696, 973, 974 and 976. (Inclusion of the founder 976 in both groups provides a convenient check on results. The other two slightly more recent founders (994 and 995) can be inferred to have had genotype MN.) For one group of founders clear inferences are obtained [Figure 3.2(a)]. Almost every founder is either substantially more likely to have carried M alleles, or to have carried N alleles. Moreover, the conclusions hold at the overall maximum likelihood estimate of founder allele frequency p for the population ($p = 0.5$), at the current population allele frequency ($p = 0.46$), and across the whole range of plausible values of the nuisance parameter p. For the other group of founders [Figure 3.2(b)] it seems no inferences are possible. However, in fact, although one can say little about these founders individually, there is substantial joint information about their collective genotypes. If certain founders contributed M alleles, then others are very likely to have contributed N alleles, and *vice versa*. The set of alleles contributed by this group of founders can be quite clearly inferred. This dependence between inferences about different founders is one of the interesting aspects of the results, and for ancestral inference in general, but will not be pursued further here.

Dependence comes up again if the question of gene extinction is considered. For any small genetic isolate, with a limited founder gene pool and/or a constricted ancestral genealogy, an immediate question is how many original founder genes survive. For example, the Tristan da Cunha population has 13 founders; 26 original genes at each autosomal locus. What is the maximum number that can have survived? The minimum? The average? Or, more generally, what are the probabilities of survival of different numbers of founder genes? Computation of extinction probabilities of given subsets of founder genes is very closely akin to computation of ancestral likelihoods (Thompson, Cannings and Skolnick, 1978). However, one of the most interesting aspects is the dependence; if genes from some founders survive, others have smaller probabilities of doing so (Thompson, 1983). The variance of the number of surviving genes at an autosomal locus is substantially smaller than the sum of the variances over individual genes, groups of genes, or even groups of founders (Geyer and Thompson, 1988).

Genetic epidemiology of West Coast Newfoundland

Where a familial disease segregates in a large genealogy, particularly the large and complex genealogy of a genetic isolate, questions of gene origins again arise. This presupposes the existence of a gene which is a major factor in the observed trait—the fitting of genetic epidemiological models is also a key component of genetic inferences from historical data in the form of a defined pedigree. An example is the genealogy of a West Coast Newfoundland population in which there was a very high incidence of lymphoreticular malignancies in the period 1955-75. This genealogy has an interesting structure, very different from that of the Tristan da Cunha population; there are over 400 founders to the current (1975) population of 1,728 individuals, but by far the largest genetic contribution is made by the original founding couple, whom we call John and Mary (J&M). This couple settled this area of the coast in 1820, with their children and grandchildren. The couple are ancestors of over 1,600 current individuals, and contribute at least 20% of the genes of over half the population (Thompson, 1981). The population was extremely isolated until around 1957.

When such a trait is observed in a genealogy, one may first ask whether there is any genetic component. This involves the familial distribution within the pedigree, and may be to some extent confounded with familial environment. The mere fact that one has a lot of related cases of course implies nothing, because this is the reason the pedigree comes to attention; it is the distribution of cases that matters. Even if there is a genetic component, there is unlikely to be a simple genetic model, but there may be apparent elements of a genetic model. For the Newfoundland study, the breakthrough came when it was recognized that, for a medically recognizable subset of the traits, there was a marked association between presence of the disease and bilateral descent from a certain set of six of the grandchildren of the original couple (Figure 3-3). J&M had 71 grandchildren, 41 of whom contribute to the current population, but only 24 have 30 or more current descendants. If J or M carried a rare allele, so would, on average, one half of their children, and one quarter of their grandchildren.

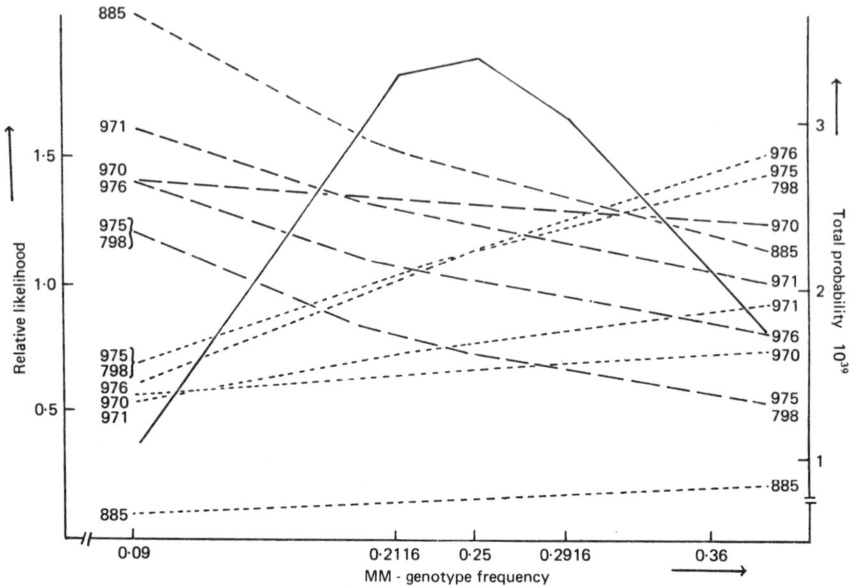

(a) MN-blood type likelihoods for each of the founders of group 1 (see Figure 3-1) as functions of the MM-genotype frequency. The solid line gives the overall likelihood for the prior probability of the MM-genotype. The broken lines give likelihoods for MM relative to MN, and the dotted lines for NN relative to MN. For further details, see text.

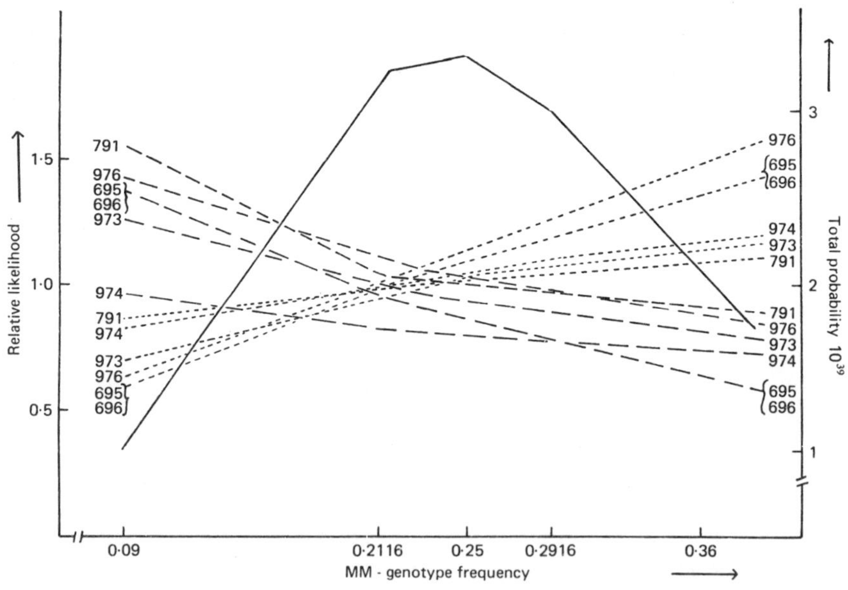

(b) MN-blood type likelihoods for the founders of group 2 (see Figure 3-1) as functions of the MM-genotype frequency. Notation as in (a).

Figure 3-2.   Tristan da Cunha ancestral likelihoods (from Thompson, 1978).

If this were a recessive predisposing allele, bilateral descent from the subset of grandchildren that carry it, would lead to potential cases. Bilateral descent only from other grandchildren would not. The finding of the strong association with bilateral descent from six grandchildren of J&M who make substantial contributions to the current population enabled us to postulate a model of recessive predisposition.

Thirdly, once a major gene component is modelled, likelihoods can be computed. The framework of a recessive predisposition does not mean that two copies of the postulated allele are either necessary or sufficient to cause disease. These uncertainties may be assessed by including penetrance parameters in the model. In particular, penetrance may be parameterised by two parameters α and β:

$$\begin{aligned}
\alpha &= \ \text{Pr(affected I two copies of the allele)} \\
\beta &= \ \text{Pr(affected I one copy of the allele)}
\end{aligned} \qquad (3\text{-}5)$$

and these penetrance parameters estimated. In fact, analysis provides the estimate $\beta = 0$; the predisposition is fully recessive. There is less information about α, but there is no strong evidence that α is not one; in homozygotes the allele may be fully penetrant. Fourthly, having fitted a genetic model, the next questions concern the ancestral origins of the relevant alleles, and about the paths of descent of those alleles to the current population. In this pedigree, the origin of the gene, if it exists, is clear; it can only be J or M. To infer paths of descent, posterior probabilities that each individual carried the gene are computed. These probabilities are conditional upon the model, the gene origin, and the current data, that each ancestral individual carried the allele. Some ancestors have high probabilities: indeed, some are obligatory carriers. However, some individuals have low posterior probabilities, even though they may be highly related to J&M and ancestors of several affected individuals; individuals 6128 and 1249 in Figure 3-3 are examples.

This question becomes of practical importance when, given evidence for a gene underlying a trait, having estimated a genetic model for the trait, and having made inferences of the pedigree origins of the gene, questions of prediction (counselling) are then addressed. For the present example, where the trait is inferred to be recessive, counselling is in principle straightforward, being essentially the inference of carrier individuals. However, in an extended and complex genealogy, joint predictions for sets of individuals are essential. All individuals are highly interdependent and, for a recessively inherited trait particularly, it is the joint carrier status of pairs of individuals that is of practical importance. It is also of interest that risk probabilities based on the trait distribution over the total pedigree may differ substantially both from those based on proximal affected relatives, and from those based solely on descent from inferred original carriers of a disease gene. Finally, many of the clearest inferences are negative rather than positive; many individuals in the population have high levels of bilateral descent from the founder couple who carried the allele, but via paths which *a posteriori* have low probability of transmitting the allele. These negative inferences may be of importance to the individuals concerned.

A final set of questions concerns the genetic mechanisms for the trait. What is the gene product and where is the gene? If there is a gene, there is presumably a gene function or product, or (since the gene appears recessive) an absence of

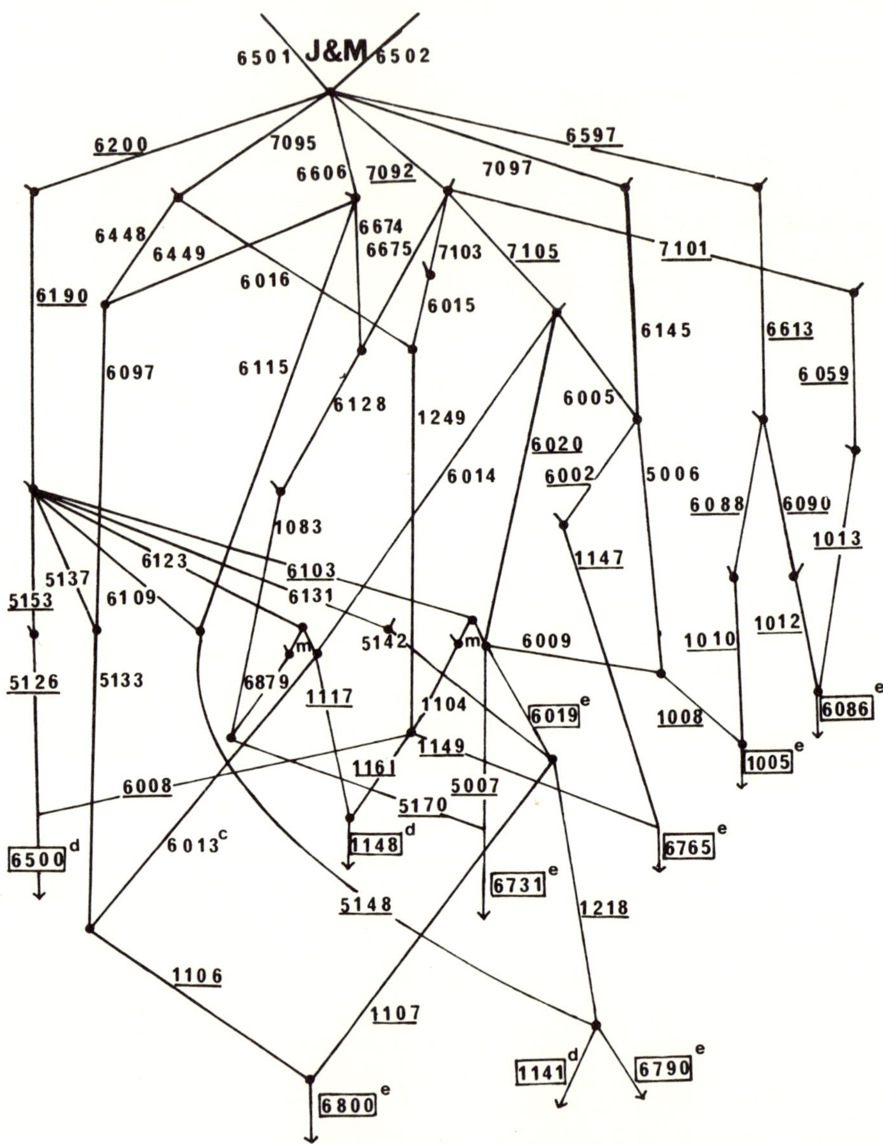

Figure 3-3. Small part of the Newfoundland genealogy, showing descent of ten affected cases ("d" = Hodgkin's disease, "e" = generalized immunodeficiency) from the founding couple J&M. Reprinted from Thompson (1981). [More recent work has necessitated minor modifications to this genealogy (see Thompson, 1986); these changes do not alter inferences.]

the product or function of the normal allele. Knowledge of this function might suggest treatment, or might suggest known enzymes or markers which may be associated with the trait, just as lipoprotein genes are studied in connection with heart disease, or the insulin gene in connection with diabetes. Linkage analysis with such known genes would provide not only evidence for the gene's location in the genome, but also more conclusive proof of its existence, and a basis for more precise counselling. In principle, successful linkage analysis would provide this even if the linkage established were with any marker gene, not necessarily one mechanistically associated with the trait. However, for this particular problem linkage analysis has yielded only negative results, and for many loci the power to detect or exclude linkage is very slight, since data are unavailable for many of the affected individuals (Goodman, 1984).

GENETIC DATA FOR HISTORICAL INFERENCES

Just as historical data, in the form of a pedigree, leads to genetic inferences, genetic data can lead to historical inferences. Normally qualitative historical information is available: without this demographic framework, genetic data are unlikely to be sufficiently informative for useful inferences to be made. Analysis of the genetic data can, however, quantify and modify that framework, providing estimates of parameters or tests of hypotheses.

Pedigree reconstruction

Normally the objective is not to reconstruct detailed pedigrees from genetic data. Such reconstruction, is, in principle, possible, if one has enough data. Genetic data have sufficient power to confirm or deny validity of putative close relationships between individuals within a small population. However, for human populations, any genetic reconstruction of ancestral genealogy is likely to be far less accurate than information available from records. For a few populations, the power to validate current genealogy may be useful; an example may be the genealogical structure within Yanomama villages (Neel, 1978a; Thompson, 1986). However, even here there are few specific conclusions that can be drawn, other than that genealogical inferences from the genetic data differ considerably from, and are often incompatible with, orally reported genealogy.

Iceland admixture

More frequently, the demographic questions to be answered from genetic data will be less specific, and on a longer time scale. For example, it is known that Iceland was colonized by Vikings, with Irish slaves, migrating by way of Scotland, but it is the genetic data that suggest the very high Celtic contribution to the population (Bjarnason *et al.*, 1973). In other admixture problems also, influences of language and culture may be dominated by a few members of the population; the genetic output provides better evidence about the likely genetic input. It is known that Iceland was colonized about 1,000 years ago by a large founder group of about 50,000 individuals, the population size remaining broadly

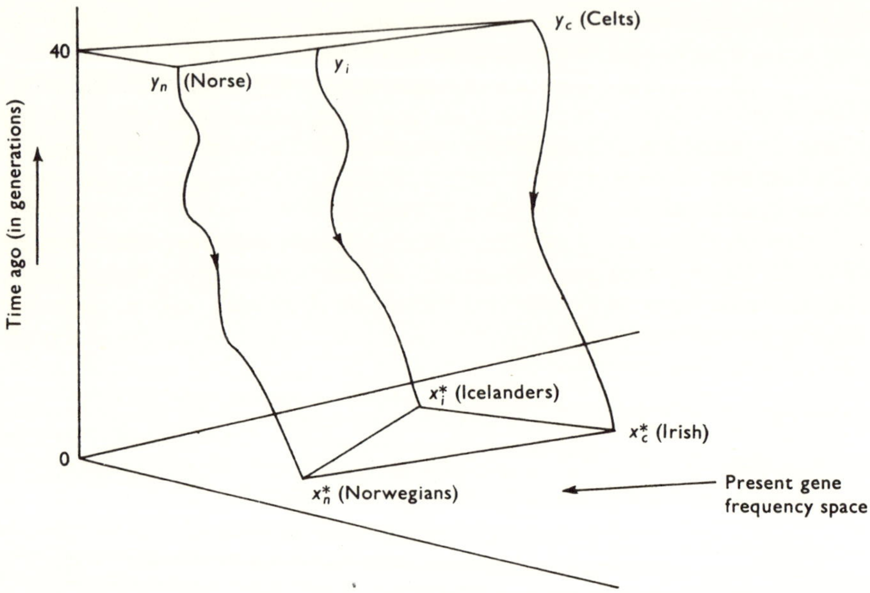

Figure 3-4.   Icelandic admixture model (from Thompson, 1973). Representation of the model for genetic changes in populations. From initial Norse, Celtic, and mixture populations (40 generations ago) current populations of Norwegians, Irish, and Icelandic populations descend, and are observed by sampling in the current gene frequency space.

constant until recent times. The founder gene pool was, presumably, a Norse and Celtic mixture. One can set up a model of genetic drift, from two founder gene pools, to the current day. This model also incorporates sampling in the current day, an initial mixture, drift in that mixture, and sampling from the current mixed population (Figure 3-4). On the basis of this model, one can estimate the original admixture proportion from current genetic data (Thompson, 1973). In this case the genetic data and previous historical information are in some disagreement, the maximum likelihood estimate of the Norse contribution being about 7%, with an upper confidence bound of about 20%. The model makes many assumptions about population sizes, and the ancestry of current populations representing the ancestral ''unmixed'' gene pools; the actual estimates should be treated with caution, but, at least, they provide some qualitative picture of the contribution of genetic information to answering the demographic question.

Inter-regional migrations of the Faroese

On a shorter time scale, more detailed inferences may be attempted. Another genetically isolated population of the Atlantic is that of the Faroe Islands, again providing interesting questions of historical demography which can be addressed via genetic analysis. This population was, like Iceland, founded about 1,000 years ago, and remained isolated and of stable size until recent times (although like many such populations it has grown rapidly in the last century). The overall time scale is thus very similar to that of Iceland, but the size (about 5,000) is an order of magnitude smaller. However, more interesting than the 1,000-year divergence from Europe, is the differentiation between the major subpopulations of the Islands; there are over 20 inhabited islands, and a natural division into seven largely endogamous regions (Harvey and Suter, 1984). To remove the effects of recent increased migration, on both genetic and demographic statistics, older members of the population may be sampled, and classified by birth-places of their grandparents.

Genetically, there are three major classes of unknowns; the genetic constitution of the original founders (Vikings), the level of continuing external migration from Denmark (over the centuries), and the level of inter-regional migration. There are some demographic knowns (or semi-knowns), such as the total population size and the relative population sizes of the separate regions, which can be built into the model. A simple model of internal migration, which at least gives some approach to analysis, is the mover-stayer model, widely used in the social sciences (Bartholemew, 1967). A proportion $(1 - K)$ of the individuals in each of L regions are assumed to be endogamous by nature, and the remaining proportion K migrate in proportion to the regional population sizes (including the possibility of "migration" to the area of origin). This gives rise to a backwards migration matrix

$$M_{ij} = \text{Pr(random parent of individual born in } i \text{ was born in } j)$$

$$= (1 - \alpha)\left[(1 - K)\delta_{ij} + K\frac{N_j}{\sum_l N_l}\right] + \alpha\delta_{j0} \quad i = 1, \ldots, L; j = 0, \ldots, L$$

$$(3\text{-}6)$$

where "0" denotes the external source, α the migration rate from that source, and $\delta_{ij} = 1$ if $i = j$, and 0 otherwise. A major advantage of phrasing the mover-stayer model in this way is that relative population sizes remain constant over the generations.

For this model, there are inferences which can be drawn from the analysis of the genetic data, again taking a likelihood approach (Thompson, 1984). It can be inferred that the genetic characteristics of the founder Vikings must have been substantially different from those of Danes, but that, nonetheless, continuing Danish immigration, at a low level, has had an effect on limiting the divergence of Faroese from other European populations. The best estimate of this continuing immigration is 0.5% per generation, which accords well with what is known of history, but external migration and genetic origins are confounded in current genetic data, and a large uncertainty surrounds the estimates. Estimates of internal migration can be made more reliably, and more precisely. The best estimate of K, effectively the inter-regional migration rate, is 16-17%, with a lower bound of about 13%, the upper bound being less sharply defined and more sensitive to assumptions about population size. Note, however, that current genetic differentiation can only estimate historical parameters relating to a period over which the rates are sufficient to establish equilibrium levels of divergence between the subpopulations. History previous to that period is no longer reflected in current data. For a value of $K = 0.165$, the period required to establish approximately equilibrium levels of divergence is 12 generations, or around 300 years. On the other hand, the estimate $K = 0.165$ accords quite remarkably well with a quite separate demographic estimate of migration levels for individuals marrying in the first years of this century. The mean proportion marrying in their region of birth was 83%, with very similar figures for most of the seven regions. So perhaps this genetic analysis does not reveal anything very new, certainly nothing very unexpected. However, it shows that there is a pattern of genetic differentiation within the population which still reflects, and can be used to infer, the levels of migration over the previous several centuries. Perhaps this is a case, not so much of convergent questions from demography and genetics, but of convergent answers, which is also an important aspect of interaction between the two subjects.

Amerindians: differentiation between and within tribes

On a longer time scale again, one may consider the South and Central American Indians, and the studies of intra- and intertribal genetic differentiation of J. V. Neel and colleagues (Neel, 1978a, b). While there are patterns of polymorphic differentiation which can be addressed by the same types of models as the more ''standard'' migration studies, the Amerindians have, as a group, been isolated for longer than other populations. On this longer time scale, and for these fragmented populations, rather different demographic processes have led to intriguing patterns of presence and absence of certain alleles in tribes, and to the existence of ''private polymorphisms''—alleles existing in appreciable numbers, but apparently confined to single tribes or regions. The patterns of presence of variants within and among tribes may reflect patterns of settlement and migration, and models of these demographic processes may help to elucidate the genetic data, although the power of this type of data to resolve specific alternative hypotheses is not large. Again it must be recognized that current data can only resolve alternative histories that are expected to have lead to substantially different current genetic characteristics of populations or of individuals.

On a longer time scale still, there are relationships between widespread populations which have not exchanged genes on any large scale for millenia—the evolutionary tree of subpopulations of the human species (see Figure 3-5). Again, random genetic drift has lead to an observable pattern of polymorphic differentiation among populations, and thus observed patterns of differentiation can be used to estimate the phylogenetic tree (Felsenstein, 1973; Thompson, 1975). This is prehistory rather than history, and the processes of evolutionary genetics rather than demography. However, the rate of genetic drift is determined by effective population size. Thus models of demographic process, at least to the extent that effective population sizes are influenced by population structure, sex ratio, and variance in reproductive success, still enter into the picture.

CONCLUSIONS

This review has spanned the range from genetic isolates of less than 300 individuals, to the megapopulations of evolutionary genetics. Throughout this range of population sizes and structures there are many fascinating historical and genetic questions that one can ask. The methods of addressing these questions play on the interaction of genetics and demography and use data from one to make

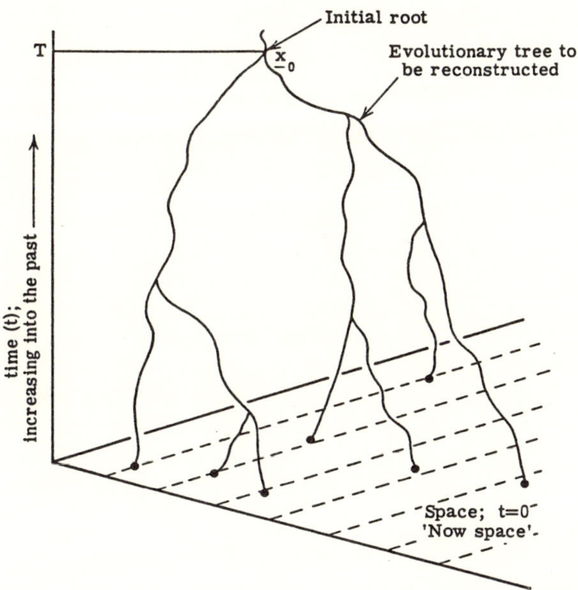

(a) Populations have genetic characteristics that change under the process of random genetic drift. Populations fission, and thereafter evolve independently.

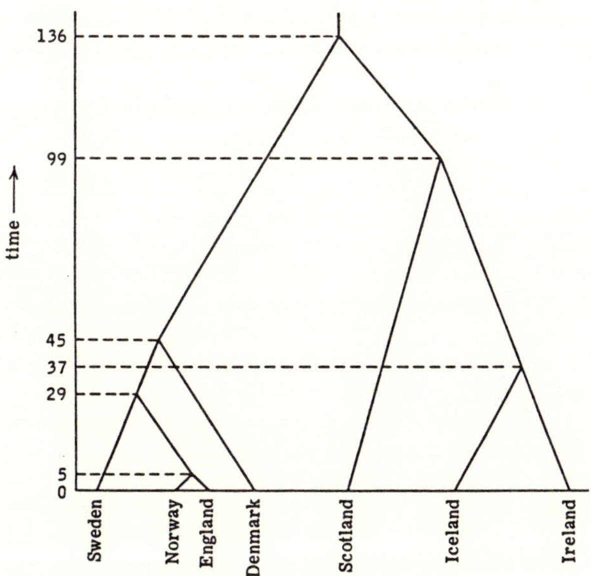

(b) Under these assumptions (not fulfilled by these European populations) the likelihood for the model can be used to estimate a tree.

Figure 3-5.   Representation of an evolutionary tree, and an estimate (based on blood type data) for some European populations (from Thompson, 1975).

inferences about the other. Genetics and demography supplement each other, to the mutual progress of both aspects of a population analysis.

ACKNOWLEDGMENTS

Work on this paper was supported in part by NSF grant BSR-8619760. I am grateful to Julian Adams, Charles Geyer, and Massimo Livi-Bacci for comments on earlier drafts.

REFERENCES

Bartholemew, D. J., 1967   *Stochastic Models for Social Processes*, (Chapter 2, Section 3.3). Wiley, London.

Bjarnason, O., V. Bjarnason, J. H. Edwards, S. Fridrikson, M. Magnusson, A. E. Mourant, and D. Tills, 1973   The blood groups of the Icelanders. Ann. Hum. Genet. 36: 425-457.

Bonné-Tamir, B., 1980   The Samaritans: an ancient living isolate. pp. 27-41 in *Population Structure and Genetic Disorders*, edited by A. W. Erikson, H. R. Forsius, H. R. Nevanlinna, P. L. Workman, and R. K. Norio. Academic Press, New York.

Bouchard, G., 1988   The population history and the gene pool of the north-eastern populations of the province of Quebec. *Proceedings of the SOREP International Symposium*, September 1987, Chicoutimi.

Felsenstein, J., 1973   Maximum likelihood estimation of evolutionary trees from continuous characters. Am. J. Hum. Genet. 25: 471-492.

Geyer, C. J., 1988   *Software for Descent Probabilities, Gene Extinction and Pedigree Drawing*. Technical Report #153, Department of Statistics, University of Washington.

Geyer, C. J., and E. A. Thompson, 1988   Gene survival in the Asian wild horse *(Equus przewalskii)*: I. Dependence in gene survival in the Calgary Breeding Group pedigree. Zoo Biol. 7: 313-327.

Goodman, D. W., 1984   *Linkage Analysis for a Recessive Trait in a Newfoundland Genetic Isolate*. Dip. Stat. Thesis, Univ. Cambridge.

Harvey, R. G., and D. Suter, 1984   Migration in the Faroe Islands. J. Hum. Evol. 13: 311-317.

Hostetler, J. A., 1974   *Hutterite Society*. The Johns Hopkins University Press, Baltimore.

Neel, J. V., 1978a   The population structure of an Amerindian tribe, the Yanomama. Annu. Rev. Genet. 12: 365-413.

Neel, J. V., 1978b   Rare variants, private polymorphisms, and locus heterozygosity in Amerindian populations. Am. J. Hum. Genet. 30: 465-490.

Roberts, D. F., 1971   The demography of Tristan da Cunha. Pop. Studies 25: 465-479.

Salmon, D., N.-F. Landre, G. R. Fraser, S. K. Buehler, J. Crumley, and W. H. Marshall, 1980   A familial aggregate of Hodgkin's disease, common variable immunodeficiency, and other malignancy cases in Newfoundland. II. Genealogical analysis and conclusions regarding hereditary determinants. Clin. Invest. Med. 2: 175-181.

Thompson, E. A., 1973   The Icelandic admixture problem. Ann. Hum. Genet. 37: 69-80.

Thompson, E. A., 1975   *Human Evolutionary Trees*. Cambridge University Press, Cambridge.

Thompson, E. A., 1978   Ancestral inference. II: The founders of Tristan da Cunha. Ann. Hum. Genet. 42: 239-253.

Thompson, E. A., 1981   Pedigree analysis of Hodgkin's disease in a Newfoundland genealogy. Ann. Hum. Genet. 45: 279-292.

Thompson, E. A., 1983   Gene extinction and allelic origins in complex genealogies. Proc. R. Soc. London B.219: 241-251.

Thompson, E. A., 1984   Inferring migration history from genetic data: application to the Faroe Islands, pp. 123-142 in *Migration and Mobility*, SSHB Symposium Volume 23, edited by A. J. Boyce.   Taylor and Francis, London.

Thompson, E. A., 1986   *Pedigree Analysis in Human Genetics*. Johns Hopkins University Press, Baltimore.

Thompson, E. A., C. Cannings, and M. H. Skolnick, 1978   Ancestral inference. I: The problem and method. Ann. Hum. Genet. 42: 95-108.

4

Migration and Genetic Structure in an Isolated Population in Argentina: Aicuña

EDUARDO E. CASTILLA AND
JULIAN ADAMS

The majority of investigations which have been published on the genetic structure and evolution of human populations involve North American or European populations. In Latin America, information available on the genetics of human populations is mainly limited to Amerindian tribal groups (e.g., see Salzano and Callegari-Jacques, 1988 and references therein), or to rare isolates which have had little impact on the genetic configuration of the present Latin American population at large.

The population of Aicuña offers the rare opportunity of having a well-documented pedigree extending for 16 generations, from its foundation in the seventeenth century until the present day. The existence of such data is unusual, and quite unique for a Latin American population. Published accounts of population genealogies, such as the Hutterites (Hostetler, 1974), the Samaritans (Bonné-Tamir, 1980), as well as for the Tristan da Cunha islanders (Roberts, 1980) cover no more than about 12 generations (Thompson, Chapter 3). Although the genealogy of the British Peerage (Hollingsworth, 1964) has a greater depth, the data do not cover a population in either the genetic or demographic sense.

In this chapter we present data concerning the history and the reproductive structure of this population that are relevant to both disciplines of genetics and demography. Changes in the mean inbreeding coefficient of the population during its history, the present breeding structure, the distribution of surnames, and the migration patterns of the historical population were analyzed in order to examine those processes of population change common to demography and genetics.

HISTORICAL BACKGROUND

The rural village of Aicuña is located in the west of the state of La Rioja in Argentina, at about 1,500 meters above sea level, on the western slopes of

45

an arid valley, bordered on the eastern side by the Famatina mountains and by the main chain of the Andes to the west. The majority of the population is agricultural; a limited commercial agricultural production of walnuts and vineyards is complemented by the rearing of small herds of goats and sheep. The frequency of illiteracy is low, limited to persons of advanced age. The high frequency of a surname (Ormeño) that otherwise is extremely rare in Argentina, and of a rare autosomal recessive trait (Oculo-cutaneous albinism), had initially brought the village to the attention of one of us (E.E.C.), as being significantly endogamous.

The village is located within the boundaries of an old *estancia* of the same name, approximately 3,565 square kilometers in extent, equivalent to an area approximately 50% larger than the state of Rhode Island in the U.S.A., or the country of Luxembourg. The estancia of Aicuña was originally bought by a Spaniard, General Pedro Nicolas de Brizuela, for his illegitimate son, Domingo de Brizuela. Due to the unusual manner (for those times) by which Domingo de Brizuela received these lands, it was not surprising that the legitimate and aristocratic descendants of the Spanish general unsuccessfully challenged the descendants of the illegitimate branch for ownership of the lands at least two times in a fifty-year period following their purchase. The first occurred around 1700, when the general's illegitimate son died in the city of La Rioja and the land was to be auctioned in order to cover the funeral expenses. The second attempt at expropriation was in 1723, by a legitimate grandson of the general, who took advantage of his position as judge and military governor of the area. Two centuries later, the third challenge occurred, in September 1955, shortly after a military coup, when unidentified individuals appeared in Aicuña claiming rights to the land. The extensive litigation surrounding the ownership of the estancia of Aicuña is significant for the existence and quality of the early genealogical records in two ways. Firstly, the legal documents from the late seventeenth and early eighteenth centuries, containing details of familial relationships, are an invaluable source of information in a period when parish records were nonexistent. Secondly, it probably stimulated the community to keep their records of their genealogical links to the founder of the population, the daughter of Domingo de Brizuela. It is unlikely however, that this stimulus (had it existed) resulted in biases in the pedigree records, as the different sources of information, even for the early generations, are remarkably concordant. In addition, the ancestral contributions of the direct descendants of Pedro Nicolas de Brizuela to the contemporary population of Aicuña are not in fact large (see Table 4-1).

A significant feature of the genealogy of Aicuña is that it includes both legitimate and illegitimate offspring, as the community conceded identical inheritance rights to both illegitimate and legitimate children. This concession was adopted by common accord, and even more surprisingly, remains in operation through to the present day. It represents a departure from the norms of the civil law in effect in Argentina since 1813, whereby only legitimate descendants are accorded inheritance rights, and is perhaps unique in Latin America. The logic underlying this attitude on the part of the inhabitants of Aicuña was perhaps that localities such as Aicuña were severely isolated from both ecclesiastical and civil centers. In those days the difference between legitimate and illegitimate children could be considered simply a reflection of the greater or lesser opportunity for the legitimization of marital bonds. The first chapel in Aicuña was only constructed in 1834, and even today a priest rarely visits the village more than twice a year. In 1833 the closest civil center for the registration of marriages was in the city of La Rioja, at least four days away by horse, though now the nearest center is located some 50 kilometers away in the town of Villa Union. However, it is also tempting to speculate that the adoption of such an accord was stimulated by the original gesture of General de Brizuela towards his illegitimate son.

A number of different sources of information exist on the genealogical history of the population. The inclusion of illegitimate children amongst the heirs meant that the community itself had to maintain genealogical records independent of civil and parish registers, since these exclude information on paternal data for illegitimate births. Church records are also available since 1724, and oral informants could provide reliable information since about 1850. These different data sources allowed us to record information on a total of 8,573 individuals, and in 1971, the date of the most recent complete census, 1,943 individuals were recorded as living within the limits of the estancia of Aicuña. Each parent-offspring link was ascertained from at least three different sources of information with very few exceptions, mainly in the eighteenth century data. The historical demographic data are obviously better for the most recent than for the more distant generations. This is particularly so for the nonbreeding individuals, namely, stillbirths, infant deaths, and early emigrants.

Aicuña was first inhabited by the Spanish general's granddaughter, her husband, and their four children in around the year 1700. A census taken some 100 years later recorded 50 people, all descendants of this couple, distributed among six different houses. The size of the village has increased considerably since then, but has never had more than 500 inhabitants. During the nineteenth and twentieth centuries, descendants from the village of Aicuña founded four other villages within the estancia. Two of these villages, Pagancillo and Los Palacios, are today considerably larger than the village of Aicuña, with twice as many inhabitants, whereas the other two, Puerto Alegre and Paso de San Isidro are much smaller, with less than a hundred inhabitants each. Due to the arid nature of this territory, human settlements were only feasible around the limited sources of water. Consequently, each one of the five villages is distinctly delineated, and interconnected by a single network of unpaved highways. The few settlements outside of the villages (known as *puestos*) rarely contain more than four families. Figure 4-1 shows the distribution of individuals within the estancia of Aicuña, together with the population sizes in 1971.

The genealogical records document a number of cases of immigration into the area and assimilation into the communities of the five villages. In the main, immigrants were isolated individuals, with few family groups, and no multifamilial groups. Significant admixture with members from the local Amerindian communities in the area occurred in the eighteenth century, but the great majority of genes can be traced back to individuals of Spanish origin. The number and temporal distribution of immigrants was different for each of the five villages, and is discussed in more detail below for those cases where it has had a significant role in determining the genetic structure of the villages.

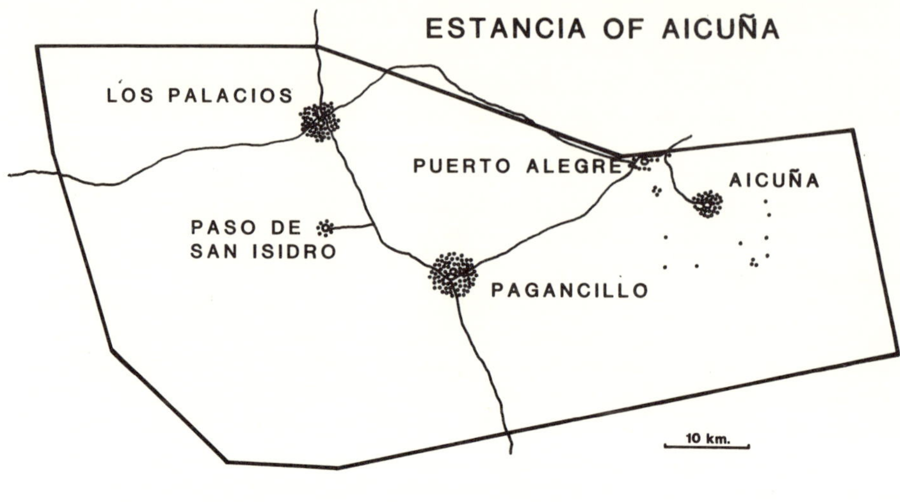

Figure 4-1. Diagrammatic map of the *estancia* of Aicuña showing the distribution of the inhabitants and the locations of the roads. Each dot represents 10 inhabitants. Geographical coordinates of the three largest villages are: Aicuña:$29°31'S, 67°44'W$ Pagancillo: $29°35'S, 68°03'W$; Los Palacios: $29°30'S, 68°12'W$.

Beyond the high frequency of albinism and of the surname Ormeño, the communities living on the estancia of Aicuña have little to distinguish them. The economy of the community is typical of other rural communities in the area. The estancia of Aicuña is crossed by federal highway 40 which runs six miles away from the village of Aicuña. This highway was built at the beginning of this century following a pre-existing path opened by the Incas in the fifteenth century. The population cannot therefore be considered to be geographically isolated, at least neither more nor less than most other rural populations in Latin America. In common with the rest of Argentina, all inhabitants of the village of Aicuña (with the exception of only one family) adhere to the Roman Catholic faith. Thus, the community cannot be considered a religious isolate. Probably, the only unique characteristic of this population is the availability of complete and detailed pedigree documentation. Since Aicuña cannot be defined as either a geographic or a cultural isolate, it seems reasonable to assume that the historical demography and genetic structure of this population are typical of most preindustrial Latin American populations of European background; a period which extends, for most of Latin America well, into the twentieth century.

PATTERNS OF INTERNAL MIGRATION

Growth of the village of Aicuña was and still is constrained by the limited quantity of water available for cultivation. Therefore, as the population expanded in size, and the quantity of food became insufficient to feed the growing population, emigrants began to occupy other oases within the estancia. In the course

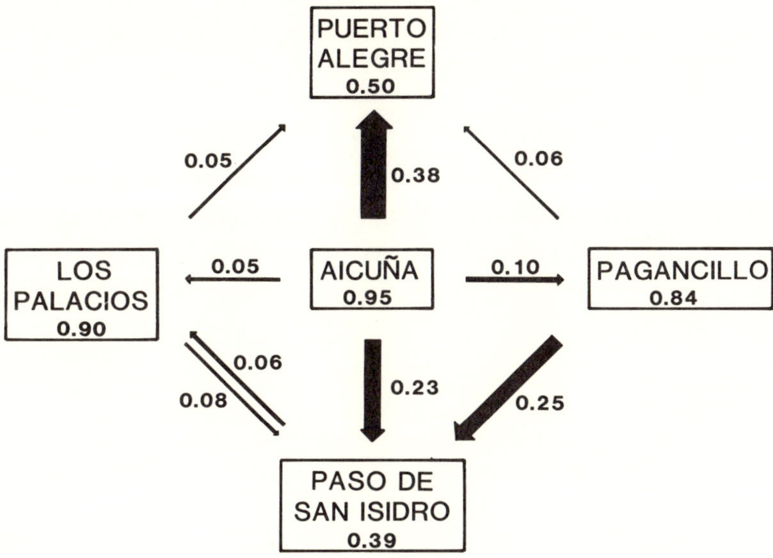

Figure 4-2. Patterns of internal migration for the estancia of Aicuña in the 20th century. Each figure is the probability that the parents of an individual born in village i were born in village j. For example, the probability that the parents of an individual born in Puerto Alegre were also born in Puerto Alegre is 0.50, whereas the probability that they were born in Aicuña is 0.38. Probabilities smaller that 0.05 are not shown.

of time these became known as the villages of Pagancillo, Los Palacios, Puerto Alegre, and Paso de San Isidro. This internal migration (that is, within the estancia of Aicuña) took place principally at the end of the nineteenth century and the beginning of the twentieth century. Migration is both a demographic as well as a genetic parameter, and can be examined from both standpoints. As will be seen below, consideration of both the demographic and genetic consequences of the diaspora from Aicuña allows us to gain a more complete understanding of the processes of population change in this community.

Demographic approach

Figure 4-2 summarizes the demographic data on internal migration for the twentieth century. The figures are elements of the average male and female backward migration matrix (Ward and Neel, 1970), and so are the probabilities that an individual in village i has parents born in village j. It can be seen that Aicuña and Los Palacios are the most endogamous of the five villages, with 95% and 90% respectively of their inhabitants in the twentieth century being born to parents from the same village. Emigrants from the village of Aicuña contributed a large portion of the populations of Puerto Alegre and Paso de San Isidro, somewhat less to the population of Pagancillo, and much less to the population of Los Palacios. In contrast to the emigration from Aicuña, there was relatively little migration between the new communities, within the exception of Paso de San Isidro and Pagancillo; Paso de San Isidro receiving many immigrants from Pagancillo.

Not surprisingly, these results are compatible with the oral histories of the villages, well known to the present day inhabitants. These accounts describe Aicuña as the original settlement, with the village of Puerto Alegre, the oldest of the other four, being founded as a simple extension of Aicuña. Pagancillo was originally populated by families from Aicuña, but soon received immigrants from outside the estancia. Families from Aicuña migrated to Paso de San Isidro, either directly or indirectly via Pagancillo with a one generation or more delay in between. Although Los Palacios is located very close to Paso de San Isidro, its history is quite distinct, receiving significant immigration from outsiders coming from the southwest, probably originating in Chile.

Genealogical approach

A quite different perspective on the relationships between the five villages can be gained by considering the genealogies of the present day inhabitants of each village and estimating the degree by which ancestry is shared. This approach differs from the preceding one in three aspects. (1) The genealogical approach takes into account immigration from outside the estancia, and considers each immigrant or immigrant family unit as unique. Immigration from the exterior was particularly important in determining the population composition of the village of Los Palacios. (2) The differences in the ancestral contributions to each village explicitly consider genetic variation within a village, whereas migration matrices treat each village as being homogeneous entities. (3) The migration probabilities describe one generation transitions only and cannot take into account secondary migration events such as those occurring between Aicuña, Pagancillo, and Paso de San Isidro.

For each village, ancestral contributions were determined by considering a one-locus genotype with two unique alleles for each of the present day inhabitants, and tracing the inheritance of the two alleles back through the genealogy, assuming identity of the alleles by descent (that is, no mutation). Table 4-1 shows the ancestral contributions for the village of Aicuña, the most endogamous of the five villages. The high level of inbreeding of the village is evident from the disproportionate contribution of a reduced set of ancestors; half of the present gene pool derives from only ten ancestors, from the eighteenth century.

Genetic divergence between related populations often has been calculated using coefficients of kinships (e.g, Morton *et al.*, 1982), derived either from gene frequencies or from genealogies. However, the coefficients of kinships determined from gene frequencies have a number of disadvantages (Jacquard, 1974), whereas those determined from genealogies involve extensive computations. An alternative approach, which is simple and involves minimal computation, utilizes the arrays of ancestral contributions (termed "origin vectors" by Jacquard, 1974). These origin vectors, which can be can be interpreted as one-locus arrays with as many alleles as the total number of ancestors, can be used to determine genetic distances between the villages. A number of distance metrics and their relative advantages and disadvantages have been described in the literature (e.g., Jacquard, 1974, Jorde, 1985 and references therein). A matrix of genetic distances between the five villages was calculated using the well-known measure of genetic identity, due to Nei (Nei, 1972). A dendrogram was then

Table 4-1 Ancestral contributions to the population of Aicuña in 1971

1. Apolinário Ormeño	0.10
2. Husband of Andrea Maria Ormeño, granddaughter of Apolinário Ormeño	0.07
3. J. Narvaez	0.06
4. Placida Flores	0.04
5. Husband of Lucia Ormeño, daughter of Apolinário Ormenõ	0.04
6. Francisco Paez de Espinosa	0.03
7. Husband of Mercedes Ormeño, granddaughter of Apolinário Ormeño	0.03
8. Pablo Reyes	0.03
9. Juana Rodriguez	0.03
10. Husband of Gregoria Narvaez, daughter of J. Narvaez	0.03
Others	0.48
Total	1.00

calculated from this matrix, and this is shown in Figure 4-3, superimposed on a map showing the geographical location of the villages, and assuming a root at the founding village of Aicuña. This dendrogram represents the best estimate of the minimum length tree describing the relationships between the villages, and was derived using the algorithm of Cavalli-Sforza and Edwards (1967) starting from an initial cluster obtained using the divisive algorithm of the same two workers (Edwards and Cavalli-Sforza, 1965). The pattern of divergence estimated using an alternative distance metric (χ^2; Jacquard, 1974) was virtually identical to that seen using Nei's metric.

Figure 4-3 shows that genetic and geographic distances correlate exceptionally well between Aicuña and the two larger villages, Pagancillo and Los Palacios, but that genetic distance exceeds geographic distance between Aicuña and Puerto Alegre, while the opposite is true for Aicuña and Paso de San Isidro. The discrepancies for these two latter villages can be explained by their small size. Both villages have less than 100 inhabitants each and are comprised of a small number of family groups. Since internal migration occurs principally in family units, random sampling variation would be expected to have a large effect on measures of genetic distance. For the large villages we may conclude that geographical distance is a surprisingly good estimator of genetic distance. For the estancia of Aicuña, cartesian measurements of distance are reasonable as the floor of the valley in which it is located is very flat, and the most common method of commuting has been and still is on foot, or on horseback following direct paths through the sparse xeric vegetation.

Inferences from the distribution of surnames

Since surnames are typically inherited paternally they can be treated as easily identifiable genetic markers, which, if carefully interpreted, can provide information bearing on the genetics and demography of human populations. For the population of Aicuña, surnames have provided important information both on the origins of the villages and their founding, as well as on the level of inbreeding (see below). Elsewhere in this volume, Piazza (Chapter 6) utilizes them in

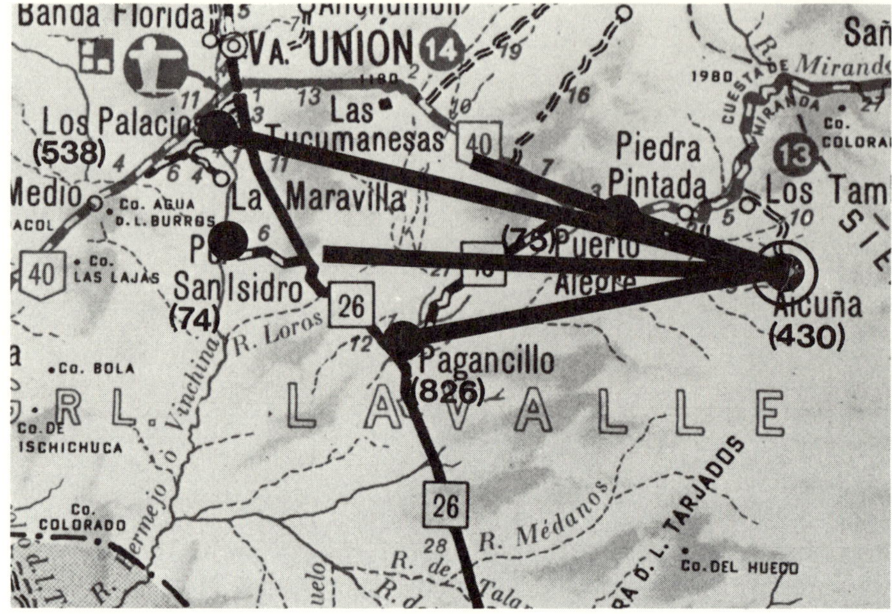

Figure 4-3. Correlation between genetic distances and geographical distance. The numbers in parentheses give the population sizes for each of the villages in 1971. See text for further details.

tracking large scale migrations within Italy. In addition, surnames have been used as indices of ethnicity (e.g., Gottlieb, 1983b).

Only six surnames account for 60% of the population of the estancia of Aicuña, of almost 2,000 inhabitants (see Table 4-2). All six were introduced into the population during the eighteenth century. In contrast to the situation in some parts of Europe (e.g., Czihak, pers. comm.), there has been virtually no modification of the lexicography of these names in the 300 or so years of the existence of the populations. The few changes that have occurred, such as the doubling of a consonant or the substitution of an ''s'' for a ''z'' are easily recognizable. It is important to consider the different modes of incorporation of these names into the population as well as their subsequent inheritance when using their frequency distributions to make genetic and demographic inferences.

The distribution of the surnames amongst the villages is not random, but illustrates their familial origins. Thus, as is well known to the inhabitants of that part of Argentina, Aicuña is known as *el pago de los Ormeños*—the ''land'' of the Ormeños. The villages of Puerto Alegre and Paso de San Isidro can also be identified in this way, while Los Palacios is known as the ''land'' of the Palacios, and Pagancillo the ''land'' of the Narvaez. In the neighboring estancia of Tambillos, legally separated from Aicuña in 1834, the two villages, Tambillos and Villa Union, are known as the ''land'' of the Paez.

Paez, being the name of the founder couple, was historically the first one in Aicuña. This couple had four offspring, only two of which, a male and a female, had descendants. The male transmitted the name to the branch of the family

Table 4-2 Surname distribution in Aicuña

Surname	Aicuña	Puerto Alegre	Pagancillo	Paso de San Isidro	Los Palacios	Total
Maldonado	0	0	1	14	2	2
Narvaez	4	0	16	0	0	7
Oliva	15	0	5	0	2	6
Ormeño	53	37	14	48	3	22
Paez	5	24	14	18	31	18
Palacios	0	0	0	0	13	4
Others	23	39	50	21	51	41
Population Size	430	75	826	74	538	1940

NOTES:

Distribution of surnames within the estancia of Aicuña, at the date of the last census in 1971. Figures are percentages of the total population size of each village.

that, as mentioned above, legally separated their land from Aicuña. The 18% of the present-day population of the estancia of Aicuña with the surname Paez originate from intermarriages between the two estancias. However, Paez most probably is not a monophyletic surname, as censuses taken in 1795 and 1810 record individuals of the same name elsewhere in the province of La Rioja, outside of the estancias of Aicuña and Tambillos. Furthermore, some of the individuals in these two censuses with the surname Paez are listed as slaves, and it is reasonable to conclude that they adopted the name from their owners.

Narvaez is historically the second name to appear in Aicuña. The original carrier of this name married the female child of the founder couple. It is probably monophyletic, as there are only a minimal number of persons in the present population who cannot be traced to the founder. However, the name Narvaez can also be found in other areas of the province.

Ormeño, the third name to appear, is no doubt the most important name in this population. It is widespread, representing almost a quarter of the present population in the estancia, and more than half in the village of Aicuña. It is undoubtedly monophyletic; every individual encountered with this name in Aicuña, in the province of La Rioja, and even in other provinces within Argentina, can be traced to the first Ormeño, who married a daughter of the first Narvaez.

The remaining three names are less important. Oliva is found mainly in Aicuña, Maldonado in Paso de San Isidro, and Palacios in the village of the same name; Los Palacios.

Therefore, even within a single and rather homogeneous population such as this, it is difficult to make generalizations for all the surnames. This sometimes introduces the need for sample stratification for proper analysis.

INBREEDING LEVELS IN AICUÑA

The average coefficient of inbreeding, estimated as the average coefficient of identity by descent, may be calculated directly from the genealogies available for the population (e.g., Crow and Kimura, 1970). Boyce (1983) describes and

Table 4-3 Estimates of inbreeding

Village	f Estimated from the Genealogy	f Estimated by Isonymy
Aicuña	0.041	0.077
Puerto Alegre	0.011	0.056
Pagancillo	0.009	0.020
Paso de San Isidro	0.011	0.072
Los Palacios	0.006	0.032
All Villages	0.016	0.023

evaluates the different approaches and algorithms that have been developed to allow automated computation of these coefficients. Using a modified version of the computer program originally written by MacLean (1969), average probabilities of identity by descent were calculated for the five villages, and these are shown in Table 4-3. As expected, the mean inbreeding coefficient, f, for the village of Aicuña, 0.0414, was the highest of the five. This means that, as directly measured by pedigree analysis, four per cent of the gene loci had two alleles which were identical by descent. Such a degree of average inbreeding is equivalent to a degree of parental relationship intermediate between first cousins and first cousins once removed.

It is well known that the average level of inbreeding can also be obtained from the frequency of isonymous pairs of individuals in the population (Gottlieb, 1983a). Since surname data is much more accessible and easily gathered than genealogical data, it is tempting to use the isonymy method to estimate inbreeding. However, as can be seen in Table 4-3, the f value obtained from the isonymy method is about twice as high as that calculated from the genealogy in the villages of Aicuña and Pagancillo, and that this difference goes up to 5 to 6 times for the other three villages. This difference seems not to be related with the absolute degree of inbreeding, since Aicuña has the highest and Pagancillo the lowest of the five values for village-mean inbreeding coefficient. Higher estimates of inbreeding using the isonymy approach have also been found by other workers (see Rogers, 1987).

The isonymy approach assumes that the surnames are monophyletic and that the variances of sibship size for male and female parents are equal. If the origin of any of the surnames is polyphyletic, or if the variance of sibship size for male parents is larger than that for female parents (Cavalli-Sforza and Bodmer, 1971), then the isonymy method will overestimate the level of inbreeding in the population. For the population of Aicuña, the assumption of a monophyletic origin of the surnames is almost certainly violated for the names Paez and Narvaez (see above). In addition the variance of sibship size for male parents is 52% larger than that for female parents.

The effect of unknown paternity on the two estimates of inbreeding, is less simple to evaluate. The overestimation of f by isonymy is greater in the three villages where the available pedigree data is less complete, that is, where the frequencies of unknown paternities are higher, namely Puerto Alegre, Paso de San Isidro and Los Palacios. It is important here to distinguish between the related phenomena of unknown paternity (where the father is recorded as unknown) and illegitimacy (where the biological father may be identified). A known, illegitimate paternity, which may be frequent when good pedigree data exist, will only affect the estimation of f by isonymy, since no paternal surname will be transmitted. The estimate of f obtained from the genealogy will obviously be unaffected. In contrast, unknown illegitimate paternities will result in incomplete genealogies thereby generating underestimates of the true level of inbreeding. It is important to bear in mind that unknown paternities are more likely to involve an outsider to the population. Consequently, most unknown paternities are outbred. They have the function of reducing the degree of genetic isolation, but at the same time, failing to bring new names into the name pool. Therefore as pedigree-estimated inbreeding is underestimated, isonymy-estimated f will be overestimated. Unknown illegitimate paternities would be expected to be more frequent in those villages where some information remains concealed because of less contact between the investigators and the population (Puerto Alegre, Paso de San Isidro, Los Palacios), resulting in a greater discrepancy between the two estimates of inbreeding.

Clearly, the degree of bias in the estimates of inbreeding will vary from population to population. However, these data suggest that estimates of inbreeding obtained from the isonymy approach are sufficiently unreliable to bring their usefulness into question. The quality of the data available appears to be an important factor determining the bias. For the estancia of Aicuña, the data are better for the village of Aicuña than for the other villages, better for recent generations than for ancient ones, and will naturally be better for nonemigrated than for emigrated family branches (though this will naturally have minimal effect on the population). In spite of their several drawbacks, names may still provide valuable insights into the genetics and demography of a population, even in Aicuña, where the pedigree data are outstanding.

GENETIC STRUCTURE AND MATE CHOICE IN AICUÑA

The average levels of inbreeding reported in Table 4-3 point to a wealth of information concerning the genetic structure of the villages. When the individual components of the mean coefficient of inbreeding for the village of Aicuña were analyzed, it can be seen that only 79 of the 430 inhabitants of Aicuña were not inbred ($f = 0$). Thus, the average level of inbreeding for the inbred sector of the population is even higher ($f = 0.051$). Within this sector of the population 22 individuals had an inbreeding coefficient greater than 0.125 (equivalent to offspring from uncle-niece and aunt-nephew matings), and 106 individuals had an inbreeding coefficient greater than 0.0625, the inbreeding coefficient of offspring of first cousins.

Such mean values, however, are not necessarily due to a high frequency of matings between close relatives, but to a very complex pattern of relationships. Figure 4-4 shows the proportional contribution of different component degrees of inbreeding, to the total inbreeding coefficient for the village of Aicuña. These are expressed as the number of paths of common ancestry with a given number of gametic steps, for each given loop going from each person in the population, through the pedigree, to a common ancestor, and back to the same person again (e.g., Crow and Kimura, 1970). The observed pattern is significantly different

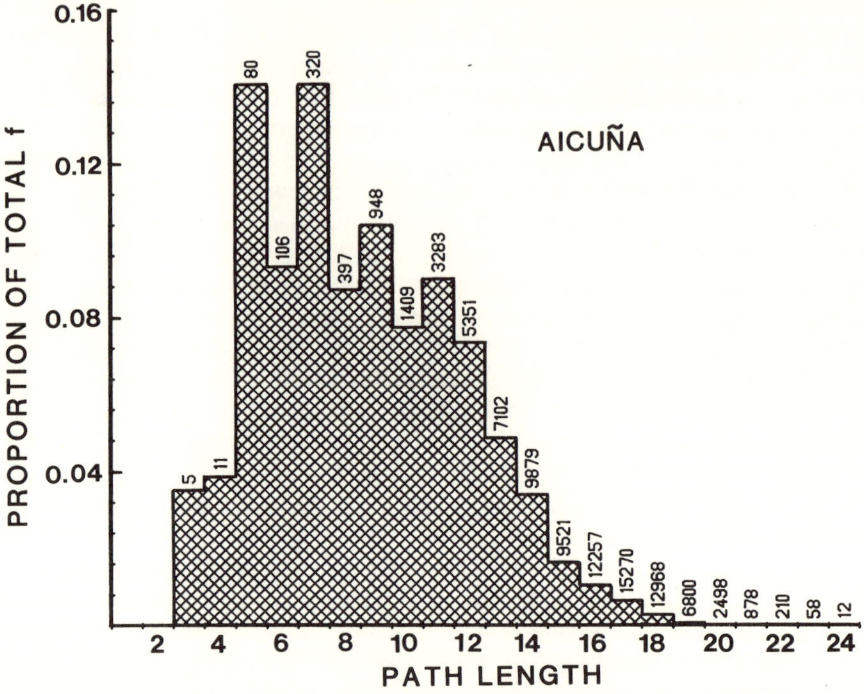

Figure 4-4. Contribution of different paths of relationship to the average inbreeding level of the village of Aicuña in 1971. Numbers on top of each histogram bar indicate the absolute number of pedigree loops of that given path length observed in the pedigree. For example, 80 instances of paths with a length of five steps (equivalent to first cousin marriage) account for approximately 14% of the total value of *f.*

from the theoretical expectation, assuming complete random mating, namely, that the proportional contribution of all paths should be similar and independent of their length (Cavalli-Sforza and Bodmer, 1971). The deviation of the results shown in Figure 4-4 from the theory is a consequence of the nonrandomness of consanguineous matings.

The complex histogram seen in Figure 4-4 can be best understood as having three defined sections. The section on the left, with paths of common ancestry of lengths three and four, correspond to traditionally discouraged incestuous and other matings, such as sib-sib, parent-child, and uncle-niece/aunt-nephew. Although products of such matings are highly inbred, overall they contribute less than 10% to the total inbreeding level. The central section, covering paths of common ancestry with lengths 5 to 12, contributes the major component of the total inbreeding, and displays a strong dentate profile. This can be explained by demographic constraints on the choice of mates within the population. It can be seen that within the range 5 to 12, the proportion of the total inbreeding contributed by paths of uneven length is greater than that for paths of even length, and that this difference becomes less for paths of greater length. Paths of uneven length correspond to ''symmetrical'' pedigrees, where there are an equal number of ancestors on the maternal and paternal sides of the pedigree which trace back to a common ancestor. For example, paths of length 5 correspond to first cousin matings, 7 to second cousin matings and so on. Thus, uneven path lengths represent consanguineous matings, where the parents can be said to be from the "same" generation. In contrast, even path lengths correspond to the transgenerational, or so called once-removed matings. The transgenerational availability of potential mates is clearly limited by age differences, making, in turn, this type of mate choice less likely (see also Chapter 5 by Leslie). As expected, this difference is less marked for more distant relationships, when the generations become more overlapping. Nevertheless, it is noteworthy that this phenomenon is strong enough to remain evident as far back as five generations, in a population such as this which practices little or no family planning. As a consequence, family sizes and reproductive spans are large, and differences between the generations would be expected to be quickly obscured.

Finally, the right part of the histogram, which corresponds to distant relationships, contributes much less to the total inbreeding. Two factors may determine this lower than expected contribution of remote consanguinity. (1) The increased probability of unknown paternities for the remote ancestors will result in many pedigree loops being undetected. (2) Immigration of unrelated individuals into the pedigree which occurred during the first generations, mainly in the eighteenth century, will reduce the contribution of remote ancestors to the total inbreeding level of the population.

The enormous complexity of this pedigree can be further illustrated by the fact that 55,081 loops with a path length of 14 steps or more, contribute approximately as much to the total inbreeding as the 16 incestuous loops, that is, about 10%. A path length of 14 corresponds to a consanguineous mating between fifth cousins once removed and, for an individual born to such parents in 1970, means that the common ancestors were born as early as in 1770. Such data are unavailable for most human populations studied, particularly those in Latin America.

The importance of remote consanguinity is more clearly seen in Figure 4-5, which shows the cumulative effect of paths of different length to the total inbreeding coefficient. For example, if only genealogical data for five generations or less were available (equivalent to a path length of eleven or less), corresponding to ancestors born around 1850, the total inbreeding coefficient would be underestimated by approximately 19%. Remote consanguinity has also been cited as an important factor contributing to the total inbreeding level in the Aland island parish of Sottunga, Finland (O'Brien *et al.*, 1988).

EVOLUTION OF INBREEDING IN THE VILLAGE OF AICUÑA

To determine changes in the level of inbreeding during the history of the population, it was necessary to estimate some missing birth, death, immigration and emigration dates, principally for individuals living during the early years of the existence of the population. Birth dates were estimated incorporating information from relatives located up to two generations removed from the propositus (individual with the missing birth date). The estimation procedure utilizes a

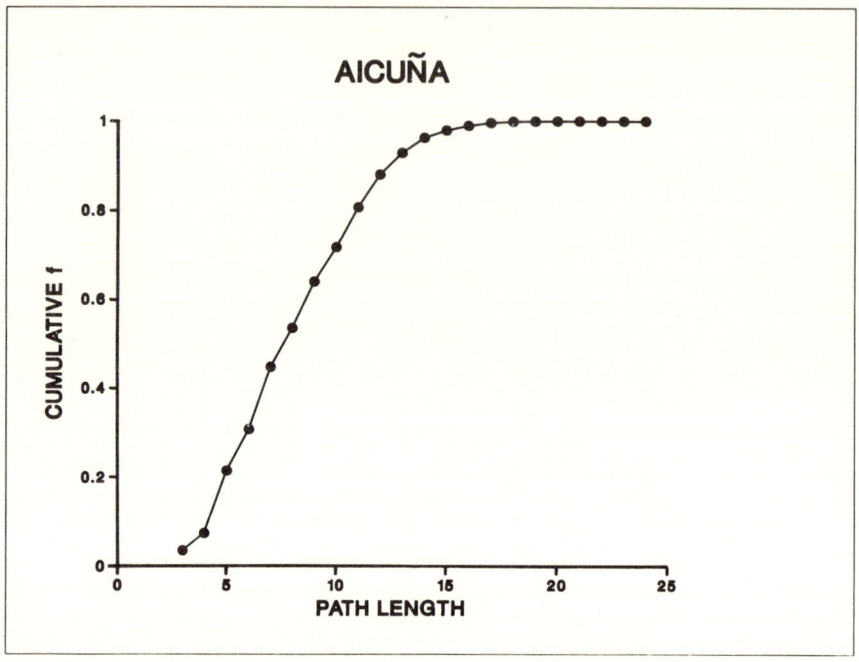

Figure 4-5.   Cumulative effect of paths of different length to the total inbreeding coefficient.

modification of a general pedigree analysis algorithm first developed by Elston and Stewart (1971) for the analysis for genetic data. This approach, which estimates birth dates using information from the entire population on the distribution of spouse age differences, the distribution of maternal age at first birth, and birth interval distributions dependent on maternal age, is described in detail in Lathrop and Adams (in prep.). Missing death dates were estimated using mortality schedules for the entire population, conditioning on information on life span given by offspring birth dates, emigration/immigration data and inclusion in censuses. Finally, missing immigration or emigration dates were estimated assuming a uniform distribution of the probability of immigration/emigration within limits set, where possible, by the incorporation of information from data on the birth places of offspring.

The estimated changes in the average inbreeding level for the population of Aicuña are shown in Figure 4-6, together with estimated changes in the population size. As expected, the mean inbreeding coefficient increased continually since the founding of the population. However, in contrast to the results for Tristan da Cunha (Roberts, 1980), the increase in the mean coefficient of inbreeding was not constant but rather displayed periods of stabilization and periods of rapid increase. For the period prior to 1900, population size was small enough that such fluctuations may be ascribed to random sampling effects. However, sampling effects cannot reasonably explain the fluctuations seen in the twentieth century, when population size was never less than 300 individuals. Of particu-

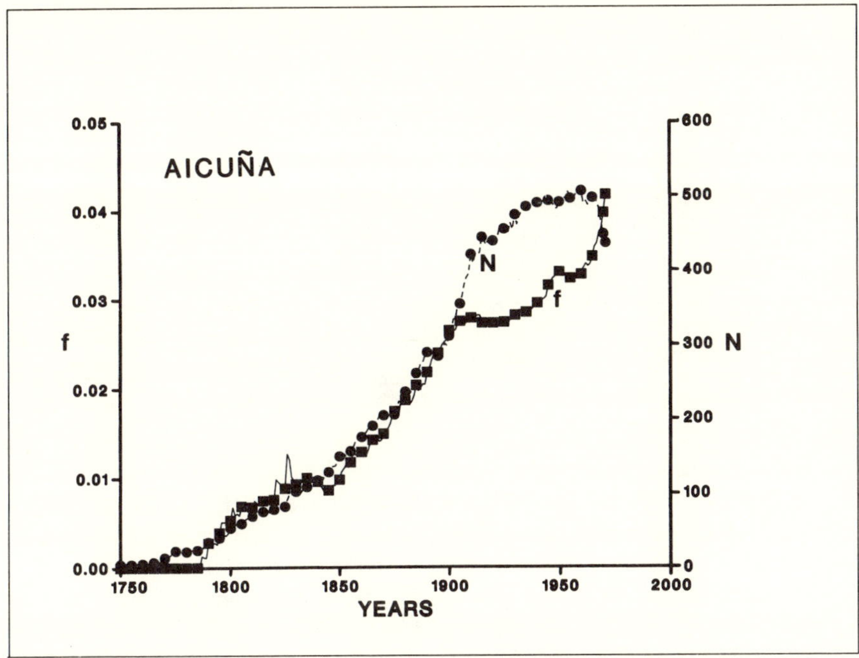

Figure 4-6. Change in the population size (*N*) and the mean inbreeding coefficient (*f*; probability of identity by descent) for the village of Aicuña.

lar interest is the relative stasis in the inbreeding levels followed by a rapid rise in the last two decades before the census in 1971.

In an attempt to explain the recent rapid rise in the inbreeding level, we have focused on the possible role of selective migration. Table 4-4 shows the average inbreeding level by decade, of non-emigrants and emigrants from the village of Aicuña during the twentieth century. It can be seen that the average inbreeding level of emigrants is significantly lower than non-emigrants for three of the seven time periods. The differences are greatest for the period since 1950, when inbreeding levels rose sharply, and the differences are nonsignificant, and the trend even reversed for the period between 1910 and 1950, during which time there was little change in the inbreeding levels of the village of Aicuña.

A higher probability of emigration of less inbred individuals may have at least two explanations. (1) If mortality and morbidity are under partial genetic control, as a number of reports have suggested (e.g., Freire-Maia and Elisbão, 1984), then a positive correlation will exist between mortality and morbidity on the one hand, and inbreeding on the other. Thus, individuals with a high inbreeding coefficient may be less likely to emigrate either by virtue of having a shorter life span or because of a higher degree of morbidity. (2) Alternatively, the results may be explained by purely sociological factors. Highly inbred individuals are more likely to be able to trace their ancestry back to more of the original founders than less or noninbred individuals, and therefore be landowners and have a guaranteed source of income and livelihood in the village of Aicuña. Consequently they will be less likely to emigrate. It is noteworthy that the period 1950-71 approximately coincides with a period of significant economic growth

Table 4-4 Selective emigration from the village of Aicuña

Period	Inbreeding Coefficient		Probability[a]
	of Emigrants	of Non-Emigrants	
1901--1910	0.0242	0.0288	0.013
1911--1920	0.0244	0.0296	0.042
1921--1930	0.0286	0.0279	NS
1931--1940	0.0331	0.0283	NS
1941--1950	0.0283	0.0340	NS
1951--1960	0.0271	0.0358	0.001
1961--1971	0.0239	0.0397	<0.001

[a]Probability of a difference between the inbreeding coefficients of emigrants and non-emigrants as determined using the Wilcoxon Mann-Whitney non-parametric test (e.g., Conover, 1971) after correcting for age and sex differences in the inbreeding levels. The tests were performed on the residuals from third degree polynomial regressions carried out separately for each sex.

in Argentina, when employment was readily available in the industrial centers such as Córdoba and Buenos Aires. Thus, there was a significant stimulus to emigrate during this period. In addition, a weekly bus service to the village was inaugurated in the early 1950s, facilitating the mechanics of emigration.

It should be emphasized that although the fluctuations in the rate of increase of inbreeding over time are consistent with selective emigration of less inbred individuals, other factors may also contribute to these changes. In addition, it may be difficult to extrapolate these findings to other populations. The nature of the relationship between the probability of emigration and inbreeding will depend on social and biological factors which can vary from population to population. Thus, the opposite relationship has been reported for the Northside population of the Caribbean Island of St. Bartholomew in 1966 (Leslie, Dyke, and Morrill, 1980). These workers explained the higher probability of emigration of more inbred individuals to the practice of avoidance of inbreeding, and the reduced number of available noninbred mates. The population size of Northside (661; Dyke, 1971) is the same order of magnitude as that of the village of Aicuña, but the estimated average inbreeding level is significantly lower.

CONCLUSIONS

Four main conclusions may be drawn from the analysis of the data presented here.

1. Analysis of migration within and between human populations can involve both genetic and demographic data. In the case of Aicuña, the village-to-village migration patterns derived from genealogical data correspond well with those based on demographic data. However, pedigree-based migration data may provide further information on long-term migration patterns, more significant than those between proximal generations, when long range historical data is being analyzed.

2. The mean inbreeding coefficient of the Aicuña population is based on a very complex pattern of multiple consanguineous matings. It is reasonable to assume that the same is also true for most inbred human populations. However, for most of them, only the closest consanguineous links may be measured due to limitations of the pedigrees available. In the case of Aicuña, the degree

of inbreeding would be significantly underestimated under those circumstances. We suggest that this will be true of other inbred populations.

3. Even in populations not practicing birth control, mating patterns are strongly conditioned by the availability of suitable mates. Therefore, the sex and age structure of a given population have to be taken into consideration whenever an observed breeding structure is to be compared with random models.

4. Finally, under certain circumstances, emigration may be conditioned by the degree of inbreeding. This observation questions the generally accepted concept that while immigration serves to break down genetic isolation, emigration does not influence it beyond the simple reduction of population size.

ACKNOWLEDGMENTS

We thank Peter Smouse for helpful criticism. This work was supported in part by grants from the National Science Foundation, the H.H. Rackham School of Graduate Studies, University of Michigan, the National Research Councils of Argentina (CONICET) and Brazil (CNPq), and the Pan American Health Office (PAHO).

REFERENCES

Bonné-Tamir, B., 1980   The Samaritans: an ancient living isolate, pp. 27-41 in *Population Structure and Genetic Disorders*, edited by A. W. Eriksson, H. R. Forsius, H. R. Nevanlinna, P. L. Wowkman, and R. K. Norio. Academic Press, New York.

Boyce, A. J., 1983   Computation of inbreeding and kinship coefficients on extended pedigrees. J. Hered. 74: 400-404.

Cavalli-Sforza, L. L., and W. F. Bodmer, 1971   *The Genetics of Human Populations*. 965 pp. Freeman, San Francisco.

Cavalli-Sforza, L. L., and A. W. F. Edwards, 1967   Phylogenetic analysis. Models and estimation procedures. Am. J. Hum. Genet. 19: 233-257.

Conover, W. J., 1971. *Practical Nonparametric Statistics*. John Wiley, New York.

Crow, J. F., and M. Kimura, 1970 *An Introduction to Population Genetics Theory*. 591 pp. Harper & Row, New York.

Dyke, B., 1971   Potential mates in a small human population. Soc. Biol. 18: 28-39.

Edwards, A. W. F., and L. L. Cavalli -Sforza, 1965   A method for cluster analysis. Biometrics 21: 362-375.

Elston, R. C., and J. Stewart, 1971   A general model for the genetic analysis of pedigree data. Hum. Hered. 21: 523-542.

Freire-Maia, N., and T. Elisbão, 1984   Inbreeding effect on morbidity. Am. J. Med. Genet. 18: 381-400.

Gottlieb, K., 1983a   Arranger. Surnames as markers of Inbreeding and Migration: A Conference. Hum. Biol. 55: 209-408.

Gottlieb, K., 1983b   Genetic demography of Denver, Colorado: Spanish surname as a marker of Mexican ancestry, in *Surnames as markers of Inbreeding and Migration*. A Conference arranged by K. Gottlieb. Hum. Biol. 55: 227-234.

Hollingsworth, T. H., 1964   The demography of the British Peerage. Pop. Studies 18, Suppl.: 1-101.

Hostetler, J. A., 1974   *Hutterite Society*. 403 pp. Johns Hopkins University Press, Baltimore.

Jacquard, A., 1974   *The Genetic Structure of Populations*. 569 pp. Springer Verlag, New York.

Jorde, L. B., 1985   Human genetic distance studies: present status and future prospects. Annu. Rev. Anthropol. 14: 343-373.

Leslie, P. W., B. Dyke, and W. T. Morrill, 1980   Celibacy, emigration and genetic structure in small populations. Hum. Biol. 52: 115-130.

MacLean, C., 1969   Computer analysis of pedigree data, pp. 82-86 in *Computer Applications in Genetics*, edited by N. E. Morton. University of Hawaii Press, Honolulu.

Morton, N. E., R. Kennett, S. Yee, and R. Lew, 1982   Bioassay of kinship in populations of middle-eastern origin and controls. Curr. Anthropol. 23: 157-167.

Nei, M., 1972.   Genetic distance between populations. Am. Nat. 106: 283-292.

O'Brien, E., L. B. Jorde, B. Rönnlöf, J. O. Fellman, and A. W. Eriksson, 1988   Inbreeding and genetic disease in Sottunga, Finland. Am. J. Phys. Anthropol. 75: 477-486.

Roberts, D. F., 1980   Genetic structure and the pathology of an isolated population, pp. 7-24 in *Population Structure and Genetic Disorders*, edited by A. W. Eriksson. Academic Press, New York.

Rogers, L.A., 1987   Concordance in isonymy and pedigree measures of inbreeding: the effects of sample composition. Hum. Biol. 59: 753-767.

Salzano, F. M., and S. M. Callegari-Jacques, 1988   *South American Indians: A Case Study in Evolution*. 259 pp. Oxford University Press, New York.

Ward, R. H., and J. V. Neel, 1970   Gene frequencies and genetic differentiation among the Makiritare Indians. IV. A comparison of a genetic network with ethnohistory and migration matrices; a new index of genetic isolation. Am. J. Hum. Genet. 22: 538-561.

5

Demographic Behavior, Mating Patterns, and the Distribution of Inbreeding

PAUL W. LESLIE

Little is known about the distribution of inbreeding in natural populations, yet this distribution may be important to some problems in genetic epidemiology and evolutionary genetics. A considerable amount of theoretical work addresses the problem of inbreeding distributions, but the models pertain to simplified mating patterns and demographic structures. One of the salient features of our species is the degree of variability among populations in genetically relevant demographic and mating behavior—that is, behavior reflected in migration, reproductive performance, or rules and preferences concerning mate choice—that has an influence on the population's genetic structure and thereby on the constitution of subsequent generations.

For most of history, humans have lived in small groups in which genealogical relationships play a profoundly important role in structuring behavior. In such populations, behavior that is nonrandom with respect to kinship, such as incest avoidance, monogamy, or kin-structured migration, may have a strong influence on genealogies and inbreeding. But the frequencies of consanguineous matings expected with random mating, and therefore the impact of mating patterns on the distribution of inbreeding, are very much functions of demographic processes. The expected distribution of inbreeding is shaped by the biological and demographic characteristics of the population (e.g., its fertility, mortality, and the timing and duration of the reproductive span). Overlaid on these expectations are the effects of mating behavior that is nonrandom with respect to kinship. The result is the observed distribution of pedigree inbreeding.

In this chapter the distribution of inbreeding as an aspect of genetic structure in human populations, emphasizing the consequences of demographic patterns and mating behavior for that distribution.

THEORETICAL BACKGROUND

Related individuals share at least one ancestor in common. The possibility arises that a gene present in a single copy in a common ancestor will be inherited by two relatives who, if they mate, may each pass on that copy to a given offspring. The offspring is then said to be "autozygous"; that is, he has inherited two alleles that are identical by descent. Inbred offspring may have a greatly increased likelihood of inheriting a double dose of an allele that is otherwise rare in the population. This is why inbreeding is associated with increased risk of recessive, often deleterious, genetic conditions.

At the population level, inbreeding statistics conveniently summarize much of the effect of mating structure and migration patterns on genotype distributions, and are consequently central to studies of genetic structure. Most studies focus on levels of inbreeding, and use mean coefficients as fundamental measures. This fails to take into account the fact that inbreeding is not uniformly distributed among individuals within a population.

The probability that an individual carries two alleles that are identical by descent at any one locus is f, the individual's inbreeding coefficient. The proportion of the population that is homozygous by descent (autozygous) at any one locus is given by the mean inbreeding coefficient, $F = \sum_i k_i f_i$, where k_i is the proportion of the population with an inbreeding coefficient of f_i.

For single loci, mean inbreeding is sufficient to predict the effect of mating structure on genotype distributions and (if the effects of the genotypes on the phenotype can be specified) on phenotypic variance. The genotype frequencies for a locus with two alleles (A and A') with frequencies p and q, in a population with mean inbreeding coefficient F, are given by

$$AA, \quad p^2 + pqF; \quad AA', \quad 2pq(1 - F); \quad A'A', \quad q^2 + pqF . \qquad (5\text{-}1)$$

Inbreeding increases the proportions of homozygotes at the expense of heterozygotes.

The probability of autozygosity is the same for all loci. In the absence of linkage, then, the probability that an individual is homozygous by descent at each of two particular loci is f^2, the joint probability of independent events. The proportion of the population that is autozygous at both loci is then

$$E(f^2) = \sum_i k_i F_i^2 = F^2 + V_f , \qquad (5\text{-}2)$$

where V_f is the variance of individual inbreeding coefficients.

Simultaneous consideration of more than two loci involves yet higher moments about the mean. For example, the proportion of the population that is autozygous at each of three particular loci is given by

$$E(f^3) = E(f - F)^2 + 3E(F)E(f - F)^2 + [E(F)]^2 , \qquad (5\text{-}3)$$

which includes the third moment about the mean as well as the variance. As the number of loci considered increases relative to the number of chromosomes, the likelihood that two or more of the loci will lie on the same chromosome also increases. Linkage relationships will then also influence the distribution of inbreeding.

It is easy to show that the two-locus genotype frequencies, for unlinked loci, take the following forms:

Double homozygotes: $Pr(AABB) = (p^2 + pqF)(r^2 + rsF) + pqrsV_f$

Double heterozygotes: $Pr(AA'BB') = 4pqrs\left[(1 - F)^2 + V_f\right]$

$$(5\text{-}4)$$

Heterozygote - homozygote combinations:

$$Pr(AA'BB) = 2pq(1 - F)(r^2 + rsF) - 2pqrsV_f \qquad (5\text{-}5)$$

where p and q are the frequencies of the alleles at the first locus (A and A'), and r and s are the frequencies of those at the second (B and B'). Thus, formulae for multilocus genotypes contain variance terms. Just as an increase in mean inbreeding elevates the proportion of individuals whose alleles at any given locus are both of the same state (homozygotes), an increase in the variance of inbreeding raises the proportion of individuals who have both of two loci in the same state (double heterozygotes as well as double homozygotes).

In a finite, randomly mating population, some matings between relatives are expected to occur by chance. The results of such matings is random inbreeding. These chance consanguineous matings also produce variation among individual inbreeding coefficients (call this V_r, the random variance; estimation of this quantity is discussed below). To assume uniform inbreeding ($V_f = 0$) in a finite population is therefore to assume less variation than would be expected by chance. When we are concerned with more than one locus at a time, models that ignore variance of inbreeding in effect assume a deviation from random mating.

If the actual frequencies of consanguineous matings do in fact differ from random expectations, there will also be a nonrandom component to the total inbreeding, and total inbreeding can be either greater or less than what is expected from random inbreeding alone. Similarly, mating patterns can also result in variation among individual inbreeding coefficients that is either greater or less than the variation expected with random mating.

If the allele frequencies refer to the founding generation, and the genealogy reaches back to that generation, then F and V_f in the equations are the mean and variance of total pedigree inbreeding. In finite populations, the change in double homozygosity is then due to both random and nonrandom inbreeding. The effect of nonrandom mating in any given parental generation can be estimated by comparing the distribution of pedigree inbreeding observed in the offspring generation with that expected if the parental generation had been paired at random with respect to kinship. The two distributions are then the products of identical population histories, differing only in the mating pattern seen in the generation of interest.

The relationships described by the genotype frequency formulae above were worked out 30 years ago by Haldane (1949), and extended by Denniston (1975). Since then, several geneticists have studied related problems, such as the effects of linkage and mating type on the variation among individuals in the proportion of genes which are identical by descent (e.g., Schnell, 1961; Franklin, 1977). Others have considered the variance of inbreeding due to pedigree differences among individuals in a population (e.g., Bennett and Binet, 1956; Cockerham and Weir, 1983; Weir, Avery, and Hill, 1980; Stam, 1980). These studies, however, deal with very simple mating systems (usually one specific type of mating), with differences arising by chance within the context of random mating, or random mating with the exclusion particular matings (selfing, sib mating). Nonuniform inbreeding in populations with complex mating structures, such as those found among humans, has not received much attention. Human mating patterns combine complete proscription of certain matings with tendencies to avoid or favor other types, defined either by specific genealogical relationship, or by broader categories such as clans or moieties.

The influence of demographic and mating behavior on the distribution of inbreeding is relevant to some problems in genetic epidemiology and evolutionary genetics because of their relationship to the variance of inbreeding, which alters multilocus genotype frequencies. For example, the efficiency of natural selection is subject to modification by the variance of inbreeding. In the case of selection for or against recessive alleles, nonrandom inbreeding alters the efficiency of selection by increasing or decreasing the proportions of recessive homozygotes that may be susceptible to selection. For a single locus, the effect of mating structure can be summarized in terms of mean inbreeding. For traits dependent on more than one locus, especially when the loci have nonadditive effects (epistasis), observed variation in the trait (phenotypic variance) will be affected not only by mean inbreeding, but also by the variance of the distribution of individual coefficients. Recessive alleles at a single locus can be protected from the action of selection by dominant alleles; the effects of homozygous loci within multilocus systems can be masked by the state of other loci. A mating structure that produces a high variance of inbreeding will increase the incidence of multiple homozygosity and may thereby increase the efficiency of selection.

Traits whose expression depends on simultaneous homozygosis (or any other specific constellation of genotypes) at several loci represent an extreme form of epistasis. Such traits tend to appear as isolated cases, and are not likely to exhibit inheritance patterns typical of either single locus traits or quantitative traits resulting from many loci with additive effects. Li (1987) presents a genetic model, based on multiple homozygosis, for the inheritance of such ''emergenic'' traits in randomly mating populations, and the model has already been shown to fit at least one human genetic disorder (vitiligo, a dermatological condition; Majumder, Das, and Li, 1988). Study of the evolution of similar traits, or of their inheritance pattern in inbred populations, will demand consideration of the inbreeding distribution.

THE DISTRIBUTION OF PEDIGREE INBREEDING

We know very little about the distribution of inbreeding in human populations. What range of values does the variance take? How much does the observed variance deviate from the random variance?

The variance of individual inbreeding coefficients may be calculated from the pedigrees of individuals born in a given generation or cohort, giving a measure of variability in total inbreeding, V_t. However, variances of individual coefficients are rarely reported along with the means, so we must resort to less direct estimates. The coefficient of kinship between two individuals is defined as the probability that two genes, one chosen at random from a given locus in each of those individuals, are identical by descent. Ignoring the effects of mutation, this measure of relationship is equivalent to the inbreeding coefficient of offspring (real or hypothetical) of the two individuals. If fertility does not vary with the consanguinity of the parents, the coefficient of kinship between mates can serve as a proxy for the inbreeding coefficient in the above equations, and the total or observed variance can be estimated from these coefficients as

$$\widehat{V}_t = \sum_i k_i \phi_i^2 - \Phi_t^2 \,, \tag{5-6}$$

where k_i is the proportion of mated pairs who are related with kinship coefficient ϕ_i, and Φ_t is the mean coefficient for all actual pairs.

The random variance can be estimated from the kinship coefficients of randomly paired potential mates (opposite-sexed people who, on demographic grounds, were appropriate mates; Leslie, 1985) rather than from actual mates. Although the frequencies of specific consanguineous marriages are commonly reported, few studies provide the information about potential mates needed to estimate the random variance as well.

Table 5-1 shows the means and variances of kinship coefficients for several populations. The values associated with total or observed inbreeding, Φ_t and V_t, are based on the actual frequencies of consanguineous matings reported in the studies. The random means and variances, Φ_r and V_r, were estimated from the frequencies of consanguineous matings expected if mates were paired at random, but the method used to estimate these frequencies differs from population to population. For the Spanish population, expected frequencies were generated by a mathematical model (discussed below). Those for the Japanese population were produced by a computer simulation by MacCluer and Schull (1970). In the remaining cases, expected frequencies were generated by randomly pairing each male who was of mating age during a specified period with a sample of his potential mates. Criteria for eligibility as mates here include appropriate sex and age, and, in some cases, geographical proximity.

The proportionate effects of observed mating patterns on the mean and variance also appear in the table. In all but one of the cases listed, the consanguinity between actual mates is less than would be expected under random mating. In all cases, the actual variation in inbreeding is less than it would be under random mating.

Thus, the direction of the effect of mating patterns appears to be consistent, but there is some variation in the absolute and relative magnitudes of the effect on the mean and variance. For example, the mean and variance for Errazu are both greatly diminished, but on the island of Sanday (early period) the observed consanguinity deviates little from that expected under random mating, while the variance is reduced by half.

For two reasons, even the apparent consistencies do not warrant too much generalization. The first reason is methodological. The figures for most of the populations in Table 5-1 are based on the frequencies of specific types of matings reported in the studies cited. Two individuals may be related in more than one way, but the primary (closest) relationship is usually the only one reported and reflected in the computations. Distant relationships (remote consanguinity)

Table 5-1 Observed and expected means and variances of kinship coefficients for several populations. Means and variances have been multiplied by 1×10^3

Population[a]	Φ_t	Φ_r	Effect[b]	V_t	V_r	Effect[b]
St. Barthélemy, F.W.I. 1960--1969	5.0	7.6	−.34	0.13	0.84	−.85
St. Barthélemy, F.W.I.[c]	9.0	8.2	.10	0.16	0.65	−.75
Sottunga, Aland Islands, 1850	2.5	4.4	−.43	0.05	0.15	−.67
Errazu, Spain	1.8	9.6	−.81	0.80	1.56	−.95
Hirado, Japan	4.1	6.0	−.32	0.21	0.87	−.76
Sanday, Orkney Is., 1880--1889	2.2	2.3	−.04	0.11	0.28	−.61
Sanday, Orkney Is. 1950--1959	0.8	2.0	−.60	0.02	0.32	−.94
St. Thomas, U.S.V.I.	7.9	10.5	−.26	0.37	1.29	−.71
Utah Mormons, 1936[d]	0.1	0.2	−.55	0.004	0.03	−.86

NOTES:

[a] Sources of data: St. Barthélemy - Leslie (1980), Leslie *et al.* (1981); Sottunga - O'Brien *et al.* (1989); Errazu - Abelson (1978); Hirado - MacCluer and Schull (1970); St. Thomas - Dyke 1971; Sanday - Brennan *et al.* (1982); Utah Mormons - L. Jorde, unpublished data.

[b] Proportionate effect of nonrandom mating in parental generation on mean and variance of inbreeding expressed as $(\Phi_t - \Phi_r)/\Phi_r$ and $(V_t - V_r)/V_r$. Values were calculated before rounding means and variances for presentation.

[c] Based on multiple relationships.

[d] Figures are averages of within-subdivision kinship and inbreeding for 22 geographic subdivisions (Mormon "Stakes").

are usually not reported at all. If remote consanguinity contributes equally to the coefficients of all pairs in a population, it will decrease the variance in relatedness between mates. This expectation has both theoretical (Leviandier and Jacquard, 1974; Cockerham and Weir, 1983) and empirical (Ward *et al.*, 1980) support, but it cannot be assumed to hold in all cases. For example, the variance of pedigree inbreeding coefficients in the population of Tristan da Cunha *increased* during the past century (Roberts, 1980). Furthermore, kin-structured migration (e.g., movement of a lineage or other group of relatives from one village to another) might prevent the attaining of a uniform distribution of remote consanguinity. So might differences in mating patterns, or asymmetries in migration between subpopulations in a partially subdivided population.

We can test the effect of more extensive genealogical information in at least one of the cases listed. The second set of entries for St. Barthélemy (St. Barth) is based on all relationships detectable in the genealogy constructed for this population, which is nearly complete for the five or six generations preceding the period of analysis (Leslie, Dyke, and Morrill, 1981). Comparison of the means and variances for the two St. Barth entries gives an indication of the effect of multiple relationships and some remote consanguinity on the inbreeding distribution. Not surprisingly, the added genealogical information increases the mean and decreases the variance of random kinship. However, it reverses the apparent effect of the population's mating patterns on mean consanguinity (the observed mean becomes greater than the expected mean), but does little to the effect on the variance. Why the ascertained level of consanguinity changes in this way is not clear. Perhaps the strong tendency in St. Barth to choose mates from one's own neighborhood, and to settle there, makes remote consanguinity between actual mates more likely than between potential mates drawn from a wider area. It may also be that two related individuals are more likely to mate if they share secondary relationships in addition to the primary one.

The second reason for caution in generalizing from these results is that the populations considered here are not at all representative, in terms of social organization, economics, or mating patterns, of the full range of human societies. Similar analysis of populations characterized by polygyny, clan exogamy, village fission along lines of descent, or other common phenomena, might yield rather different results.

PROXIMITY AND POPULATION BOUNDARIES

The genetic structure of subdivided populations is often analyzed in terms of Wright's (1965) hierarchical F-statistics model, by comparing observed genotype distributions within and among subdivisions to panmictic expectations. Estimates of F_{IS} (a measure of nonrandom inbreeding within a subdivision) and F_{ST} (a measure of the effects of subdivision that can be equated with random genetic drift; see also, Piazza, Chapter 6) are sensitive to the demarcation of subdivisions (Neel and Ward, 1972; Jorde, 1980; Long, 1986). A genealogical approach to population structure to a certain extent obviates this problem. The observed pedigree is the product of past mating structure and is not subject to the effects of inadvertently lumping or splitting breeding units. However, the definition of population boundaries remains a problem if we wish to determine the consequences of mating structure by comparing the observed distribution of kinship coefficients with that expected under random mating. The latter requires defining pools of potential mates, which obviously entails establishing population boundaries.

Table 5-2 gives some idea of the effect of changing the boundaries used in defining potential mate pools for the St. Barth population. The figures also illustrate the effect on the inbreeding distribution of the tendency for people to choose mates who were born nearby. The values for St. Barth in Table 5-1 are the result of imposing the criterion that potential pairs must have been born within 2,600 meters of one another. This distance was chosen because nearly 95% of all married couples on St. Barth were born less than this distance apart. Despite the fact that St. Barth is small (about 25 square kilometers), restricting the boundaries of potential mate pools to 2,600 m greatly increases both the mean and variance of expected consanguinity. The island is divided into a dozen neighborhoods or quarters. More than 90% of marriages involve people born in the same or adjacent quarters; nearly 75% of all mates were born in the same quarter. The consequences of choosing mates from these increasingly restricted pools are clear in Table 5-2. Comparison of the means expected for potential mates with that for actual mates shows that, if observed matings were interpreted as having been drawn from island-wide potential mate pools, actual consanguinity would be greater than random consanguinity (that is, nonrandom inbreeding would be positive). With potential mate pools defined to reflect actual marital movement more realistically, nonrandom inbreeding is only slightly positive, or even negative, indicating overall consanguinity avoidance. On the other hand, the variance of kinship coefficients among potential mates is in all cases greater than the observed variance.

Table 5-2 Mean and variance of kinship coefficients between potential mates on the island of St. Barthélemy, F.W.I.

| | Birthplace of Potential Pair | | | | |
	Anywhere on Island	Within 2600 Meters	Same or Adjacent Quarter	Same Quarter	Actual Mates
Mean	0.0049	0.0082	0.0116	0.0308	0.0090
Variance	0.0004	0.0007	0.0010	0.0034	0.0002

DEMOGRAPHIC EFFECTS

Mating that is nonrandom with respect to kinship is by no means the sole determinant of the inbreeding distribution. This is clear from the range of values for V_r, even in the culturally rather homogeneous sample in Table 5-1. Differences among these populations are due at least in part to demographic differences, past and present. It is of interest to probe the effects of demographic structure and behavior on the underlying expectations. The demographic factors that should most strongly influence the expected distribution of inbreeding are, first, those that affect the numbers of the various types of relatives present (e.g., mortality, migration), and second, those that affect the probability that a particular pair of relatives will in fact mate.

We can explore the effects of various demographic characteristics on the expected distribution of inbreeding with a mathematical model that predicts the frequencies of consanguineous matings in age-structured populations, where the chance of mating between a couple depends on their age difference. Equivalent versions of this model were developed by Hajnal (1963) and Cavalli-Sforza, Kimura, and Barrai (1966). The model has been extended in several ways (Leslie, 1983), the most important of which here allows the generation of probabilities of the various mating types when mating is random with respect to age, constrained only by the limits of the reproductive spans. This allows examination of the consequences of age correlation between mates (see below).

At the core of the model is the relationship between two distributions: that of age differences between mates, and that of age differences between relatives of a given type. The latter is a function of several other distributions: those of age at maternity, age at paternity, and age difference between siblings. The model is based on the notion that the greater the overlap between the distribution of age difference between mates and the distribution of age difference between relatives of a given type, the greater the probability of that type of mating. For example, if all males and females tend to reproduce at about the same age (that is, there is a low variance in the distributions of ages at maternity and paternity), first cousins will tend to be close to one another in age. This will make first cousin matings more likely than would be the case if reproductive behavior were more variable, which would result in many cousins being quite different in age. The same is true for other even degree relationships (those with equal numbers of generations separating each of the related pair from their common ancestor, as with second or third cousins). On the other hand, relatives of odd degree (for example, uncle-niece or first cousins once removed) tend to be less similar in age, and are therefore less likely to be mates. The likelihood of each type of mating must then be adjusted to account for the prevalence of relatives of each type, which is itself a function of several demographic parameters (for example, sex-specific migration rates).

This model has several deficiencies when applied to problems concerning the genetic structure of small populations. One is that it provides for people being related in only one way. That is, we are limited to the analysis of primary relationships, and must assume that remote consanguinity contributes equally to the degree of consanguinity of all relatives (see above). Another is that it is based on the assumptions of stable population theory, and ignores the effects of stochastic fluctuations that may be important to the demographic and kinship structures of small populations (see MacCluer and Dyke, 1976). Nevertheless, the model is convenient and suitable for generating qualitative predictions about the effect of certain demographic variables on the distribution of inbreeding.

Except where noted, parameter values for the crucial distributions in the model were set as follows: mean and variance of age at paternity, 35 and 70, respectively; of age at maternity, 30 and 40; of age difference between spouses, 5 and 30; and the variance of intervals between births of all pairs of siblings, 55. These values were suggested by Hajnal (1963) as being representative of a number of natural fertility populations, and are adequate for our purposes here. We take as a starting point a stationary population with a moderately high level of inbreeding, where people are related on average as second or third cousins. We then vary selected demographic characteristics and note the consequences.

Migration

Here, migration is modeled as a certain proportion of each generation being replaced, before mating age, with unrelated people of the same age and sex. Figure 5-1 shows the effect of migration on the inbreeding distribution expected if mates are chosen at random with respect to kinship. The mean and variance are expressed as proportions of the values predicted for a similar population with no migration.

Migration decreases mean inbreeding because it introduces unrelated potential mates into the population, and because it separates related pairs. The variance also decreases with the migration rate, but the explanation is perhaps less obvious. Migration reduces the probability of all types of consanguineous matings, but it affects more distant relationships much more strongly than close ones (the more generations that are needed in the pedigree to establish the relationship, the greater the chance that the genealogical link will be broken by migration). In the case of migration with replacement, the increase in unrelated potential mates is by far the strongest influence, and the variance decreases. If, however, migration is modeled as emigration without replacement, the greater reduction of more distant relationships makes matings at the extremes of the distribution (close relatives and unrelated pairs) proportionately more likely. The variance may then increase.

In many populations, male and female migration patterns differ markedly. Differential sex-specific migration rates do affect the frequency of some specific relationships more than others (depending on the numbers of males and females in the genealogy defining the relationship). However, the distribution

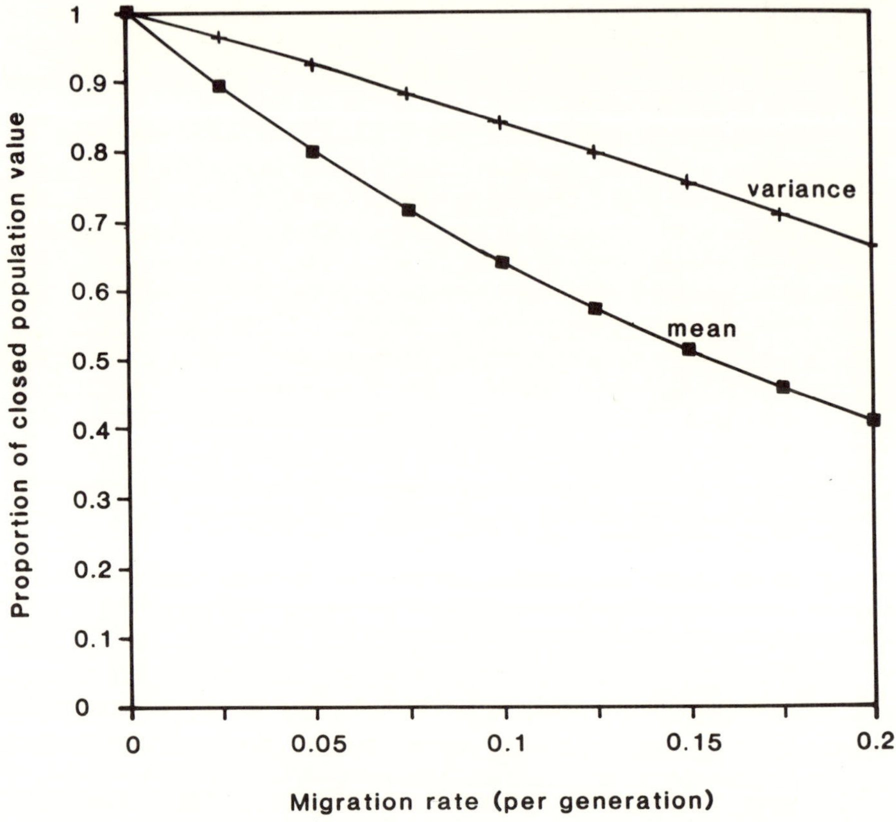

Figure 5-1. Effect of migration on the mean and variance of inbreeding.

of expected consanguinity predicted by the model when males and females are assigned different migration rates is virtually identical to that predicted when the mean of the two rates is applied to both sexes.

There are several factors that may complicate the effects of migration on genotype distributions. For example, different allele frequencies in male and female parents result in an excess of heterozygotes. Unequal sex-specific migration rates, or asymmetrical migration among subpopulations with allele frequency differences, may produce such excesses (Smouse, Neel, and Liu, 1983), which would complement the effect of reduced inbreeding associated with migration. On the other hand, migration is often (perhaps usually, historically) not random with respect to kinship. Migrants may consist of groups of related individuals (Neel and Salzano, 1967; Fix, 1978, Castilla and Adams, Chapter 4), or may have more relatives in the sending population than do nonmigrants (Leslie, Dyke, and Morrill, 1981; Leslie, 1980). Migration may thus work either to increase or decrease the level of inbreeding and rates of genetic differentiation, relative to what would be expected with random migration. The effect of migration on the inbreeding distribution may be qualitatively different when it is structured by kinship.

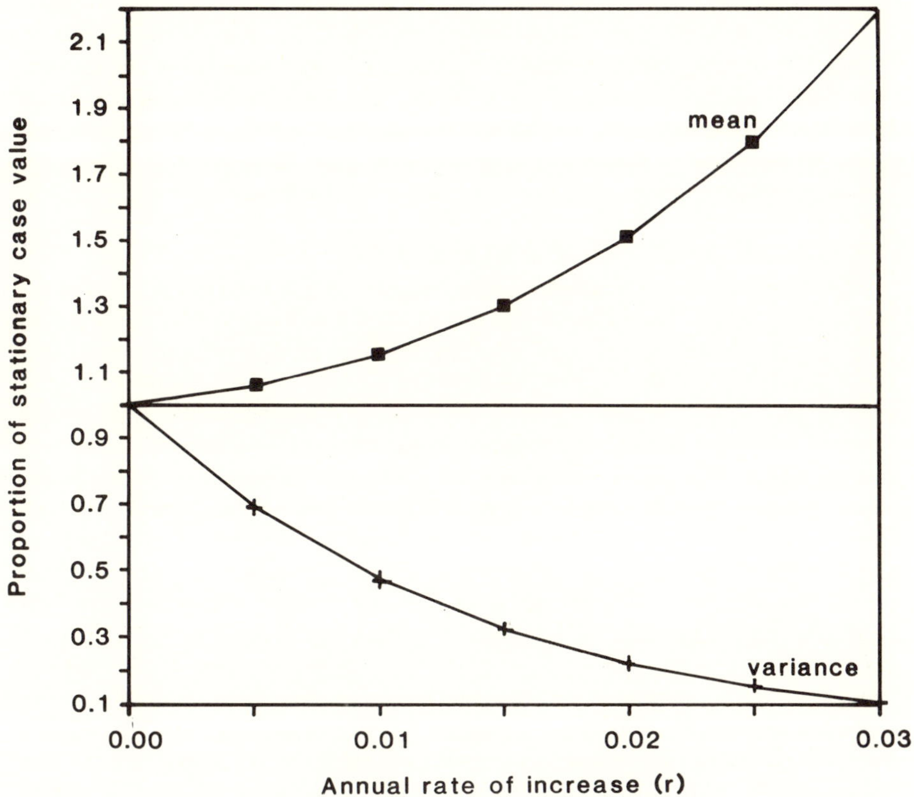

Figure 5-2. Effect of population growth on the mean and variance of inbreeding.

Population growth

It is well known that the rate of increase of random inbreeding (or random genetic drift) declines as a population grows. Put another way, there will be fewer close consanguineous matings expected in generation t if the population grows from time 0 to time t than if it remained at the initial size. But often we wish to interpret the genetic structure of a contemporary population, without having full knowledge of its history. What is the effect on the expected inbreeding distribution of assuming that a population has been at its current size for some generations, rather than having attained that size through natural increase (not through immigration)? As shown in Figure 5-2, population growth increases the mean and decreases the variance of inbreeding. This is because, with growth, each generation is descended from a smaller previous generation. Consequently, the overall frequency of consanguineous matings increases. But the more distant relationships, which are defined by common descent from smaller, earlier generations, are affected more than the closer relationships. The distribution becomes dominated by the more distant relationships, and the variance is reduced.

Age correlation between mates

Population growth and migration can also have some interesting interactions with other aspects of demographic behavior. One example is the effect of age correlation between mates on the inbreeding distribution. The tendency for mates to be close in age has a dramatic effect on the likelihood of specific consanguineous matings. For example, first cousin matings may be more than twice as frequent as would be the case if there were no age correlation between mates (Cavalli-Sforza, Kimura, and Barrai, 1966).

Figure 5-3 shows the overall effect of age correlation between mates on mean inbreeding. The effect is expressed as the ratio of the mean expected with age correlation to that expected if mating were random with respect to age, constrained only by the limits of male and female reproductive spans (set here as ages 15 and 45 for females, 15 and 70 for males). The parameters in the model were set as above, but the mean age difference between mates was varied over a wide range. The effect varies with the mean age difference between mates, being greatest when the mean is zero and falling off in both directions. With larger age differences, odd-degree matings become more frequent, but do not make up for the loss of even-degree matings. This is because the probability that odd-degree relatives are both within the limits of their reproductive spans (and therefore eligible as mates) is less than the probability that even-degree relatives are both within those spans (see Castilla and Adams, Chapter 4). By way of illustration, first cousins twice removed have the same kinship coefficient as second cousins. When mating is random with respect to age in a population with the demographic parameters noted, matings between second cousins are expected to be more than three times as frequent as matings between first cousins twice removed. The difference is accounted for by the difference in overlap of mating spans—first cousins twice removed are likely to have been born many years apart. Imposing age preference on the mating pattern (mean and variance of age difference between mates equal to 5 and 30, respectively) nearly doubles the expected frequency of matings between second cousins, but reduces that for first cousins twice removed by nearly two orders of magnitude (see Leslie, 1983).

The compensatory effects on even- and odd-degree matings are not complete, but the overall effect of age correlation on mean inbreeding tends to be small or moderate. The effect of age correlation on the variance of inbreeding (not shown) follows a similar pattern, but is a bit greater; the variance may be increased by 50% or more. Note, though, that the magnitude of the effect of age correlation differs depending on whether there is population growth or migration (curves G and M in Figure 5-3). Even though migration with replacement reduces inbreeding (see above), the effect of age correlation on mean inbreeding may be greater when there is migration. This is because there is a variance attached to the ages at parenthood of each ancestor, so the more distant the relationship, the greater the variance in the age difference between the couple. Migration has a disproportionate effect on more distant relationships, and therefore tends to eliminate those relationships that obscure the effect of age correlation.

Thus far we have considered the effect of certain demographic variables on both the mean and variance of inbreeding. However, the focus has been on

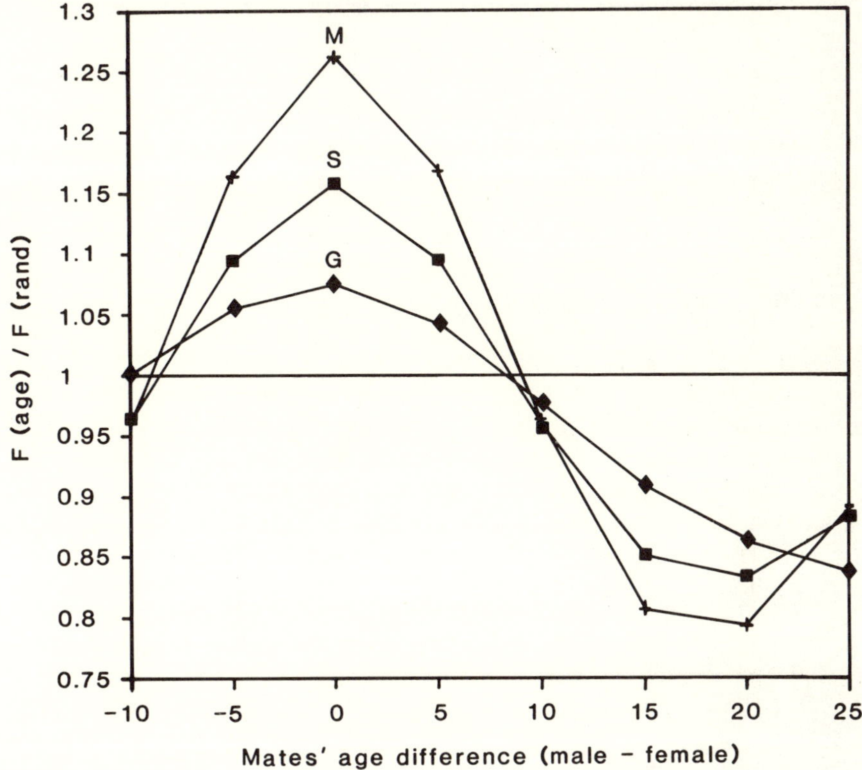

Figure 5-3. Effect of age correlation between mates on mean inbreeding. S - stationary population. M - 20% migration per generation. G - growing population, annual rate of increase $r = .025$.

the *level* of those demographic variables. It is appropriate here to consider also consequences of *variability* in demographic behavior. For example, it is expected that the more uniform reproductive behavior is in a population (that is, the smaller the variances of age at marriage, age at maternity, number of offspring, etc.), the smaller will be the variance of age difference between relatives of a given type. Thus, the effect of age correlation between mates on the inbreeding distribution depends not only on the mean age difference, discussed above, but also on the variance of that difference. Figure 5-4 and Figure 5-5 show how the impact of age correlation between mates changes, in a stationary population, if the variance of age difference between mates is higher or lower than the value (30) set for Figure 5-3. Values of all other parameters were unchanged. As expected, the less variability there is in this aspect of mating behavior, the greater the effect of age correlation. For example, Figure 5-5 shows that if the mean age difference is zero and the variance is 10, then the variance of inbreeding expected with age correlation between mates is 70% greater than it would be with random mating; if the variance of age differences between mates is greater (90), then the effect of age correlation on the variance of inbreeding is only 30%. Again, for the same reasons discussed above, the effect depends on the mean age difference, and for some values of the mean difference, the effect is very small.

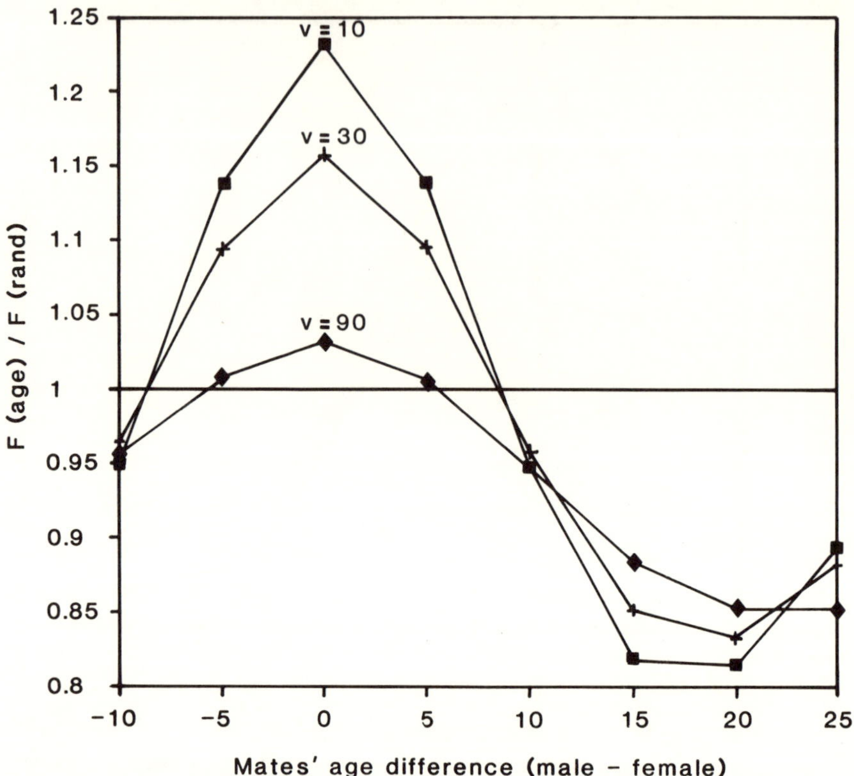

Figure 5.4. Effect of variability in age difference between mates on mean inbreeding. Curve labels v = 10, 30, and 90 refer to the variance of age differences.

CONCLUSIONS

We are not yet in a position to draw firm conclusions about the likely importance of the distribution of pedigree inbreeding to microevolutionary processes, or its relevance to other problems. It may turn out that the absolute magnitude of effects attributable to the variance of inbreeding distributions will, as a rule, be negligible. But for some problems, such as assessing the genetic implications of various patterns of mating or demographic behavior, it is not just the magnitude of the observed distribution that is important. It is the deviation of the observed distribution from that expected in the absence of particular factors (consanguinity avoidance, age preference, etc.) that must be assessed.

The analysis presented in this paper is only a beginning; the problem warrants investigation of other aspects of demographic and mating behavior, and a broader consideration of the consequences of population structure (for example, subdivided populations with subpopulations of unequal sizes). It would be particularly interesting to look at migration that is not random with respect to kinship, because kin-structured migration has the potential for large and diverse effects on the distribution of inbreeding, and has probably been characteristic of humans during most of our evolution. And, of course, more information is needed about the observed and expected distributions of inbreeding in a wider range of populations.

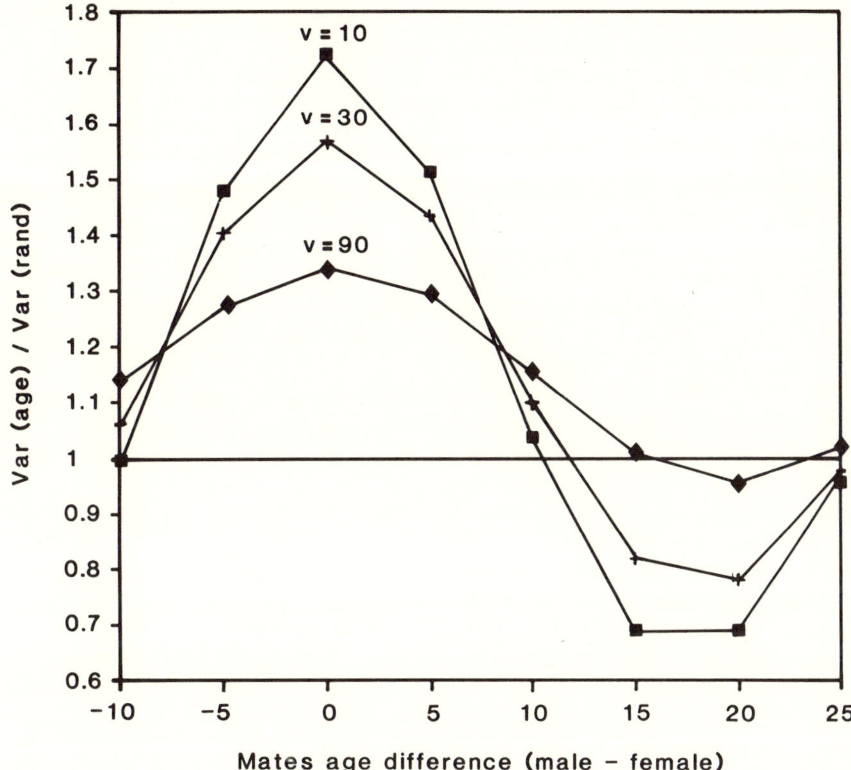

Figure 5-5. Effect of variability in age difference between mates on the variance of inbreeding. Curve labels v = 10, 30, and 90 refer to the variance of age differences between mates.

More generally, there is much of interest to geneticists in the details of demographic behavior, and in the distributions as well as the levels of demographic variables.

ACKNOWLEDGMENTS

James Crow, Bennett Dyke, and Jean MacCluer reviewed an earlier version of this paper, and made numerous useful suggestions. I have also benefitted from discussions with Julian Adams and Peter Smouse, and from the comments of two nearly anonymous reviewers. Lynn Jorde and Elizabeth O'Brien kindly provided some unpublished data on the distribution of consanguinity in the Utah Mormons and on Sottunga, Aland Islands. I am grateful to all of these people.

REFERENCES

Abelson, A., 1978   Population structure in the Western Pyrenees: social class, migration and the frequency of consanguineous marriage, 1850 to 1910. Ann. Hum. Biol. 5: 165-178.

Bennett, J. H., and F. E. Binet, 1956   Association between Mendelian factors with mixed selfing and random mating. Heredity 10: 51-55.

Brennan, E. R., P. W. Leslie, and B. Dyke, 1982   Mate choice and genetic structure: Sanday, Orkney Islands, Scotland. Hum. Biol. 54: 477-489.

Cavalli-Sforza, L. L., M. Kimura, and I. Barrai, 1966   The probability of consanguineous marriages. Genetics 54: 37-60.

Cockerham, C. C., and B. S. Weir, 1983   Variance of actual inbreeding. Theor. Pop. Biol. 23: 85-109.

Denniston, C. 1975   Probability and genetic relationship: two loci. Ann. Hum. Genet. 39: 89-104.

Dyke, B., 1971   Potential mates in a small human population. Soc. Biol. 18: 28-39.

Fix, A. G., 1978   The role of kin-structured migration in genetic micro-differentiation. Ann. Hum. Genet. 41: 329-339.

Franklin, I. R., 1977   The distribution of the proportion of the genome which is homozygous by descent in inbred individuals. Theor. Pop. Biol. 11: 60-80.

Hajnal, J., 1963   Concepts of random mating and the frequency of consanguineous marriages. Proc. R. Soc. London B 159: 125-177.

Haldane, J. B. S., 1949   The association of characters as a result of inbreeding and linkage. Ann. Eugenics 15: 15-23.

Jorde, L. B., 1980   The genetic structure of subdivided human populations: a review, pp. 135-208 in *Current Developments in Anthropological Genetics, Vol 1. Theory and Methods*, edited by J. H. Mielke and M. H. Crawford. Plenum, New York.

Leslie, P. W., 1980   Internal migration and genetic differentiation in St. Barthé-lemy, French West Indies, pp. 167-177 in *Genealogical Demography*, edited by B. Dyke and W. T. Morrill. Academic Press, New York.

Leslie, P. W., 1983   Age correlation between mates and average consanguinity in age-structured human populations. Am. J. Hum. Genet. 35: 962-977.

Leslie, P. W., 1985   Potential mates analysis and the study of human population structure. Yearbk. Phys. Anthropol. 28: 53-78.

Leslie, P. W., B. Dyke, and W. T. Morrill, 1981   Genetic implications of mating structure in a Caribbean isolate. Am. J. Hum. Genet. 33: 90-104.

Leviandier, T., and A. Jacquard, 1974   Homogeneisation of kinship in a small population, pp. 285-290 in *Genealogical Mathematics*, edited by P. A. Ballonoff. Mouton, Paris.

Li, C. C., 1987   A genetical model for emergenesis. Am. J. Hum. Genet. 41: 517-523.

Long, J. C., 1986   The allelic correlation structure of Gainj- and Kalam-speaking people. I. The estimation and interpretation of Wright's F-statistics. Genetics 112: 629-647.

MacCluer, J. W., and B. Dyke, 1976   On the minimum size of endogamous populations. Soc. Biol. 23: 1-12.

MacCluer, J. W., and W. J. Schull, 1970   Frequencies of consanguineous marriage and the accumulation of inbreeding in an artificial population. Am. J. Hum. Genet. 22: 160-175.

Majumder, P. P., S. K. Das, and C. C. Li, 1988   A genetical model for vitiligo. Am. J. Hum. Genet. 43: 119-125.

Neel, J. V., and F. M. Salzano, 1967   Further studies on the Xavante Indians. X. Some hypotheses-generalizations resulting from these studies. Am. J. Hum. Genet. 19: 555-573.

Neel, J. V., and R. H. Ward, 1972   The genetic structure of a tribal population, the Yanomama Indians. VI. Analysis by F-statistics. Genetics 72: 639-666.

O'Brien, E., L. B. Jorde, B. Rönnlöf, J. O. Fellman, and A. W. Eriksson, 1989   Consanguinity avoidance and mate choice in Sottunga, Finland. Am. J. Phys. Anthropol. 79: 235-246.

Roberts, D. F., 1980   Genetic structure and the pathology of an isolated population, pp. 7-26 in *Population Structure and Genetic Disorders*, edited by A. W. Eriksson, H. R. Forsius,   H. R. Nevanlinna, P. L. Workman, and R. K. Norio. Academic Press, New York.

Schnell, R. W., 1961   Some general formulations of linkage effects in inbreeding. Genetics 46: 947-957.

Smouse, P. E., J. V. Neel, and W. Liu, 1983   Multiple-locus departures from panmictic equilibrium within and between village gene pools of Amerindian tribes at different stages of agglomeration. Genetics 104: 133-153.

Stam, P., 1980   The distribution of the fraction of the genome identical by descent in finite random mating populations. Genet. Res. 35: 131-155.

Ward, R. H., P. Raspe, M. Ramirez, R. Kirk, and I. Prior, 1980   Genetic structure and epidemiology: the Tokelau study, pp. 301-325 in *Population Structure and Genetic Disorders*, edited by A. Erikkson, H. R. Forsius, H. R. Nevanlinna, P. L. Workman, and R. K. Norio. Academic Press, New York.

Weir, B. S., P. J. Avery, and W. G. Hill, 1980   Effect of mating structure on variation in inbreeding. Theor. Pop. Biol. 18: 396-429.

Wright, S., 1965   The interpretation of population structure by F-statistics with special regard to systems of mating. Evolution 19: 39-420.

6

Migration and Genetic Differentiation in Italy

ALBERTO PIAZZA

Migration plays an important role in the biological evolution of human populations. The purpose of this paper is to discuss two examples on how human genetic differentiation is related to migration.

Genetic differentiation is usually expressed in terms of gene frequencies that may differ not only among populations but also among different samples of the same population. If we bypass the nontrivial problem of defining a population unit, especially when only a sample of it is available for analysis, two relevant problems have to be addressed: (1) how migration can change gene frequencies; and (2) how to distinguish migration from other evolutionary processes once gene frequency changes have been observed.

MIGRATION AND GENETIC DIFFERENCES

The effect of migration upon gene frequencies depends on two related factors: the type of migration and the structure of the population involved in the migrational process (for reviews see, for instance, Wijsman and Cavalli-Sforza, 1984; Karlin, 1983).

Generally speaking the effect of migration is similar to that of mutation: it introduces new genes into a population. The change in gene frequency is proportional to the difference in frequency between the recipient population and the average of the donor populations. Unlike the mutation rate, which is known to be very low in humans (from 10^{-7} to 10^{-5} per genetic locus per generation), the migration rate can be large, so the change in frequency can be substantial. Obviously all new genetic variation derives ultimately from gene mutations: what migration does is simply to convert genetic differences between populations into genetic variation within a population. In other words the introduction of a new gene from a donor population increases the genetic variability of the recipient population at the expense of the variability between the donor and the recipient populations.

The simplest way to make the previous statement more quantitative is to express the genetic variabilities within and between populations in the standard statistical framework of the analysis of variance. In fact, the ratio of the variance between populations to the total variance (between and within) turns out to have a biological meaning: it represents the proportional reduction in the frequency of heterozygous individuals due to structuring of a population in substrata without genetic contacts among each other, rather than consisting of only one pooled population without restrictions to random mating. This ratio of variances is usually known as F_{ST} (Wright, 1965; Cockerham, 1973; Cockerham and Weir, 1987). Thus, the greater the rate of migration into one population, the lower the F_{ST} between the donor and the recipient populations.

It is convenient to distinguish migrational processes according to their extent in time and in space. More specifically, we will distinguish: (1) *small scale* migrations of individuals from (2) *large scale*, massive migrations involving large movements of populations (Wijsman and Cavalli-Sforza, 1984). The two kinds of migration differ also in the demographic data usually available in either case: in the former, written records provide information on movements of individuals over the span of a few generations or on the distances between mates' birthplaces or residences; in the latter, information on the relative sizes of the moving and the settled populations are mostly scanty or unreliable, historical and linguistic data being sometimes the only sources of knowledge.

Another kind of migration pattern has to be mentioned even if its discussion may be beyond the purpose of this paper: it was called (3) *demic diffusion* (Ammerman and Cavalli-Sforza, 1984 and references therein) which associates large-scale migrations with population growth. If increase in population numbers coincides with migratory activity that is random in direction, a wave of population expansion will arise and progress at a constant radial rate. It involves slow, continuous expansion with movements over short distances. This manner of expansion is appropriate to describe the diffusion of a technology that may increase the growth-rate of the population which adopts it. It is not caused by people motivated by a desire to explore distant lands, nor is it necessary that people have to be moving always in the same direction. The average distance moved per generation is rather modest, and whatever direction has been chosen, the overall outcome will still be the spread of the technology outwards from the area already settled, at a relatively steady rate.

This model of demic diffusion was specifically developed to study the spread of early farming techniques in the early neolithic. There it is assumed that the population density goes up to an average of five persons per square kilometer, and that the population grows in the early stages of occupation of an area at such a rate that it doubles every 18 years. The distance of the local migratory activity—that is the distance of the movement of settlements which is random in direction—is taken to be 18 kilometers for each generation of 25 years. Under these conditions, the rate of movements of the wave of advance turns out to be 1 kilometer per year which agrees with the earliest sites of farming settlements in Europe, as determined by radiocarbon dates.

MIGRATION AND OTHER EVOLUTIONARY PROCESSES

Migration is not the only factor which can change gene frequencies between and within populations. In addition to mutation, other forces mold our genes in

Table 6-1 The effect of the forces of evolution which increase (+) or decrease (−) variation within and between populations and their influence on gene frequencies

Evolutionary Force	Variation		Genes Affected	
	Within	Between	Some	All
Random genetic drift (or inbreeding)	−	+	NO	YES
Mutation	+	−	NO	YES
Migration	+	−	NO	YES
Natural Selection				
-directional	−	+ or −	YES	NO
-balancing	+	−	YES	NO
-incompatibility	−	+	YES	NO

NOTE: Adapted from Lewontin, 1982.

time and space. These forces are listed in Table 6-1 with some of their relevant properties.

The forces that increase or maintain variation within populations prevent the differentiation of the populations from each other, whereas divergence between populations is a result of forces that reduce the proportion of heterozygotes in each population (that is, making the population more homogeneous, F_{ST} increases). Thus, random genetic drift produces a decrease of heterozygosity while causing different populations to diverge. This divergence is counteracted by the constant flux of mutation and migration, which introduce variation into the population again and tend to make them more like each other. This equilibrium situation can be described in more quantitative terms: if m is the migration rate per generation into a given population, and N is the number of reproducing individuals in that population, then (to an order of magnitude) the population will maintain its heterozygosity if $Nm > 1$, i.e., if more than one single individual per generation migrates, irrespective of population size. Many human populations undergo more migration than this minimal value.

One major problem in the analysis of gene frequency data is to discriminate between the variation due to natural selection and that due to migration (counterbalanced by drift). This is of the greatest practical importance when insufficient demographic data offer no support for choosing either interpretation: this is the case for most of the massive movements that have marked our history. An inspection of Table 6-1 suggests an approach which was originally proposed by Cavalli-Sforza (1965). The two processes of migration and natural selection can be distinguished by the pattern of the last two columns of the table. Natural selection depends on the particular environment in which the population is placed, so that it is likely to affect only those genes which happen to be advantageous or disadvantageous to their carriers in that environment. In contrast, migration as well as random genetic drift affect the population structure, not the single genes. Thus, all gene frequencies are expected to show the same variation, and methods for detecting this common variation can supply evidence for genetic differentiation due to these processes.

AN ANALYSIS OF SMALL-SCALE MIGRATIONS IN ITALY

Over the last century, with the industrial revolution and the corresponding concentration of people in very large urban settlements, there have been documented migrations and changes in the population structure of most European countries. Italy, in particular, has changed its population structure over the last 30 years, as the concentration of industries in the north induced an internal south-north migration of remarkable proportions (Golini, 1974).

The slow pace of genetic changes and the relatively gross measurements of the traditional blood groups cannot resolve this process. Depth in time is too small for a genetic difference to be singled out from sampling noise when using blood group markers for which data are freely available in this context. The power of these genes is limited by the number of ''variant'' types (in technical terms the number of ''alleles'' or ''polymorphism'') they can exhibit. The higher the number of alleles of a gene, the higher the potential number of heterozygous individuals and the more powerful this gene can be for resolving small differences in the geographical distribution of its frequencies. A very polymorphic HLA genetic system in man carries some hundred alleles. Its function is to control some immunological reactions, for instance in organ transplants. It has, however, the disadvantage of involving an expensive test which not all blood-typing laboratories are able to do.

A system which is inexpensive, simple to test, and sensitive enough to detect small-scale migrations, is the collection of surnames. Surnames, considered as alleles of a gene transmitted only by the male line, can be assumed to be genetic markers unaffected by natural selection (Zei *et al.*, 1983) and therefore satisfy the expectations of a theory of evolution based only on random genetic drift, mutation and migration (Karlin and McGregor, 1967, Ewens, 1972). By making use of this theory we compared the estimates of migration rates in Italy, as inferred by the surname distribution found in the telephone directories, with the corresponding estimates from official demographic sources (Piazza *et al.*, 1987). It was a pleasant surprise to find that at least with this source of data, the ratio of surnames to individuals makes it possible to obtain reliable estimates of migration rates.

The model is the following. Each of N individuals carrying one of S different surnames is subjected to a process of death at random. Dead individuals are replaced by new ones carrying the same surname or externally from immigration occurring with rate v. When the number of individuals N is much higher than the number of surnames S, the following system of non-linear equations:

$$N = \frac{\alpha(1 - v)}{v} \tag{6-1}$$

$$S = -\alpha \, log \, v \tag{6-2}$$

allows v, the immigration rate per generation, and the parameter α to be estimated given N and S (Fisher, 1943; Zei *et al.*, 1983). The quantity α, a parameter of the model measuring the ''richness'' of surnames in a given sample of individuals, is not a function of its size N: values of α from different samples of surnames collected in the *same* population are expected to be equal. On this assumption, α can be estimated by equations (6-1) and (6-2) from any collection of surnames, for instance our Italian telephone directories; then equation (6-1) can be used to evaluate the immigration rate v by substituting the estimates for the parameter α and the sample size of individuals, N, referred to for the rate v (in our application this is the census size which differs from the size of telephone users).

Surname data were collected from the telephone directories of 91 Italian provinces after elimination of commercial subscribers. Our sample has 10,473,727 registered telephone users with 59,961,000 inhabitants, that is, 1 telephone every 5.44 residents; therefore, we assume that most families are represented. Immigration was estimated from (6-1), rewritten as:

$$v = \frac{\alpha}{P_m + \alpha} \cdot \tag{6-3}$$

The parameter α was estimated from the solution of equations (6-1) and (6-2) simultaneously with S and N being the number of surnames and the number of telephone users provided for each province of Italy by the telephone directories. Therefore different α's are found, one for each province. P_m is the male census size from the same provinces; this was obtained by halving the number of people officially registered in each province of Italy at the end of 1978, the year in which the telephone company's computerized file was available. The immigration rates v, as estimated by equation (6-2), have been compared with actual immigration rates, calculated by averaging the yearly ratios 1967-80 of newly registered individuals to total residents in the 91 Italian provinces, as published by the Italian registration offices (Istat, 1967-80).

Figure 6-1 shows the correlation between "observed" and "estimated" migration coefficients. The Pearson correlation coefficient is 0.596 ± 0.083, significantly different from zero. Special statistical techniques have been used to detect possible outliers. Of the putative outliers displayed in the figure, the most extreme (Trieste) was only recently annexed to Italy; history and/or tourism may be influential for Venice and Genoa (which deviate in the same direction as Trieste) and may be responsible for an anomalous pattern of migration. The interpretation of the other outliers corresponding to the towns of Nuoro, Caserta, and Avellino (all above the line in Figure 6-1) is still to be investigated; all three belong to regions (Sardinia and Campania) where emigration was remarkably higher than immigration in the considered period of time, and this may introduce a bias in the estimate of their α. Obviously the exclusion of these outliers increases the correlation ($r = 0.741$).

In addition, we analysed a second body of surnames originating from a list of about 540,000 consanguineous marriages for which dispensations were required by the Catholic church in Italy. The surnames refer to the years 1936-1964, the census data to the year 1961. The correlation $r = 0.721$ between the two sets of data, higher but not significantly different from that obtained with the telephone directories, has been calculated on 59 geographical areas and is shown in Figure 6-2.

Our findings show that the ratio of surnames to individuals in the sample makes it possible to calculate an estimate of the male immigration rate. The number of individuals is very high when compared to the usual size of test

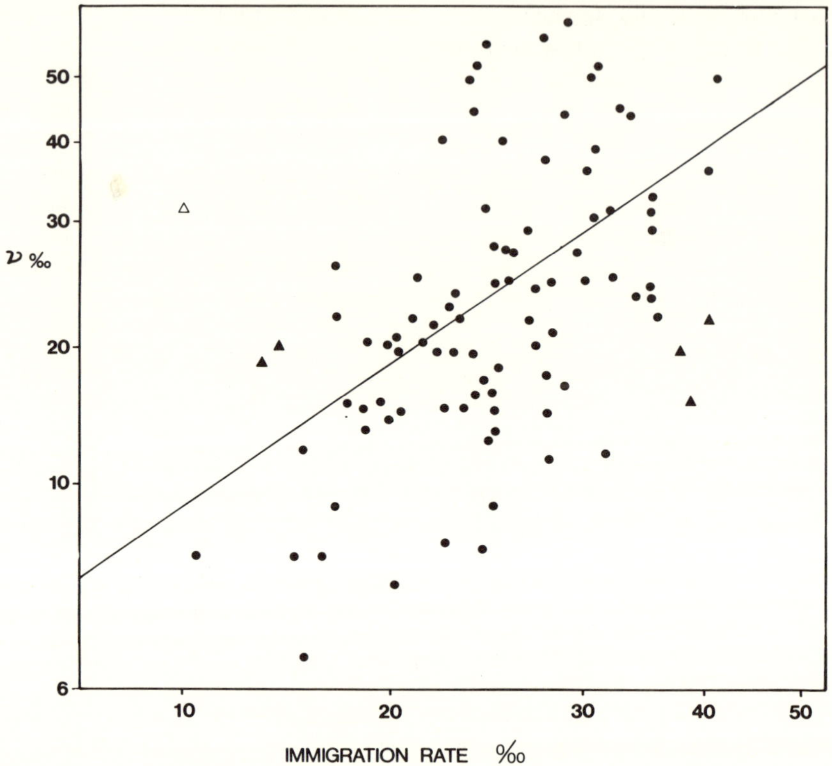

Figure 6-1.   Immigration rates of 91 Italian provinces estimated from telephone directories using surname data, as explained in the text, are plotted against immigration rates from Italian official demographic data (years 1967-80). Both axes are represented on a logarithmic scale. The statistically significant regression line is also shown. Open triangles: outlier data points are defined by the analysis of standardized residuals. The most significant of them (filled triangle) has been excluded from the regression analysis (from Piazza *et al.*, 1987).

samples, taken from blood groups for example, and this is one of the potential advantages of the technique when other demographic sources are not available.

A limitation on the application of the method, however, is that the migration of the males is the only one considered. Furthermore the introduction of surnames is too recent for tracing old, large-scale movements and settlements of populations: more appropriate, albeit less direct approaches must be developed.

AN ANALYSIS OF LARGE-SCALE MIGRATIONS IN ITALY

An alternative analysis for the identification of putative migrations into Italy, possibly older than those detected by the evolution of surnames, has already been used to interpret the geographical distribution of the gene frequencies in Europe (Menozzi, Piazza, and Cavalli-Sforza, 1978) and in the world (Piazza, Menozzi, and Cavalli-Sforza, 1981a). The procedure of principal components

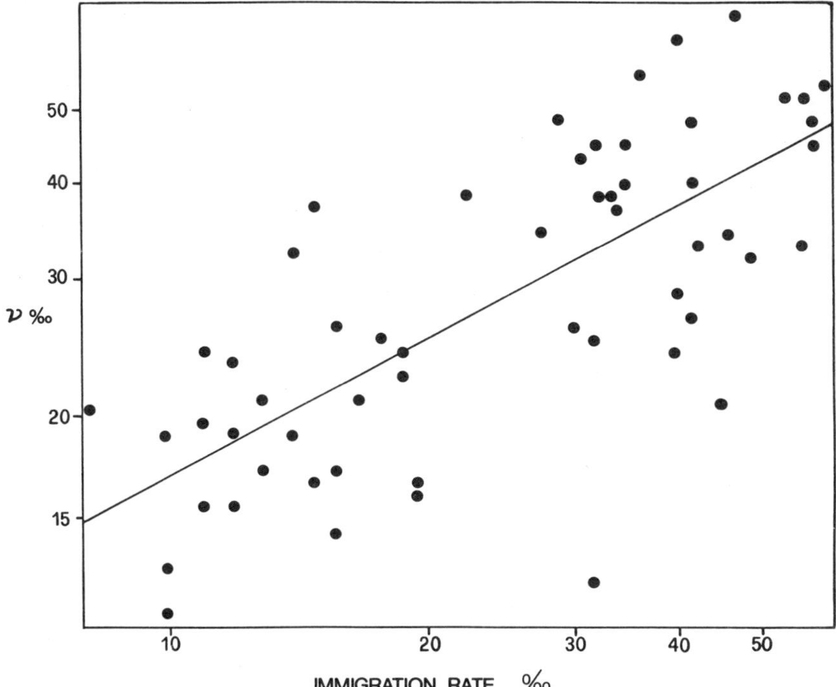

Figure 6-2. Plot of immigration rates of 59 Italian geographic areas estimated from 1936-1964 consanguinity dispensations of the Catholic Church, against immigration rates from Italian demographic data (year 1961). Both axes are represented on a logarithmic scale. The statistically significant regression line is also shown.

has been applied to display, in a few synthetic images, that portion of genetic differentiation which is common to all considered genes. For this purpose a complete set of 34 variables (gene frequencies) calculated at 28 geographical locations was selected. The resulting data matrix of 28 cases and 34 variables has been rotated so as to condense the information of the original data matrix into fewer new variables by minimizing the amount of information lost in such variable reduction. The new variables, linear transformations of the original 34 variables, are ordered in such a way that the information provided by the first principal component is larger than that one provided by the second principal component, and so on.

Therefore a contour surface plot of the *first* principal component represents the most informative display of the genetic differentiation over the geographical area covered by the original data points. Without going into further details (see also Piazza, Menozzi, and Cavalli-Sforza, 1981b), the idea justifying the use of this multivariate analysis is that the migrational component of genetic differentiation must be shared equally by all the gene frequencies (see discussion of Table 6-1). In fact, it has been shown that the synthetic displays provided by plotting the principal component surfaces are particularly useful in detecting gradients (clines) of genetic differentiation associated with movements of populations like those accompanying the Neolithic expansion of farmers from the Near East

Figure 6-3.   Contour plot of the first principal component coordinates of genes frequencies in Italy from 34 independent alleles at the human loci: ABO, Rhesus, MNS, Kell, Haptoglobin, HLA-A, HLA-B. Shades indicate different values of the principal component scores (from Piazza *et. al*, 1988).

(Rendine, Piazza, and Cavalli-Sforza, 1986) or the putative diffusion of Indo-European speaking populations (Ammerman and Cavalli-Sforza, 1984; Renfrew, 1987).

The results of the analysis can be summarized in synthetic images (Piazza *et al.*, 1988) showing clines of genetic variation. One of these is represented in Figure 6-3. It is the contour plot of the first principal component of all genetic data and it explains 35% of the total variation. The island of Sardinia is not included in the map because its great genetic distance from the peninsula would flatten the genetic differentiation within Italy to a scale too small to be resolved by the human eye. The genetic isolation of Sardinia, is supported by many other pieces of evidence (Menozzi, Piazza, and Cavalli-Sforza, 1978; Olivetti *et al.*, 1986; Piazza *et al.*, 1985).

The Italic world, with its local cultures, seems to be more or less clearly defined at the beginning of the Iron Age, that is, since the ninth-eighth century B.C., when an approximate correspondence between archeologically defined cultures and linguistic areas has been established through epigraphic records and literary sources (Pallottino, 1984). Figure 6-4 which shows the geograph-

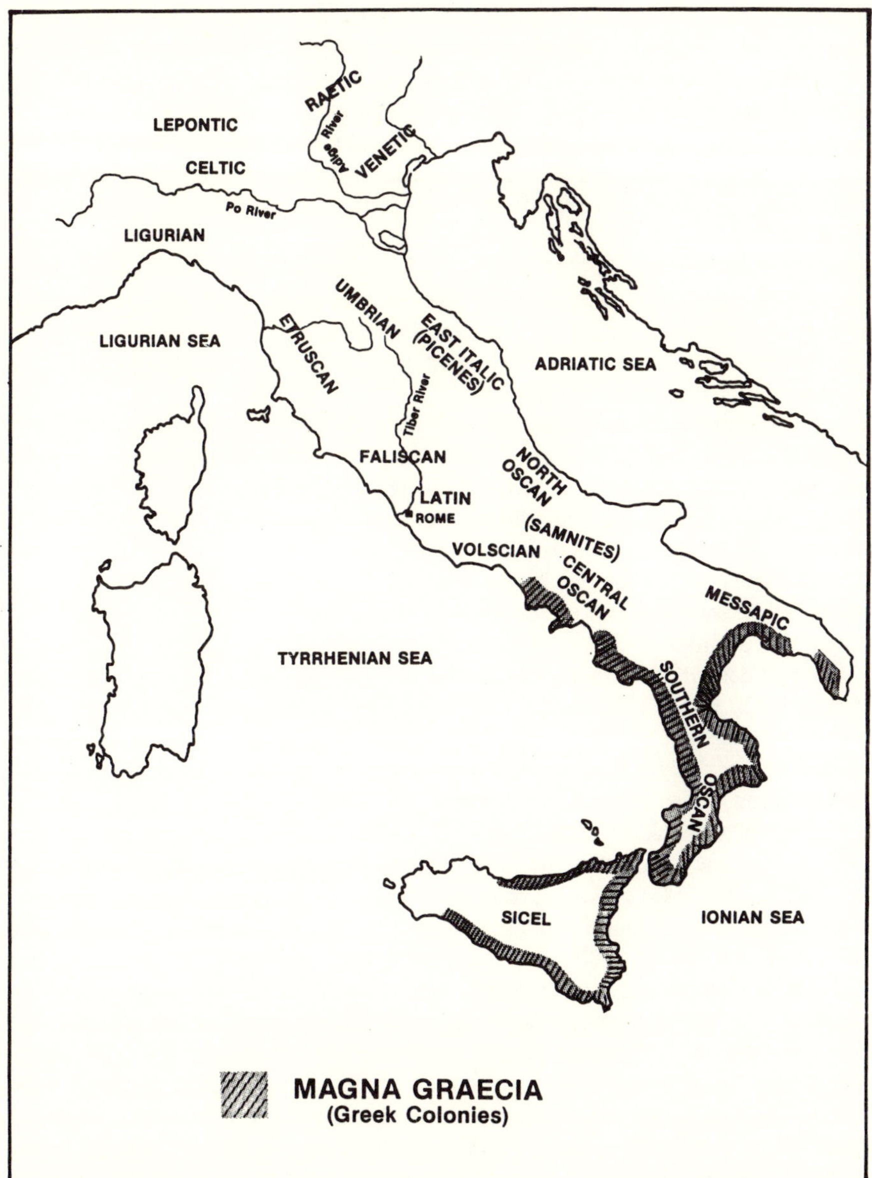

Figure 6-4. The languages of ancient Italy at the end of the sixth century B.C. (adapted from Pulgram, 1958). The areas of the Greek colonization ("Magna Graecia") are also shown.

ical distribution of languages in Italy as documented in the fifth century B.C. (Pulgram, 1958) can be used as a reference to describe the various ethnic groups which formed Italy at the beginning of its history.

Literary tradition and archeological evidence report a sharp increase in population size in the eastern Mediterranean in the late part of the second millennium B.C. (McEvedy and Jones, 1978); a logical consequence of this was the foundation of new cities overseas ("colonies"). During the eighth century B.C. the western Mediterranean was the site of a complex network of colonies founded mainly by Phoenicians and Greeks (Sherrat, 1980). Whereas the Phoenicians directed their main colonizing efforts towards the coasts of North Africa, Spain, Malta, Sardinia, and the western triangle of Sicily, the Greeks settled mainly along the southern and western shores of the mainland and also along the fertile coastal belt of Sicily (excluding the Phoenician western triangle). More than 40 towns are known whose size and architectural magnificence underline the importance of what came to be known as *Magna Graecia*; its geographical area of influence is shown in Figure 6-4.

An interesting feature of our genetic map (Figure 6-3) displaying the first principal component of 34 genes, is a clear North-South gradient showing a remarkable differentiation of *Magna Graecia* from the rest of the mainland. The genetic similarity of the Italian southern regions with modern Greece as shown by their similar gene frequencies, the genetic difference between the two parts of Sicily which appear in this map (it is noteworthy that the Phoenician colony of Motya in the western area of the island is very well defined) and finally the definition of a geographical area around the delta of the Po river which was described as the possible Adriatic site (Spina, Adria) of later (fourth century B.C.) Greek settlements, (Bérard, 1957) all suggest a Greek gene flow as the most relevant factor which could explain the genetic differentiation shown in this map of Italy. A nontrivial question to be raised in order to make this interpretation more plausible is whether the Greek colonies were of such size as to justify a diffusion of their genes. Because of the geographical position of Greece as the door from the Near East to the Mediterranean, by the end of the Bronze age (1000 B.C.) the average density of the population was higher in Greece than in Europe (3.7 inhabitants per square kilometer) by a factor of 3 (McEvedy and Jones, 1978). Between 1000 B.C. and 400 B.C. the population doubled in Europe, increasing from 10 to 20 million; in the same period the population trebled in Greece, reaching a total of 3 million (McEvedy and Jones, 1978). By 400 B.C. Italy, the second most densely populated country in Europe after Greece, had about 4 million people (McEvedy and Jones, 1978; Beloch, 1886). The Greek colonies of Sicily alone accounted for 1.5 million people, of which more than 10% (about 200,000) were of Greek origin (Beloch, 1886). To these Greek inhabitants of Sicily may be added at least another 100,000 Greek colonizers in the Italian peninsula, so that before the Roman period, one out of every 10-13 inhabitants in Italy was Greek (Beloch, 1886). Even if these numbers must be taken with caution, their order of magnitude does not contradict the idea of a possible introduction in Sicily and in the South of Italy, of Greek genes whose dilution in the autochthonous gene pool is still apparent as a gradual genetic change from south to north.

The lack of relevant archaeological records makes dating prehistoric movements of peoples and ascertaining their extent in Italy difficult. Linguistic records are sometimes the only source of useful information for identifying the different ethnic groups of prehistoric times. A main finding of our study is that languages and genes, at least as far as Italy is concerned, have a similar geographical distribution and therefore a probable common history. Comparison of linguistic records and archeological cultures (mostly burial patterns) makes it possible to date the presence of the ancestors of the peoples mentioned in Figure 6-4 after the diffusion of the Indo-European languages and before or during the early Iron Age (Peroni, 1979).

It seems plausible that the Indo-European languages were not brought to Italy through massive migrations, but rather by a moderate number of carriers originating roughly in the Danube basin, who reached Italy by a transalpine and/or an Adriatic route (Pulgram, 1958; Devoto, 1962).

The use of a language by a population is a cultural trait which can be imposed by a few politically powerful people without any substantial effect on the genetic structure of the population itself. However when it is associated with a specific genetic identity, then language and population may share a common origin. In the case of Italy, we suggest the pre-Roman times as those of the main settlement and movement of peoples determining the present pattern of genetic differentiation. Such early times are not as surprising as it might appear. In fact it has been shown that the genetic structure of the present European populations is likely to be the result of the expansion of Neolithic farmers from the Near East which occurred 10,000 years ago (Menozzi, Piazza, and Cavalli-Sforza, 1978; Ammerman and Cavalli-Sforza, 1984). Moreover a computer simulation of the same process (Rendine, Piazza, and Cavalli-Sforza, 1986) suggested that the clines of gene frequencies generated by this expansion are not so easily dissolved by successive migrations between contiguous populations once the expansion is over; it can also be shown theoretically that a linear gradient is particularly stable over long periods of time.

The romanization of Italy and Europe, which was obviously of the greatest importance for many other aspects of our history, is not likely to have changed the genetic individuality of the conquered populations in a substantial way. Colonization produced changes in the political, administrative, urban, and commercial systems rather than a massive substitution of people such as occurred in the European colonization of the Americas or in the genocides of our times. A unique language, Latin, was also imposed by the Roman conquerors, but its adoption was not complete since even today each region in Italy speaks a different dialect, revealing possible traces of the ancient language spoken by the ancestors of its inhabitants. A parallel analysis of the geographical distribution of gene frequencies in France (Piazza, 1986) seems to lead to similar conclusions as if the genetic roots of France were to be found before the settling of the Franks.

Our analysis provides information on the evolutionary history of the population—or rather the populations—of Italy. It is clear, however, that genetic similarities among populations living in different geographical areas might be due to environmental similarity or common origin, the corresponding differences being the outcome of natural selection or genetic drift balanced by migration. The study of the gene frequency maps does not by itself allow a clear-cut choice between these possibilities, but natural selection as the main cause of the genetic gradients shown above could reasonably be excluded because of the following reasons.

1. The representation of the principal components is by itself a technique which cannot easily accommodate selective agents: in fact it displays the genetic differences shared by *all* genes, and it is very difficult to imagine a selective factor affecting a large fraction of genes at the same time and in the same proportion. On the contrary this holds in the case of genetic drift and migration; they depend on the population structure, that is, on demographic parameters like size, rate of migration, growth, etc., but they do not discriminate among genes.

2. Genes clearly correlated with environmental or pathological factors, such as thalassemia, G-6-PD, etc., were excluded from the analysis.

3. The small geographical area of Italy and its temperate climate make significant changes in the physical environment implausible.

On the basis of these arguments, we suggest that the genetic history of Italy is not a classification of different kinds of environment, but more likely the result of the ancient history of peoples with their settlements and their movements. Genetic drift was the probable cause of differentiation among the many different ethnic groups who settled in that country. The migrational network established among them over the centuries contributed to create the genetic gradient still discernible in our synthetic map of modern Italy.

REFERENCES

Ammerman, A. J., and L. L. Cavalli-Sforza, 1984   *The Neolithic Transition and the Genetics of Populations in Europe*. Princeton University Press, Princeton, NJ.

Beloch, J., 1886   *Die Bevölkerung der grieschisch-römischen Welt*. Leipzig.

Bérard, J., 1957   *La colonisation grecque de l'Italie méridionale et de la Sicile dans l'antiquité*. Presses Universitaires de France, Paris.

Cavalli-Sforza, L. L., 1965   Population structure and human evolution. Proc. R. Soc. London B 164: 362-379.

Cockerham, C. C., 1973   Analyses of gene frequencies. Genetics 74: 679-700.

Cockerham, C. C., and B. S. Weir, 1987   Correlations, descent measures: Drift with migration and mutation. Proc. Natl. Acad. Sci., USA 84: 8512-8514.

Devoto, G., 1962   *Origini Indoeuropee*. Sansoni, Firenze.

Ewens, W. J., 1972   The sampling theory of selectively neutral alleles. Theor. Pop. Biol. 3: 87-112.

Fisher, R. A., 1943   The relation between the number of species and the number of individuals in a random sample of an animal population. J. Anim. Ecol. 12: 42-58.

Golini, A., 1974   *Distribuzione della Popolazione, Migrazioni Interne ed Urbanizzazione in Italia*.   Istituto Demografia della Università. Roma.

Istat, 1967-1980   *Popolazione e Movimento Anagrafico dei Comuni. Vols. 1967-1970, 1972-1974, 1976, 1978, 1980*. Istituto Poligrafico dello Stato, Roma.

Karlin, S., 1983   Classification of selection-migration structures and conditions for a protected polymorphism. Evol. Biol. 14: 61-204.

Karlin, S., and J. McGregor, 1967   The number of mutant forms maintained in a population. Proc. Fifth Berkeley Symp. Math. Stat. Prob. 4: 415-438.

Lewontin, R. C., 1982   *Human Diversity*. Freeman, San Francisco.

McEvedy, C., and R. Jones, 1978   *Atlas of World Population History*. Harmondsworth: Penguin.

Menozzi, P., A. Piazza, and L. L. Cavalli-Sforza, 1978   Synthetic maps of human gene frequencies in Europeans. Science 201: 786-792.

Olivetti, E., S. Rendine, N. Cappello, E. S. Curtoni, and A. Piazza, 1986   The HLA system in Italy. Hum. Hered. 36: 357-372.

Pallottino, M., 1984   *Storia della prima Italia.* Rusconi, Milano.

Peroni, R., 1979   The Iron Age, Orientalizing and Etruscan Periods. pp. 7-30 in *Italy before the Romans,* edited by D. Ridgway and F. R. Ridgway. Academic Press, London.

Piazza, A., 1986   The genetic data from the French Provinces: a tentative summary. pp. 345-352 in *Human Population Genetics*, edited by E. Ohayon and A. Cambon-Thomsen. Inserm, Paris.

Piazza, A., N. Cappello, E. Olivetti, and S. Rendine, 1988   A genetic history of Italy. Ann. Hum. Genet. 52: 203-213.

Piazza, A., W. R. Mayr, L. Contu, A. Amoroso, I. Borelli, E. S. Curtoni, C. Marcello, A. Moroni, E. Olivetti, P. Richiardi, and R. Ceppellini, 1985   Genetic and population structure of four Sardinian villages. Ann. Hum. Genet. 49: 47-63.

Piazza, A., P. Menozzi, and L. L. Cavalli-Sforza, 1981a   Synthetic gene frequency maps of man and selective effects of climate. Proc. Natl. Acad. Sci., USA 78: 2638-2642.

Piazza, A., P. Menozzi, and L. L. Cavalli-Sforza, 1981b   The making and testing of geographic gene frequency maps. Biometrics 37: 635-659.

Piazza, A., S. Rendine, G. Zei, A. Moroni, and L. L. Cavalli-Sforza, 1987   Migration rates of human populations from surname distributions. Nature 329: 714-716.

Pulgram, E., 1958   *The Tongues of Italy.* Harvard University Press, Cambridge, MA.

Rendine, S., A. Piazza,and L. L. Cavalli-Sforza, 1986   Simulation and separation by principal components of multiple demic expansions in Europe. Am. Nat. 128: 681-806.

Renfrew, C., 1987   *Archaeology and Language.   The Puzzle of Indo-European Origins.* Cambridge University Press, Cambridge.

Sherrat, A., (ed.), 1980   *The Cambridge Encyclopedia of Archaeology.* Chs. 33, 34. Cambridge University Press, New York.

Wijsman, E. M., and L. L. Cavalli-Sforza, 1984   Migration and genetic population structure with special reference to humans. Annu. Rev. Ecol. Syst. 15: 279-301.

Wright, S., 1965   The interpretation of population structure by F-statistics with special regard to system of mating. Evolution 19: 395-420.

Zei, G., R. G. Matessi, E. Siri, A. Moroni, and L. L. Cavalli-Sforza, 1983   Surnames in Sardinia. I. Fit of frequency distributions for neutral alleles and genetic population structure. Ann. Hum. Genet. 47: 329-352.

II

HETEROGENEITY, PHENOTYPIC VARIATION, AND FRAILTY

7

Heterogeneous Frailty Analysis in Demography and Genetics

DAVID LAM AND PETER E. SMOUSE

Heterogeneity is the term used in the demographic literature to describe the existence of significant *phenotypic* (see below for a distinction between phenotypic and genotypic variation) variation among individuals or sub-populations for demographic and socioeconomic traits (principally vital rates). Although sometimes viewed as a nuisance by demographers, it is both the basis of fundamental analytical problems in both disciplines, and at the same time provides the foundation for understanding important biological and social phenomena.

A fundamental purpose of demography is analyzing the factors which determine life table distributions of different population strata in some coherent way that leads to inferential generalization. Heterogeneity (genetic variation) is the *sine qua non* for the study of genetics and is the cornerstone of population genetics.

Analysis of heterogeneity is an important area of intersection of genetics and demography, but these convergent threads have received only limited attention in existing literature. The papers in this section provide a variety of approaches to the study of heterogeneity, and point out some important interactions between the two disciplines.

OBSERVABLE VERSUS UNOBSERVABLE HETEROGENEITY

For analytical purposes, we need to make a distinction between those sources of heterogeneity in vital rates that we can identify (observables) and those sources of heterogeneity that we cannot identify (unobservables). Observable heterogeneity exists wherever one can identify correlated variables, variables covering the range from mere correlates to causative factors. The key is identification of which individuals belong to which frailty class. Given identification, numerous regression-type analytic approaches exist. Unobservable heterogeneity probably

97

always exists, inasmuch as we can never identify all of the relevant variables. The extent to which unobservable heterogeneity is a problem depends on the nature of that heterogeneity and its relationship to the observable variables under investigation. In an ordinary linear regression context, the question is whether unobservable (omitted) variables are correlated with observable (included) variables. In more general models, the issue of when unobserved heterogeneity creates problems of statistical inference can be more complex. The nature of the problems caused by unobserved heterogeneity and their sensitivity to alternative statistical methodologies has been the focus of an extensive literature in recent years in a variety of disciplines.

As the papers in this section demonstrate, considerable progress has been made in understanding and dealing with unobserved heterogeneity. One of the more important thoughts to emerge is that much demographic heterogeneity has a genetic component. Such heterogeneity is the basis of adaptive evolution of populations, a major theme in population genetics. In a demographic vein, there is always unobservable variation in vital rates that can be ascribed to genetic variation within the population. Exploiting the links between genetics and demography in modeling heterogeneity would appear to provide substantial returns in identifying functional forms for the distribution of heterogeneity, in understanding the determinants of heterogeneity, and in most effectively using information on related individuals and repeated events.

HETEROGENEITY AS A NUISANCE IN DEMOGRAPHIC RESEARCH

In demographic research, heterogeneity is routinely considered a nuisance, complicating empirical research and potentially creating seriously misleading statistical inferences. Although omitted variables are a potential nuisance in any kind of statistical analysis, concern with unobserved heterogeneity has been closely associated in demographic research with analysis of duration models. The popularity of hazard-rate methodologies throughout demography in recent years has led to increased sensitivity to the problems created by unobserved heterogeneity. As noted by Trussell and Rodríguez (Chapter 8), unobserved heterogeneity can cause problems in duration models under conditions that would not cause problems in a simple cross-sectional regression.

Heterogeneous frailty

One of the most natural meeting grounds for genetics and demography is analysis of frailty, by which we mean susceptibility or vulnerability to morbidity and/or mortality. As discussed in the section on epidemiology below, many human diseases have an age-dependent (developmental) component that can be modeled in the form of a survivorship curve (hazard function). The various genotypes in a population differ in their frailty, yielding different parameters for their hazard functions for the disease in question, and proper genetic analysis requires that we account for that heterogeneity (see Chapter 12 by Weiss and Chapter 16 by Ott).

Some implications of heterogeneous frailty can be illustrated with a simple example. Consider analysis of the mortality effects on a population caused by some event such as famine or exposure to a hazardous substance, with mortality data collected for a period of time following the event. Unobserved heterogeneity could create several kinds of misleading inference. The first relates to "duration-dependence" in the data, one of the standard examples of the problems of unobserved heterogeneity. Suppose the passage of time *per se* has no effect on the probability that an event occurs, but the population from which event data are collected contains two subpopulations, one with a higher hazard rate than the other, with the data not allowing identification of the separate subpopulations. The population at risk of death will change over time as deaths occur, creating the appearance that the hazard rate decreases with duration (see Chapter 8 by Trussell and Rodríguez).

The issue of duration-dependence demonstrates one of the fundamental features of heterogeneity, and one of the reasons why it is especially important in duration models. Heterogeneous frailty will lead to systematic changes in the composition of the population over time, changes which may create the illusion of effects that are not really present. The systematic relationship between unobserved frailty and the composition of the population means that any life table analysis will be potentially misleading. Omitted variables will be correlated with the probability of inclusion in the sample due to heterogeneity in the rate of exit from the population at risk. This is not simply a problem in estimation of the effects of time *per se*, but can affect identification of effects of other covariates. Trussell and Rodríguez (Chapter 8) provide an instructive example of the effects of race and gender on mortality.

A second kind of empirical problem caused by heterogeneity is based on the distribution of frailty. Returning to the example of the mortality effects of some catastrophic event, suppose that the mortality crisis under investigation initially killed the most frail portion of the population. In comparison with a control population, the population under study might actually appear to have *lower* mortality after the crisis. As in the example of duration-dependence, the systematic effects of heterogeneous frailty on the composition of the population at risk would lead to misleading statistical inference. If we could actually observe frailty we would be able to test whether the affected population had higher mortality, *controlling for frailty*. As long as frailty is unobservable, we are faced with observing a confusing combination of true mortality differences and the effects of mortality on the composition of the population.

These misleading inferences are simple examples of the kinds of issues that have been the subject of extensive analysis in recent literature. Trussell and Rodríguez (Chapter 8) discuss many of the potential nuisances created by heterogeneity in analysis of duration data. Vaupel, Manton, and Stallard (1979) and Vaupel and Yashin (1985a, 1985b) point out a number of potentially misleading inferences that can occur in populations with heterogeneous frailty.

Frailty, in its literal application to morbidity and mortality, can affect analysis of a wide variety of issues, including behavioral issues that do not arise directly from interest in mortality. An interesting economic application, for example, in which biological structure imposed on models may help answer behavioral questions, relates to changing labor force behavior of the elderly in recent decades. Significant declines in the labor force participation rates of older men have been observed in the United States and other industrialized countries (see, for example, Bound and Waidmann, 1989), with an increasing proportion of these men reporting that physical disabilities prevent them from working. Two competing hypotheses, one biological and one behavioral, with very different policy implications, have been suggested. These explanations are based on two trends that have coincided with the decline in labor force participation, an increase in the benefit levels of retirement and disability programs, and a significant decline in mortality. The behavioral explanation is that the decline is a response to increasingly generous retirement and disability programs that make it possible for workers to withdraw from the labor force who in the past would have continued to work. The second, biological, explanation, is that declining mortality has left a group of "frail" workers alive who in the past would have died. The two explanations have very different policy implications, and both appear plausible in terms of orders of magnitude. Convincing tests of the behavioral hypothesis require careful attention to the role of heterogeneous frailty, with the fundamental identification problem being the same as in the mortality crisis example above.

Frailty more broadly defined

In addition to literally referring to morbidity and mortality, the distribution of "frailty" may be interpreted more broadly. Duration or failure time models (Kalbfleisch and Prentice, 1980) have been applied extensively in both behavioral and natural sciences. Just as any duration model can be interpreted as an analysis of "failure times," whether the actual issue is the lifetimes of light bulbs, the lifetimes of cancer patients, the spacing between births, or the duration of unemployment spells, the population under analysis can be thought of as having an underlying distribution of "frailty" with respect to risk of experiencing the event under analysis. Whatever the application, the implications of heterogeneous frailty for empirical analysis are similar, and virtually every branch of demography and genetics has well-known examples of misleading inferences resulting from such heterogeneity.

Concern over unobserved heterogeneity has been important in demographers' analysis of fertility, especially as duration models have been applied to fertility research in recent years. One important effect of heterogeneity on analysis of reproduction is demonstrated by Wood and Weinstein (Chapter 11). In a model of a homogeneous population with a constant risk of conception, the unconditional probability of conceiving in a given month will be a simple exponential decline. In a population in which there is heterogeneous fecundability, however, the most fecund will conceive first and leave the population at risk. The density of waiting times to conception will thus be shifted to the right.

Another area of demographic research in which heterogeneity has come to play an important role is in the analysis of marriage and divorce. Although there would seem to be no particular genetic component to the story, the relationship between cohabitation and the probability of divorce, analyzed in Hoem's chapter in this volume, is an instructive example with wide analytical implications. Hoem finds, as have others, that Swedish couples who cohabit before marriage have a *higher* probability of divorce after marriage than do couples who do not cohabit, but few observers of this empirical regularity would take it at face value. Rather than interpreting this simple correlation as implying that cohabitation has a negative effect on the viability of a marriage, most would provide some kind of a heterogeneity explanation. Couples who cohabit are different than couples who do not, and the way that they are different makes them more likely to divorce as well as more likely to cohabit. Most observers, whether statistician or lay person, would take the view that (1) cohabitation does not occur randomly among couples, and that (2) cohabitation is unlikely to actually increase the probability of a subsequent divorce. The underlying heterogeneity might be thought of as the ''frailty'' of marriage, with the empirical problem being that this frailty is likely to be highly correlated (due to some common determinant, loosely identified as independence or liberal values) with the probability of cohabitation prior to marriage.

Issues such as these can be found in empirical research throughout the social sciences. An example that has been analyzed extensively in econometric treatment of duration models is the analysis of data on unemployment spells in labor economics (see, for example, Heckman and Singer, 1984a,b). Does an observation that the hazard rate for leaving the pool of unemployed rises with duration of the spell mean that there is a ''duration dependence'' in the probability of finding a new job, possibly indicating a stigma of unemployment? Or, alternatively, is it explained by worker heterogeneity, with more employable workers leaving the unemployed pool earlier, making the population of unemployed increasingly made up of those with the lowest hazard rates? ''Frailty'' in this case is the propensity to find a job, but appropriately translated, the example is identical to the mortality example given earlier.

In most social science research of this type there is only limited interest in unobservable heterogeneity *per se*. It is primarily a nuisance, complicating the identification of some more direct empirical issue. It is important from a public health perspective to know whether the mortality effects of a health crisis are declining over time. Interest in learning the distribution of underlying frailty in the population derives from the fact that this heterogeneity may cause misleading inferences from mortality data. An indication of the fact that the underlying heterogeneity is of little direct interest is that a number of empirical solutions to the heterogeneity problem, such as ''fixed effect'' (dummy variable) models based on repeated events, circumvent the problem without directly estimating the distribution of heterogeneity (see Vaupel, Chapter 10).

GENOTYPIC AND PHENOTYPIC VARIATION

In many of the areas of empirical research in which heterogeneity has received attention, there is little to guide researchers on the nature of the distribution of unobserved heterogeneity. One of the intriguing aspects of the intersection of genetics and demography in the area of heterogeneity is that genetics may provide a basis for imposing structure on the form of the heterogeneity. In contrast to the demographic research discussed above, in which unobserved heterogeneity is a nuisance, geneticists are directly interested in the distribution of heterogeneity itself, and in the proportion of the total (phenotypic) variation ascribable to genetic differences (genotypic variation).

Weiss (Chapter 12) provides an interesting example of such a possibility. Weiss describes a curious result in the epidemiological modeling of cancer incidence. Most cancers for which good population data are available show age-incidence patterns that can all be described by the same family of hazard functions, a "multi-hit" survival model, representing a multi-stage or multi-event developmental process. While that model is general enough to fit most data sets, given some flexibility of the parameters, the two parameters of the model are very highly correlated across cancers, geographic settings, and perturbations of environment and genotype.

Those who are genetically predisposed to a particular cancer, having inherited one or more of the "hits," show a general leftward shift and steepening of the hazard function. Different cancers have different intercepts, but the same slope. There is nothing intrinsic to the mathematics or statistics that would produce this result, since there are some striking exceptions; Weiss (Chapter 12) argues that this regularity has its roots in the biology of cancer, involving (as it does) sets of related genes and developmental processes.

Weiss searches for an explanation in the genetic biology of these various cancers, beginning with the observation that many cancers arise from deregulated growth behavior at the cellular level. Many cancers are now known to have at least a partial genetic etiology, arising from genetic defects in growth regulation in different tissues, leading to perturbations in development. Several of the genes involved are now thought to be evolutionarily homologous, representing ancient duplications of single ancestral genes. Different developmental processes in different tissues are controlled by a family of related genes. Weiss argues that the parallelisms of the onset pattern of different cancers and the high correlation of the two parameters of the hazard function are natural consequences of this evolutionary history.

While Weiss' interpretation of the deeper biological and evolutionary meaning to be attached to a generic two-parameter, multi-hit hazard model and its evolutionary implications is perhaps controversial, the fact that one can describe a whole class of diseases with a single family of two-parameter hazard functions, and then reduce the problem to a single dimension (the two parameters being almost perfectly correlated) is intriguing. Given the difficulties expressed elsewhere in this volume over the question of how to choose a biologically reasonable frailty distribution, the regularities described by Weiss should prove empirically useful, whatever their ultimate meaning.

DEALING WITH HETEROGENEITY

Significant advances have been made in dealing with heterogeneity in empirical research. It might be argued that one of the most important contributions has simply been increased understanding of the potential for misinterpretation in the presence of unobserved heterogeneity. Systematic modeling of heterogeneity not only provides a warning of the problems caused by heterogeneity; it also suggests solutions, at least in principle, to those problems. Trussell and Rodríguez (Chapter 8) discuss many of the important contributions to this literature, including a number of tests of the robustness of various techniques to assumptions imposed on both the distribution of heterogeneity and the form of the underlying hazard function.

Approaches to modeling heterogeneity

A great deal has been learned from analytical modeling of heterogeneity. Trussell and Rodríguez (Chapter 8) explain the use of three powerful analytic techniques (hazard regression, binomial regression, and Poisson regression), the choice being made on the basis of data form and mathematical convenience, and show that heterogeneity modeling can be used with each. They provide a somewhat pessimistic picture about the ability to identify structural relationships in the data in the presence of unobserved heterogeneity. Trussell and Rodríguez show that one can fit either a simple functional model (with heterogeneity) or a more elaborate model (without) to the same set of data with equal facility. That choice is inevitably a matter of modeling taste.

One way to analyze the effect of heterogeneity on observed empirical patterns is by simulation models. Heckman and Singer (1984a,b) and Trussell and Richards (1985), for example, use Monte Carlo simulations to analyze the robustness of alternative estimation strategies to alternative forms of both the underlying hazard and the distribution of unobservables.

Biological and genetic models

Given unobserved heterogeneity, empirical analysis requires either some functional form assumptions or a nonparametric analogue (see Hoem, Chapter 9). In demographic research, the frailty distributions are chosen primarily for mathematical convenience, with some guidance being provided by Monte Carlo simulations to test the sensitivity of estimates to functional form assumptions. Heckman and Singer (1984a,b) provide a nonparametric alternative, an example of which is presented by Hoem (Chapter 9).

In genetics, there are a variety of well-developed functional forms for hidden heterogeneity, deriving from standard models of Mendelian inheritance or its quantitative genetic analogue. There are thus some constraints on the forms that heterogeneity can take and some population theory against which to compare the outcome of analysis.

An attraction of modeling heterogeneity with obvious biological components is that genetics may provide a basis for identifying the distribution of that heterogeneity. In addition to using simulation models to explore the consequences of heterogeneity and potential treatments of it, models can also be instructive regarding the determinants of heterogeneity.

Wood and Weinstein (Chapter 11) simulate heterogeneity in fetal loss probability, and show that such variation can be a major determinant of heterogeneity in fecundability. Relying on previous empirical work, they point out that there is considerable variation in fecundability (adjusted for age) in most populations, translating into a right-skewed distribution of interbirth intervals. They begin with the (genetically plausible) premise that there is heterogeneity in the probability of fetal loss (adjusted for age), and query the extent to which the distribution of interbirth intervals can be traced to fetal loss rate heterogeneity.

Assuming a β-distribution for fetal loss probability within an age class, they examine the impact of increasing the mean probability of loss with age. They also compare the impact of increasing variance for this distribution (with age), relative to the impact of a constant variance. They discover that even with a constant variance, the imposition of a heterogeneous frailty distribution within an age class yields a minimal change in the mean fecundability but a dramatic increase in the variance.

This same heterogeneity of fetal loss probability shifts the waiting time distribution to the right, doubling the expected waiting time (relative to the fetal loss homogeneity model). With a variance that increases with the mean, for which there is some empirical justification extraneous to their data sets, the right-shift in the waiting time is drastic and the heterogeneity in fecundability itself becomes substantial.

Wood and Weinstein's objective is not so much to measure the magnitude of hidden heterogeneity in fecundability as it is to trace the source of that heterogeneity to fetal loss probability. The effort is dependent of the particular model they have used, but is instructive nonetheless. There is a growing suspicion that some of the fetal loss rate heterogeneity may have a genetic component, and they interpret their results as indicating that physiological factors may be substantially more important than behavioral factors in determining heterogeneity in fecundability. In addition to its direct importance to our understanding of reproductive physiology, this conclusion has implications for research on fertility more generally. Because of the potentially confounding role of heterogeneity in fecundability in all fertility research, the source of that heterogeneity can have important implications. If it is behavioral, resulting, for example, from variations in coital frequency, there may be correlations between that heterogeneity and other observable determinants of fertility that need to be accounted for in empirical research. If it is primarily biological, then assumptions that unobserved fecundability is uncorrelated with observables under investigation may be more defensible.

TURNING UNOBSERVABLES INTO OBSERVABLES

The philosophical difficulties of modeling unobservables are almost insurmountable; the ideal solution would be to turn a set of unobservable z's into a set of observable x's. This can be done in two different ways. The first is illustrated by Vaupel (Chapter 10), who develops a methodology for using monozygotic twins as replicated observations on the unobservable vector z. The second strategy is to define and characterize candidate measures for the z's.

Related individuals and repeated events

An approach that has been used for some time in both demography and genetics to deal with unobserved heterogeneity is the use of repeated events. A standard demographic technique, for example, is a "fixed effect" model, in which individuals or families are assumed to differ by some additive effect that remains constant over time or across family members. Given repeated observations of the same individual, or multiple observations from the same family, this "fixed effect" can be factored out, leaving only the systematic relationships of concern to the researcher.

Vaupel (Chapter 10) provides an approach to frailty modeling in the case where there are repeated and identifiable observations of each frailty class, and gives an example of how this can be used for twin analysis, where the genetically identical members of each twin pair represent replicate observations on the unobservable frailty distribution. The twin example is just an illustration, but the strategy is reminiscent of the traditional quantitative genetic methodology, where the covariate affecting frailty is genotype.

In the terminology of quantitative genetics, while one still cannot observe the covariate, its effect is replicated, and that provides useful information on the distribution of frailty. We can imagine a number of genetic applications, where the trait of interest behaves in failure time mode: age of onset patterns for multifactorial diseases (see Schull, Chapter 14, and Ott, Chapter 16), incidence patterns for carcinoma development (see Weiss, Chapter 12). As we have pointed out (Smouse and Teitelbaum, Chapter 13), this is a standard strategy in human population genetics and one that has been widely efficacious. One need not use monozygotic twins; other sorts of relatives can be used as well (Vaupel, Chapter 10). One need not use exact replicates, as long as one can specify the degree of correlation among relatives of any kind. This strategy is widely used in genetic epidemiology and could be more widely deployed in demographic work to advantage.

Measured genotypes

A second strategy for turning unobservable z's into observable x's is to define and characterize candidate measures for the z's. In genetic work, this is called the "measured genotype" approach, and consists of identifying candidate genetic markers that might be involved in (or predictive of) the disease process itself. There is no guarantee that any given candidate will be predictive. We effectively obtain an expanded x-vector, while reducing the uncertainty encompassed by the unobservable.

Examples of common diseases are emerging that have yielded increased resolution with a small number of parameters for "measured genotypes." The trick, of course, is to choose reasonable candidate genes to assay, genetic markers that can plausibly be expected to be involved in the relevant developmental processes (see Smouse and Teitelbaum, Chapter 13, and Schull, Chapter 14).

Choice among models: The need for prior restrictions

Trussell and Rodríguez describe the difficulties of comparing simple models containing heterogeneity with more complicated models without. Given enough parameters, one can fit any set of data to perfection, and beyond a certain point, the choice of those parameters is irrelevant and uninformative. One can describe the data with simple models using observed covariates, sweeping most of the unexplained variation into a convenient unobservable, or one can fit a very detailed specification of the observables, leaving out the unobservables; in either case, the result comes out about the same. One needs some extra information to resolve the situation.

Put another way, we need to impose some prior information in order to draw meaningful inferences. The chapters by Hoem (Chapter 9), Vaupel (Chapter 10), Wood and Weinstein (Chapter 11), and Weiss (Chapter 12), can all be thought of as imposing priors of some kind, but the nature of those priors and the ways that they are applied are quite different. Taken together, these papers provide an instructive set of exercises in attempting to deal with unobserved heterogeneity. On the one hand, they are the source of some optimism about the possibility of using biology and genetics to impose structure on heterogeneity. On the other hand, they demonstrate that even when a strong case can be made for appealing to genetic variation, we have a considerable distance to travel before there is a clear "right way" to deal with heterogeneity. They also provide a reminder that, since not all heterogeneity is genetically based, such solutions will never provide a universal solution to all the heterogeneity problems which plague social science research. When it is heterogeneity in tastes or values that lead to misleading inference, as in the case of Hoem's chapter on cohabitation and divorce, an appeal to genetic heterogeneity would require a major leap of faith, no less extreme than that required to adopt any other particular assumption about the distribution of unobserved heterogeneity.

DISCUSSION

Statistical treatments of heterogeneity have become increasingly sophisticated. Formal analysis of heterogeneity has clarified the problems it can create in empirical research, and has helped motivate a variety of statistical techniques to deal with heterogeneity. A great deal has been learned from Monte Carlo simulations about the robustness of these techniques and their ability to distinguish between alternative interpretations of the data.

In spite of the considerable progress that has been made in dealing with heterogeneity, however, empirical researchers are far from having a simple recipe for success. Trussell and Rodríguez (Chapter 8) provide a somewhat pessimistic view, pointing out that simple structural models containing heterogeneity may be indistinguishable from more complicated models without heterogeneity.

An instructive example is provided by Hoem's analysis of conjugal union disruptions in Sweden. As pointed out above, this is a case where there is a strong presumption of heterogeneity. But while much demographic research may have enough of a biological foundation to provide a hope for guidance in modeling unobserved heterogeneity, Hoem's example is an important reminder that a large part of demographic research cannot reasonably appeal to genetics. The distribution of "progressiveness" may or may not behave like the distribution of genetic frailty.

Hoem, in a sense, confirms some of the pessimism of Trussell and Rodríguez. Even with a strong prior that heterogeneity must be present, Hoem is not able to identify it well enough to estimate the model he believes to be appropriate. Hoem demonstrates that even with a fairly large data base, it is easy to get into the unfortunate situation of asking too much of the data; there is not enough information to identify a large number of parameters with any confidence. The only solution is to model with some care and to use some common sense.

A more optimistic view might be drawn from the chapters by Vaupel (Chapter 10), Wood and Weinstein (Chapter 11), and Weiss (Chapter 12). These chapters all offer some hope that in areas where there is a clear genetic or biological process to appeal to, it may be possible to impose some structure on unobserved heterogeneity. Even if one is willing to draw optimism from Weiss' view that genetic models may provide the basis for modeling unobservable heterogeneity in analyzing mortality, or from Wood and Weinstein's instructive modeling of heterogeneity in fecundability, these biological links will never be able to solve all problems of unobserved heterogeneity in the behavioral side of demography. In the case of marriage, for example, a fundamental demographic process, there is little biological data to appeal to in modeling the distribution of heterogeneous preferences that lead to results such as the the apparent negative effect of cohabitation on marital stability.

The extent to which demographic processes are dominated by behavior versus biology is, of course, a matter of degree. In the case of mortality and morbidity, biology is of fundamental importance in determining the distribution of frailty. It seems likely that it is in these areas where a convergence of demography and genetics is likely to offer the greatest rewards in dealing with unobserved heterogeneity. Behavior remains important, however, even in mortality and morbidity. Even if they provide quantitatively a relatively small part of the variance, behavioral effects may be more critical in creating the potential for misleading inference. Thus, for example, as Fisher (1957; 1958a; 1958b; 1958c) had suggested a number of years ago, the correlation between an observable such as smoking and a variable that might be unobservable, such as income, could be an important issue in interpretation of mortality data. There are even interactions between observable behavioral factors (smoking) and unobservable genetic ones (such as the genotype for the α-1 anti-trypsin locus, involved in emphysema; Marx, 1989) that could impinge on morbidity and mortality.

Analysis of fertility provides an intriguing middle ground between mortality and marriage, with both biology and behavior clearly playing important roles in determining reproduction (Fisher, 1958d). Wood and Weinstein (Chapter 11) have shown that a model grounded primarily in reproductive physiology can improve our understanding of both the behavioral and biological determinants of fecundability. As pointed out by Trussell and Rodríguez (Chapter 8), unobserved heterogeneity has become an important issue in fertility analysis, with much of this heterogeneity having a biological foundation. It seems likely that this is an area where there is a good deal more to be gained from exploring the intersection of genetics and demography.

Analysis of heterogeneous frailty provides an intriguing area of convergence for genetics and demography. From one perspective, it is an area in which there may be a valuable complementarity in the two areas of research. A major nuisance to one set of researchers is ''bread and butter'' for the other. At the same time, geneticists are likely to benefit from the powerful set of statistical tools that demographers have developed while struggling to deal with the problems unobserved heterogeneity creates for statistical inference. It would be overly simplistic and overly optimistic to suggest that genetics offers the solution to the problems of unobserved heterogeneity in demographic research. Nonetheless, the analysis of heterogeneous frailty is an exciting area of convergence between genetics and demography, and an area in which there are likely to be significant returns from cross-disciplinary interaction.

REFERENCES

Bound, J., and T. Waidmann, 1989   Disability transfers and the labor force attachment of older men: evidence from the historical record. Discussion Paper No. 899-89, University of Wisconsin Institute for Research on Poverty.

Fisher, R. A., 1957   Dangers of cigarette smoking. British Medical Journal 2: 297-298. Reprinted in the collected works of R. A. Fisher 1972, Volume V, edited by J. H. Bennett. University of Adelaide Press.

Fisher, R. A., 1958a   Cigarettes, cancer and statistics. Centennial Rev. 2: 151-166. Reprinted in the collected works of R. A. Fisher 1972, Volume V, edited by J. H. Bennett. University of Adelaide Press.

Fisher, R. A., 1958b   Lung cancer and cigarettes? Nature 182: 108. Reprinted in the collected works of R. A. Fisher 1972, Volume V, edited by J. H. Bennett. University of Adelaide Press.

Fisher, R. A., 1958c   Cancer and smoking. Nature 182: 596. Reprinted in the collected works of R. A. Fisher 1972, Volume V, edited by J. H. Bennett. University of Adelaide Press.

Fisher, R. A., 1958d   *The Genetical Theory of Natural Selection.* Second Revised Edition. Dover, New York.

Heckman, J., and B. Singer, B., 1984a   A method for minimizing the impact of distributional assumptions in economic models for duration data. Econometrica 52: 271-320.

Heckman, J., and B. Singer, 1984b   Econometric duration analysis. J. Econometrics 24: 63-132.

Kalbfleisch, J. D., and R. L. Prentice, 1980   *The Statistical Analysis of Failure Time Data.* John Wiley, New York.

Marx, J. L., 1989   Getting to the heart of genetic disease. Science 243: 315-316.

Vaupel, J. W., K. G. Manton, and E. Stallard, 1979   The impact of heterogeneity in individual frailty on the dynamics of mortality. Demography 16: 439-454.

Vaupel, J. and A. Yashin, 1985a   The deviant dynamics of death in heterogeneous populations, pp. 179-211 in *Sociological Methodology 1985*, edited by N. Tuma. Jossey-Bass, San Francisco.

Vaupel, J., and A. Yashin, 1985b   Heterogeneity's ruses: some surprising effects of selection on population dynamics. Am. Statistician 39: 176-185.

Trussell, J., and T. Richards, 1985   Correcting for unmeasured heterogeneity in hazard models using the Heckman-Singer procedure, pp. 242-276 in *Sociological Methodology 1985*, edited by N. Tuma. Jossey-Bass, San Francisco.

8

Heterogeneity in Demographic Research

JAMES TRUSSELL AND
GERMÁN RODRÍGUEZ

Concern with heterogeneity in demographic research is as old as the discipline itself. The very first life tables constructed by John Graunt in the seventeenth century explicitly incorporated heterogeneity in the propensity to die at different ages. The literature since that time is filled with studies of fertility, nuptiality, mortality, and migration differentials across observable characteristics such as race, ethnicity, gender, age, period, cohort, and region. The search for differentials is simply an attempt to stratify the population into distinct groups, with the groups being as demographically homogeneous as possible. A more recent strand of demographic research concerns fairly general models of mortality, nuptiality, fertility and migration (Coale and Demeny, 1966; Coale, Demeny and Vaughan, 1983; Coale and Trussell, 1974; Rogers and Castro, 1981). Whereas those concerned with demographic variation explicitly look for differences in behavior among strata, model builders search for commonalities. The two strategies might at first appear to be contradictory, but in fact they are complementary. The goal in either case is to separate the population into the minimum number of groups sufficient to capture the important sources of variation.

Both approaches, one involving the search for empirical regularities and the other the search for differentials, are essentially descriptive in nature. Yet the ultimate goal of demographic research is prediction, not description. Demographers have long been concerned with measuring the effects of predictor variables on outcome variables. For example, the study of the demographic transition and the evaluation of intervention programs to reduce infant mortality and teenage fertility are (partly) aimed at discovering effective policy instruments for achieving policy objectives. The problem in the social sciences of drawing causal inferences from observational (as contrasted with experimental) data is both old and important; we will not explore it here, but instead refer the interested reader to the excellent discussion by Marini and Singer (1988).

111

This chapter will focus on three *statistical* models appropriate for the analysis of demographic data: hazard regression or multivariate life tables (suitable for event history analysis), Poisson regression (appropriate for rates or counts of events, conditional on exposure to risk), and binomial regression (suitable for current status data, with logistic regression as a special case). Incorporation of observed predictors (which capture *observed* heterogeneity) is by now relatively well understood, at least in the context of logistic regression and hazard regression. Although binomial regression on current status data and Poisson regression have seldom been used in demographic analysis, principles of estimation are relatively straightforward, as we will demonstrate. There has been increasing concern lately with the effect of *unmeasured* or *unmeasurable* predictors (hereafter in either case denoted *unobservables*) in demographic research. One source of such unobserved heterogeneity is genetic variability, which may be particularly important in the analysis of mortality. We explore ways in which unobservables may be incorporated in each statistical model. The discussion is heavily weighted toward hazard models, simply because the literature is richer and we ourselves have worked more extensively on this case. Before turning to the statistical models, we explore briefly variations in modelling philosophy in different strands of demographic research.

Because demographers are trained in a variety of disciplines, their philosophies of modelling differ. Economists, for example, often refer to the *true* model and explore the consequences of model misspecification caused by omitting relevant predictor variables (Judge *et al.*, 1980). Modelling unobserved heterogeneity is often viewed as a way of correcting for model misspecification, thereby allowing one to draw causal inferences. Our own approach to modelling is essentially descriptive, although we seek ultimately to understand demographic behavior. We examine confounding effects by analyzing changes in estimated effects of a covariate as other covariates are added to a model. We search for parsimonious models that faithfully represent the data, in the sense of yielding fitted values of the outcome variable that are close to the observed values. Our selection of covariates is generally based more on data availability and a set of specific research questions than on all-encompassing theories of behavior. We proceed inductively, by placing models at the intersection between data and theory, with each informing the other through empirical tests. We model unobserved heterogeneity when doing so would yield a better-fitting or more parsimonious model, but we are cautious in interpreting the results when identification rests on untestable assumptions.

HAZARD REGRESSION

Suppose that we wish to analyze the determinants of failure times, where failures simply refer to events that signal the end of a state and do not imply literal lack of success; for example, first marriage marks the end of the single state. We assume that T_i, the failure time for the ith individual, has a distribution with density $f(t_i/\Omega_i)$, where Ω_i is a vector of parameters, some or all of which may vary across individuals. Associated with each individual is an observed vector of predictor variables x_i and a vector of parameters or coefficients β. Both of these vectors could be functions of time, to allow for predictor variables whose values or effects changed over time; for notational convenience, we confine attention here to fixed covariates with fixed effects.

There are several possible ways to proceed. For example, we might specify that $\log(T_i) = x_i'\beta + \epsilon_i$, where ϵ_i is a random error with density $f(\epsilon)$. This specification leads to the *accelerated failure time model*, in which the effect of covariates is to alter the speed at which individuals move along the time axis (Kalbfleisch and Prentice, 1980). For convenience, we will consider instead modelling the hazard function:

$$h(t_i|\Omega_i) = f(t_i|\Omega_i)/S(t_i|\Omega_i) , \qquad (8\text{-}1)$$

where $S(t_i|\Omega_i)$, the survivor function, is the complement of the distribution function $F(t_i|\Omega_i)$. Suppose we assume that the log-hazard for each individual can be expressed as:

$$\log[h(t_i)|\psi, x_i, \beta, z_i, \gamma] = \alpha(t_i|\psi) + x_i'\beta + z_i'\gamma , \qquad (8\text{-}2)$$

where ψ is a vector of parameters (a subset of Ω), common to all individuals, governing the shape of the underlying hazard [whose functional form is given by $\alpha(t)$] and z_i is a vector of unobserved characteristics. For example, when studying the determinants of child mortality, the analyst cannot observe the underlying *frailty* of each individual (Vaupel, Manton and Stallard, 1979). Likewise, the *fecundity* of a woman is not observable to the investigator analyzing waiting time to conception (Sheps and Menken, 1973).

Suppose that the unobserved variables z_i are not included in the model. Then the inevitable consequence is that individuals with high risk values of z will fail earlier than those with low-risk values of z. Therefore, as time goes by, the apparent risk (not controlling for the effects of z_i) either will not rise as fast or will decline more rapidly than it would have, had the effects of z_i been controlled. This result is well-known in the demographic literature, and extensive discussions can be found in Sheps and Menken (1973), Keyfitz (1985), Cohen (1986), and in a series of articles by Vaupel and his colleagues (Vaupel, Manton and Stallard, 1979; Vaupel and Yashin, 1985a,b). What happens to the estimates of ψ and β when the analyst fails to include z_i in the model? We might initially conclude on the basis of well-known results for the linear regression model that these parameter estimates would not be contaminated so long as the z_i are independent of the x_i. In fact, however, even this result does not hold for survival analysis, as a simple example will show. Suppose that the log-hazard is given by

$$\log[h(t_i)|\alpha, \beta, \gamma, X_i, Z_i] = \alpha + \beta X_i + \gamma Z_i . \qquad (8\text{-}3)$$

where X_i is a dummy variable that takes the value 0 unless the individual is black, in which case it is 1, and Z_i is another dummy variable that takes the value 0 unless the individual is a female, when it takes the value 1. Suppose further that the values of α, β, and γ are -2.30, 0.69, and 0.91, respectively. Then the risk for white males is $\exp(-2.3) = 0.10$, the risk for white females is $\exp(-2.3 + 0.91) = 0.25$, the risk for black males is $\exp(-2.3 + 0.69) = 0.20$, and the risk for black females is $\exp(-2.3 + 0.91 + 0.69) = 0.50$. Suppose that initially (at $t = 0$) race and gender are independently distributed; for example, for example, let there be 50 white males, 50 white females, 100 black males and 100 black females. The problem is that at any time after $t = 0$, the distributions cannot remain independent. To see that this statement is true, note that at time $t = 5$ there would be $50\exp(-0.5) = 30.3$ white males, $50\exp(-1.25) = 14.3$ white females, $100\exp(-1.0) = 36.8$ black males and $100\exp(-2.5) = 8.2$ black females. Therefore, whereas males constituted initially 50% of both the black and white groups, after five years the group of blacks is 8% male but the group of whites is only 68% male. Because there are different rates of attrition, the composition of the sample cannot remain fixed.

Had we omitted the factor gender and examined the hazards for blacks and whites, we would find that both hazards decline, because the females, who have higher risks than the males, "die faster." Among blacks, for example, the risk of death is initially 0.5 for the 100 females and 0.2 for the 100 males, so that the observed risk would be 0.35. After five years the observed risk will have fallen to 0.25 [$=0.82(0.2)+0.18(0.5)$]. Among whites, the initial risk is 0.175; after 5 years this risk falls to 0.15. Although the net effect (net of gender) of being black is to double the risk of death at all ages, the observed risk will be double only at $t = 0$. After that time the observed (gross) effect declines as the black females are quickly selected out; later, when both white and black females have been virtually eliminated, the relative risk approaches 2.0 once more. As a description of the sample (and the population from which it was drawn), these gross risks for blacks and whites have a straightforward interpretation. Furthermore, predictions of the mortality risks for blacks and whites would be valid, so long as the distributions of race and gender in the population for which the prediction is made are roughly the same as the sample from which estimates were generated. Of course, if model (8-3), including the effect of gender, had been examined, then the more refined net effects would be obtained; these could be used for prediction in a population with different distributions of race and gender, because calculations could be made for each of the four relevant groups.

The omitted variables problem introduced above has no real solution, in the sense that if the model is given by equation (8-2) and z_i and x_i are jointly dependent, then β cannot be recovered without a lot more knowledge about the joint distribution of the omitted and included variables than will ever be available. The developing literature on omitted variables in proportional hazards models seems to suggest that an "analysis based on the proportional hazards model will reflect the relative importance of the {included} covariates when the true model is accelerated failure time or proportional hazards with omitted covariates. In the latter case {estimated regression coefficients will be} asymptotically biased toward zero. The size of the bias, however, will be small unless {the effect of the omitted covariate is large}" (Struthers and Kalbfleisch, 1986). Omitting a dichotomous covariate has fairly modest effects on efficiency when that covariate splits the sample into a very large group and a very small group. For more equal splits, the loss of efficiency can be as much as 50% or more when the omitted covariate has a large effect on survival (Lagakos and Schoenfeld, 1984; Morgan, 1986; Lagakos, 1988).

If a model that includes only observables does not fit the data well (cf. Heckman and Walker, 1987), then one may attempt to improve the fit by explicitly incorporating a source of variation independent of the x_i. Two primary strategies have been adopted. In the first strategy, the analyst assumes a distributional form for the unobservable and integrates it out of the likelihood function before maximizing. This procedure is discussed at length by Sheps and Menken (1973) in their excellent book on modelling conception and birth. It was applied recently in a study of the effects of covariates on birth interval length in Costa Rica by Newman and McCulloch (1984). The second strategy is a variant of the first, proposed by Heckman and Singer (1982); a discrete approximation to the distribution of the unobservable is estimated to avoid the necessity of choosing the correct functional form for the distribution. In the next two sections, we discuss these alternatives.

Assuming a functional form for the distribution of the unobservable

Consider an alternative version of the hazard model (8-2) given by

$$\log\,[h(t_i)\psi, x_i, \beta, \theta_i] = \alpha(t_i|\psi) + x_i'\beta + \theta_i \, , \tag{8-4}$$

where θ_i is an unobservable, independent of x_i. The density of T_i, conditional on θ_i is given by $f(t_i|\theta_i) = S(t_i|\theta_i)h(t_i|\theta_i)$, where

$$S(t) = \exp\left[-\int_0^t h(u)du\right] . \tag{8-5}$$

If we could observe θ, then we could proceed directly to maximize the log-likelihood, subject to some imposed functional form for $\alpha(t|\psi)$. Since, however, θ is not available to the analyst, another strategy is required. To eliminate the θ_i, we multiply $f(t_i|\theta_i)$ by $f(\theta_i)$, the *marginal* distribution of the θ_i, to obtain the joint distribution of the T_i and the θ_i; finally, we integrate over the θ_i to obtain the marginal density of the T_i.

To illustrate this approach, suppose in equation (8-4) that there are no observed covariates relevant to the model ($\beta = 0$) and that $\alpha(t) = 0$. In this case, the underlying hazard $[\exp(\theta_i)]$ is constant for each individual but individual identifiers are unobservable. Then, if we replace $\exp(\theta_i)$ by γ_i, we see that the conditional density of T_i simplifies to

$$f(t_i|\gamma_i) = \gamma_i \exp(-\gamma_i t_i) . \tag{8-6}$$

Next, assume that the γ_i are independently and identically gamma distributed with mean η/τ and variance η/τ^2:

$$f(\gamma_i|\tau, \eta) = \tau^\eta \gamma_i^{\eta-1} \exp(-\tau\gamma_i)/\Gamma(\eta) \, , \tag{8-7}$$

where $\Gamma(\eta)$ is the gamma function. The joint density for each pair of T_i and γ_i is the product of equations (8-6) and (8-7):

$$f(t_i, \gamma_i) = \gamma_i^\eta \exp[-\gamma_i(\tau + t_i)]\tau^\eta/\Gamma(\eta) . \tag{8-8}$$

To integrate over γ_i in equation (8-8), we use the conventional trick of multiplying and dividing by $(\tau + t_i)^{1+\eta}/\Gamma(1+\eta)$ to obtain:

$$f(t_i|\tau,\eta) = [\tau^\eta \Gamma(1+\eta)]/[(\tau+t_i)^{1+\eta}\Gamma(\eta)] \int f(\gamma_i|\tau+t_i, 1+\eta)d\gamma_i$$

$$= [\tau^\eta \Gamma(1+\eta)]/[\tau+t_i)^{1+\eta}\Gamma(\eta)] = \eta\tau^\eta/(\tau+t_i)^{1+\eta} \qquad (8\text{-}9)$$

where the integral disappears because the integrand is the density of a gamma with parameters $\tau + t_i$ and $1 + \eta$, and where $\Gamma(1+\eta)/\Gamma(\eta)$ equals η. Equation (8-9) is the *unconditional* density for each of the T_i. Note (with hindsight) that the gamma distribution was chosen for $\gamma_i = \exp(\theta_i)$ because there is a closed form for the integral. In general, however, there will be no closed form solution to the integral, so that numerical techniques must be used for computation.

Adding the covariates x_i and the associated coefficients β to the model is not difficult now that we have worked out the algebra for the simpler case. The hazard for each individual would be constant at $\exp(x_i'\beta + \theta_i) = \gamma_i\exp(x_i'\beta)$. We assume, as before, that γ_i has a gamma distribution with parameters η and τ. Then it follows that $\gamma_i\exp(x_i'\beta)$ has a gamma distribution with parameters η and $\tau\exp(-x_i'\beta)$. Therefore, to obtain the unconditional density of T_i with covariates, we simply replace τ with $\tau\exp(-x_i'\beta)$ in equation (8-9). Before doing so, we note that it is not possible in this model to identify separately the parameters τ, η and the constant term β_1 in the linear form $x_i'\beta$; hence, we normalize the gamma to have mean 1 and variance $\sigma^2 = \eta^{-1}$ to obtain:

$$f(t_i|\eta,\beta,x_i) = \eta[\eta\exp(-x_i'\beta)]^\eta/[\eta\exp(-x_i'\beta) + t_i]^{1+\eta} . \qquad (8\text{-}10)$$

How well does this approach work? In principle, it ought to work well if the model specification is consistent with the data. Therefore, the answer is an empirical one that cannot be known until one estimates the parameters and examines goodness of fit.

The Heckman–Singer solution

Heckman and Singer (1982, 1984b), having observed that estimates of β are sensitive to the distribution assumed for the unobservable, reasoned that a more robust approach was needed. Following Laird (1978), they proposed estimating the mixing distribution using a nonparametric maximum likelihood approach. Specifically, the procedure involves expressing the unknown density for θ as a discrete probability function with mass π_j concentrated at relatively few support points ν_j:

$$Pr(\theta_i = \nu_j) = \pi_j \;, j = 1\ldots, J . \qquad (8\text{-}11)$$

In this formulation, the π_j must sum to one, but there is no restriction on the values of ν_j. This same probability function applies to all θ_i. The unconditional density of T_i becomes:

$$f(t_i|\psi,x_i,\beta,\nu,\pi) = \Sigma_j [(S(t_i)h(t_i)\pi_j] , \qquad (8\text{-}12)$$

where $h(t_i|\psi,x_i,\beta,\nu_j) = \exp[\alpha(t_i|\psi) + x_i'\beta + \nu_j]$ and $S(t_i)$ is given by equation (8-5). The log-likelihood must be maximized numerically. One problem with this approach is that the analyst cannot in general be certain that a unique global maximum exists. Therefore, one must generate several sets of starting values and check that the optimization algorithm converges to the same point. The cost of applying the Heckman-Singer correction can therefore be very high. (Software, called CTM, to analyze such models and a manual are available at nominal charge from James Heckman, Department of Economics, Yale University).

Heckman and Singer have found in many applications that only a very small number of support points, ν_j, two or three, or perhaps four, are needed. Adding more fails to improve the likelihood. In several applications, two support points will suffice, and either ν_1 or ν_2 drifts toward $-\infty$; such models in the analysis of migration or mobility are called mover-stayer models, because part of the sample has essentially zero risk of moving (Blumen, Kogan, and McCarthy, 1955).

Trussell and Richards (1985) and Montgomery and Trussell (1986) have carried this analysis further. They reasoned that models estimated with the Heckman-Singer correction for heterogeneity could be sensitive to the functional form for the shape of the hazard governed by $\alpha(t)$ in equation (8-4), just as Heckman and Singer had found that results can be sensitive to the choice of distribution for the unobservable. Those suspicions were confirmed in analyses of the determinants of birth interval length and of child mortality in Korea.

Two other analyses did not find the same instability. Marini (1985) found no sensitivity in parameter estimates when she examined models of marital disruption and specified the distribution of the unobservable as (alternatively) a gamma or a log-normal; however, these two distributions are very similar, so that one might not expect to find much sensitivity. Likewise, Ridder and Verbakel (Ridder and W. Verbakel, unpublished results, 1983) did not find the same sensitivity to the specification of the distribution of the unobservable as did Heckman and Singer. There was, however, an important difference in their approach. Heckman and Singer analyzed their alternative models (each with the same functional form for the hazard) on unemployment data. Ridder and Verbakel, on the other hand, used Monte Carlo techniques to simulate data with a known hazard and a known distribution of the unobservable. Their results show that as long as the hazard is correctly specified, then results are not sensitive to the choice of the distribution of the unobservable. Misspecification of the hazard, however, biases the estimates of all parameters. They conclude that the solution is to specify a flexible functional form for the hazard, such as a piecewise constant hazard [so that $\alpha(t)$ in equation (8-4) would be a step-function], and then apply any convenient distribution for the unobservable. Their approach is also the one more recently adopted by Heckman and Singer (1984a), who suggest a general flexible hazard where $\alpha(t)$ in equation (8-4) is given by

$$\alpha(t) = \alpha + \xi(t^{\psi} - 1)/\psi + \theta(t^{\omega} - 1)/\omega . \tag{8-13}$$

This general form reduces to a Weibull when $\psi = \theta = 0$, to a Gompertz when $\psi = 1$ and $\theta = 0$, to a log-quadratic when $\psi = 1$ and $\omega = 2$, and to an exponential when $\xi = \theta = 0$.

We draw the following conclusions from these varied results; those interested in this topic should also consult the fine discussion in Manton, Singer, and

Woodbury (1988). First, the Heckman-Singer and Trussell-Richards-Montgomery instability findings illustrate the point that an analyst may achieve nothing but a false sense of security by examining a first model with observed covariates and a second model with an unobservable assumed to have a specific functional form, and then concluding that the second model allows better inference because it has corrected for unobserved heterogeneity. The instability and Marini's stability results do not generalize much further, because there was no test of goodness of fit of the models. Such tests might reveal that none of the models fit. Alternatively, the tests might reveal that several models fit the data well, although the estimated coefficients appear dissimilar. Even if such tests did not reveal the latter outcome, *it is always possible to find many models incorporating unobserved heterogeneity and one model not incorporating unobserved heterogeneity that fit equally well because they yield identical unconditional distributions of the outcome variables.* Only external information (perhaps biological or genetic, as contrasted with social or cultural) seldom available to the analyst could be used to distinguish among these alternative models. We explore this identification issue in greater detail in the next section.

Identification

We first consider single spell models (such as the spell from birth to death or the spell from birth to first marriage) and then sketch briefly some results for multiple spell models (such as the spells from one birth to the next).

Single spell models

Let us suppose, without loss of generality, that someone has fitted model (8-4) to data. While there may be some "theory" that motivates the choice of the baseline hazard $\exp[\alpha(\cdot)]$ in some situations, there is currently not sufficient theory to pin down $\alpha(\cdot)$ very precisely in demographic applications. The analyst must assume some distribution $g(\theta|\sigma)$ for θ_i, such as a gamma or the nonparametric representation of Heckman and Singer (1984a,b). To proceed, the analyst must write down the likelihood for individual i and then integrate over (or sum over) the unobservable θ. Then the maximum likelihood estimates of ψ, β, and σ are obtained. This analyst, whom we hereafter dub the "first analyst," explicitly modelled an unobservable.

However, given model (8-4), we could (as above) derive the marginal survival and density functions of T_i by integrating over θ, name the resulting distribution the $\pounds$ distribution, and give it to a second analyst who would obtain the same MLEs of ψ, β, and σ. This analyst would not have modelled any unobservable explicitly, and would (probably) be unaware that such a model could be interpreted (or motivated) in such a manner. The two models would fit identically, no matter what definition of fit is used, although the two analysts may interpret their models quite differently.

To illustrate this point, suppose that in model (8-4), $h(t)$ is constant and the distribution of $\exp(\theta_i)$ is a gamma. Earlier, in equation (8-10), we derived the unconditional distribution of survival time T_i. It is identical to the distribution that would be obtained if T_i were assumed to have a Pareto distribution of the second kind

$$f(t) = \kappa\delta / \left[(t + \xi)^{\delta+1} \right] , \qquad (8\text{-}14)$$

with parameters $\delta = \eta, \xi = \eta\exp(-\beta'x_i)$ and $\kappa = [\eta\exp(-\beta'x_i)]^\eta$. The second analyst, using the Pareto, will have found a declining hazard and nonproportional effects of the covariates on the hazard, whereas the first will have discovered a gamma mixture of exponentials. Their results might appear to the casual observer to be different, but their models are observationally identical.

This result can be generalized quite nicely using equations derived by Vaupel, Manton and Stallard (1979) and Hougaard (1984). Let $h_i(\cdot)$ and $h_p(\cdot)$ denote the hazards for an individual and for the population, respectively. Then

$$h_p(t|x) = E[h_i(t|x)|T > t] , \qquad (8\text{-}15)$$

The hazard for the population with covariates x at time t is the average of the individual hazards, where the average is taken over the distribution of the unobservable among those still alive at duration t. Suppose we assume that $h_i(t, x) = \gamma_i\lambda(t, x)$, where $\lambda(t, x)$ is the baseline hazard at time t for persons with characteristics x and γ has a gamma distribution with mean 1 and variance σ^2 *at birth*. It can be shown that

$$h_p(t, x) = \lambda(t, x) / [1 + \sigma^2\Lambda(t, x)] , \qquad (8\text{-}16)$$

where $\Lambda(t, x) = \int_0^t \lambda(u, x)du$.

It is also possible to obtain the inverse of this relation by expressing $h_i(t, x)$ in terms of $h_p(t, x)$. Let the integral of $h_p(u, x)$ from 0 to t be $H_p(t, x)$. Then

$$h_i(t, x) = \gamma_i h_p(t, x)\exp[\sigma^2 H_p(t)] . \qquad (8\text{-}17)$$

To obtain this result, integrate $\lambda(u, x) = h_i(u, x)/\gamma_i$ in equation (8-17) from 0 to t to obtain $\Lambda(t, x) = \{\exp[\sigma^2 H_p(t, x)] - 1\}/\sigma^2$. Substitute this expression into equation (8-16) to obtain $h_p(t, x) = h_p(t, x)$, as required.

Suppose that we estimated a model with only observed covariates x and that the estimated hazard is given by $h_p(t, x)$. Then given any value of σ^2, the resulting $h_i(t)$ can be derived, and this model with an unobservable distributed as a gamma with mean 1 and variance σ^2 will fit exactly the same as the model $h_p(t, x)$ without an unobservable. Suppose, however, that γ was distributed not as a gamma but as an inverse Gaussian with mean 1 and variance σ^2. Then it can be shown that

$$h_i(t, x) = \gamma_i h_p(t, x)[1 + \sigma^2 H_p(t)] . \qquad (8\text{-}18)$$

We have shown that any hazard has one representation without an unobservable and infinitely many representations as a gamma mixture of unobservables and infinitely many representations as an inverse Gaussian mixture of unobservables. Yet other representations as mixtures of unobservables could be found. These would all have the same fitted unconditional hazard. *Nevertheless*, the interpretations of the structural (conditional) models would differ greatly. Therefore, we conclude that one can never separately identify the baseline hazard and the effects of the covariates on that hazard. Identifiability in the sense of estimability is obtained by imposing some structure, such as assuming that the model for individuals takes the form of equation (8-4) with a specific functional form for $\alpha(\cdot)$. It is difficult to imagine, however, that one could convincingly defend such an assumption.

These observations lead us to the following strategy for selecting a model. First, fit a model with no interactions of the covariates with the baseline hazard. If it fits well, then stop. One can then find many equivalent representations as a model with a mixture of unobservables, should one desire to do so. If the model does not fit well, then one can test whether adding a mixture of unobservables adds to the fit by using a procedure outlined in Rodríguez (1988). This procedure, which can be implemented in the framework of generalized linear models (and therefore estimated with the software package GLIM) should be considerably less costly than the approach of Heckman. An alternative would be to specify the baseline hazard $\alpha(\cdot)$ as a semi-parametric step-function and search for interactions between the observed covariates and the hazard in models employing only observed predictors.

Multiple spell models

Suppose that the process we wish to study has multiple spells. If there is unobserved heterogeneity, but its value in one spell is independent of its value in another spell, then we are back in the single spell situation discussed above. If the unobservable is persistent, however, then one cannot analyze spells as independent observations. Nevertheless, one can model each spell conditional on the outcomes of all previous spells. We do not claim that one can simply enter the length of the first spell as an additive main effect in a model of the length of the second spell, but we do claim that we can find a model of second spell that incorporates the length of the first as a covariate *and* that fits as well as a model that explicitly incorporates persistent unobservable heterogeneity.

All we need to do to validate this claim is to note that the joint distribution of the first two spells $f_{12}(t_1, t_2)$ can always be factored as the marginal of the first spell $f_1(t_1)$ times the conditional of the second spell given the first $f_{2.1}(t_2|t_1)$. Given some model with persistent heterogeneity, we derive the joint distribution $f_{12}(\cdot)$ as a function of the parameters of the two baseline hazards, the parameters of the distribution of the unobservable, and the two covariate parameter vectors as we did for single spells by integrating over the unobservable. We then derive the marginal distribution $f_1(\cdot)$ as a function of the heterogeneity parameters, the parameters of the baseline hazard of T_1, and the covariate parameter vector for T_1 by integrating out T_2. The conditional distribution $f_{2.1}(\cdot)$ is then simply the ratio of $f_{12}(\cdot)$ to $f_1(\cdot)$, and will in general contain the length of the first spell t_1, the parameters of the distribution of the unobservable, the parameters of both baseline hazards, and both covariate parameter vectors.

If the second analyst employs $f_1(\cdot)$ and $f_{2.1}(\cdot)$, then this analyst must do as well as the first analyst. This time, however, the second analyst will have to estimate more parameters, because the parameters of the distribution of the unobservable will be estimated twice, in $f_1(\cdot)$ and in $f_{2.1}(\cdot)$, and $f_{2.1}(\cdot)$ will generally contain all of the parameters in $f_1(\cdot)$ plus the ones that pertain to the second spell only. Hence, this second analyst will lose on grounds of parsimony, but only on those grounds. The second analyst could, however, achieve a better fit, because his unconstrained model (which consists of two sub-models, estimated completely separately) could never fit less well than the model of the first analyst.

In practice, the second analyst could examine separate models for the two spells using semiparametric step-function hazards, and could include the length of the first spell as an additive main effect in the model for the second spell. We claim that this analyst could always match the performance of a model that explicitly embodies persistent heterogeneity, by incorporating enough duration categories and by allowing for duration-dependent covariate effects. While the interpretations of such a model and an alternative model explicitly incorporating persistent heterogeneity might differ radically, no empirical goodness-of-fit test could ever distinguish between them.

We hasten to emphasize, before concluding, that it is possible that some theory might push one either to model an unobservable explicitly or to omit it entirely. Unfortunately, we are unaware of any demographic applications where the theory is anywhere near so specific. For example, Heckman and Walker (1989) empirically discovered two models with different behavioral implications when they analyzed birth intervals among an anabaptist sect called the Hutterites. Extant theory does not enable them to choose between the two models.

POISSON REGRESSION

Demographic data often come in the form of rates or counts of events. Examples abound in the fields of fertility, nuptiality, mortality and migration analysis. Indeed, most demographic data come in this form. Although appropriate statistical methods for the analysis of rates or counts of events (notably Poisson regression analysis) have been developed by statisticians, they are seldom applied by demographers in this context. Poisson models are used in demographic research for the analysis of event histories (hazard models with piecewise constant hazards) and contingency tables. Poisson regression analysis is described in the context of generalized linear models by Nelder and Wedderburn (1972) and McCullagh and Nelder (1983). The application of these models to the analysis of rates is noted, for example, in Holford (1980) and Frome (1983). For an excellent discussion of the role that the Poisson process plays in the analysis of vital rates and associated statistics, see Brillinger (1986).

Rodríguez and Cleland have used Poisson regression to study socioeconomic determinants of marital fertility rates in developing countries (Rodríguez and Cleland, 1988; Cleland and Rodríguez, 1988). The data they analyze are typical of information collected in fertility surveys based on a sample of women in the reproductive ages, and include information on (1) the number of births in a given period, which could be a fixed number of years before the survey or the woman's entire reproductive career, and (2) one or more covariates such as age, marital duration, education, occupation and work status, believed from general social or biological theory to affect fertility in the period in question. The challenge is to develop a model that makes statistical as well as demographic sense.

Rodríguez and Cleland (1988) propose treating the birth counts B_i in a recent period before the survey as realizations of independent Poisson random variables with means μ_i equal to the product of exposure time in the period E_i and a theoretical fertility rate $f(x_i)$ that depends on a vector of covariates x_i.

The key statistical feature of the model is the assumption of Poisson variation. This assumption is consistent with the discrete nature of the data, which for a five-year period consist mostly of zeroes, ones, and the occasional woman with two or three births. More importantly, the model allows the variance of the observations to be proportional to the mean, in agreement with empirical observations and in contrast to the usual linear regression assumption of homoscedasticity. It remains to show via convincing applications that these two features do make an important difference in practice.

On the demographic side of the model, Rodríguez and Cleland (1988) develop an expression for the theoretical fertility rate $f(x_i)$ that builds on earlier work of Henry (1961), Coale (1971), Coale and Trussell (1974), and Page (1977). Specifically, they partition the vector of covariates into its demographic and social components, writing $x_i = (a_i, d_i, x_i^*)$, where a_i denotes age, d_i denotes duration of marriage, and x_i^* the remaining covariates of interest, always for the ith woman. They then write

$$\log(\mu_i) = \log(E_i) + \log[n(a_i)] + \alpha' x_i^* + \beta' x_i^* d_i \,, \qquad (8\text{-}19)$$

where $n(a_i)$ is the age-pattern of natural fertility first described by Henry (1961), α is a vector of parameters affecting the level of natural fertility, and β is a vector of parameters affecting the degree of control of marital fertility. Note that both sets of parameters are allowed to depend on the covariates. For a similar extension of the Coale (1971) model, but where only the level of natural fertility is allowed to depend on the covariates, see Broström (1985).

A key feature of this model is that it falls in the class of generalized linear models described by Nelder and Wedderburn (1972), representing the special case where (1) the stochastic or error structure is Poisson, (2) the link between the expectation of the response (here the number of births) and the predictors is the logarithm, and (3) the linear predictor itself has a known part or offset. This fact implies that maximum likelihood estimates of the parameters are readily available, together with estimates of standard errors and a likelihood ratio goodness of fit chi-square statistic. Moreover, all calculations may be done using the software package GLIM described by Baker and Nelder (1978), now available for personal computers.

We turn next to the issue of unobserved heterogeneity. To do so, we return to the analysis of fertility rates described earlier. Specifically, let B_i denote the number of births and E_i the amount of exposure contributed by the ith woman. We have assumed in our prior work that the B_i are realizations of independent Poisson random variables with means μ_i depending on age, duration and other characteristics of the individual via a log-linear model. Specifically, we assume that $\log(\mu_i) = x_i'\beta$. Implicitly, we assume that all individuals with the same characteristics x_i have a Poisson distribution with the same mean. Suppose, however, that there is further individual variation, and that in fact

$$\log(\mu_i) = x_i'\beta + \epsilon_i \,, \qquad (8\text{-}20)$$

where the ϵ_i are independent perturbations representing other sources of variation not included in x (or strictly speaking, the part of those variables lying in a space orthogonal to x). If the ϵ_i were observed, we could work with the *conditional* distribution of B_i given ϵ_i, which is Poisson with mean μ_i as given above. Unfortunately the ϵ_i are not observed (indeed, they may not even be observable), and we are forced to consider the *unconditional* distribution of the B_i.

One way to proceed is to assume a distribution for the unobservables. Suppose that $\exp(\epsilon_i)$ has a gamma distribution with mean 1 and variance σ^2, so that μ_i (which is now random) has a gamma distribution with mean $\exp(x_i'\beta)$ and variance $\sigma^2\exp(2x_i'\beta)$. This mixture leads to the negative binomial distribution. In other words, B_i now has a negative binomial distribution with mean $\exp(x_i'\beta)$ and variance $\exp(x_i'\beta)[1 + \sigma^2\exp](x_i'\beta)$ (McCullagh and Nelder, 1983). There are three key points we would like to make about this result. First, note that under the assumptions we have made, an investigator who had decided to treat the data as negative binomial (say on the basis of an examination of the mean and variance, which would have led to the binomial if $\sigma^2 < \mu$, the Poisson if $\sigma^2 = \mu$, and the negative binomial if $\sigma^2 > \mu$) would in fact be fitting the "correct" model and would obtain a consistent estimate of the vector of parameters β, even though he had not "corrected for unobservables". Put in a different way, correcting for unobservables is not a universal recipe for success. What is always required, instead, is to fit a model consistent with the data (i.e., do not fit Poisson if $\sigma^2 > \mu$). Second, McCullagh and Nelder (1983) note that estimates of β based on the negative binomial and the Poisson likelihoods will in general differ, but only by a term that is of order σ^2/μ^2 in probability. They therefore conclude that "for modest amounts of over-dispersion this difference may be neglected." The lesson here is that introduction of unobservables in Poisson regression problems is not likely to make much of a difference in results unless the amount of unmeasured heterogeneity is substantial. Note also that a rough indication of the extent of heterogeneity may be obtained by comparing the variance against the mean. Third, Cox (1983) has shown that maximum likelihood estimation (MLE) of Poisson (or binomial) models retains high efficiency in the presence of modest amounts of over-dispersion.

McCullagh and Nelder (1983) discuss fitting the negative binomial distribution in the context of generalized linear models, indicating that the MLE is inconvenient to use, because it requires the log-gamma function and its derivative, although these may be approximated numerically. Fitting the model by maximum likelihood as well as by weighted least squares to determine how close the two procedures come in actual applications would be a useful exercise. Breslow (1984) has described an alternative method of accommodating extra-Poisson variation in log-linear models that has two advantages: (1) it makes no distributional assumptions about the unobservables ϵ_i other than the fact that they have mean 0 and variance σ^2, and (2) the parameters β and σ^2 may be estimated using a simple iterative procedure. Under these assumptions, conditional on ϵ_i, the response B_i is Poisson distributed with mean $\mu_i = \exp(x_i'\beta + \epsilon_i)$; to a first order approximation, the *unconditional* moments are $E[B_i] = \exp(x_i'\beta)$ and $\mathrm{Var}[B_i] = \exp(x_i'\beta)[1 + \sigma^2\exp(x_i'\beta)]$. If one assumes that $\exp(\epsilon_i)$ has a distribution with mean 1 and variance σ^2, then this result is exact; note also that these moments are identical to those obtained above for the negative binomial.

For fixed σ^2, the regression coefficients β are estimated by quasi-likelihood (Wedderburn, 1974), and the resulting estimate of β is used to estimate σ^2 by setting the Pearson χ^2 statistic equal to its degrees of freedom. The procedure is easily set up in GLIM. Large sample properties of the estimates have been studied by Moore (1986) For a similar approach in the context of generalized linear models, see Nelder (1985).

We think that Breslow's procedure is potentially very useful, but we are unaware of any demographic applications. Only an extensive test using actual data will reveal whether the estimates of the regression coefficients obtained under this procedure (which by definition is robust to the choice of a distribution for the unobservable) are similar to estimates obtained under suitable parametric models such as the negative binomial (where appropriate) or other mixture models. Brillinger (1986) suggests that the likelihood approach may be anticipated to be more efficient when the assumed distribution is reasonable, and adds that it may well be close to a physical description of the situation, but he recognizes that study of the robustness of the likelihood approach is needed.

REGRESSION ANALYSIS OF CURRENT STATUS DATA

When demographic data do not come in the form of rates, they often come in the form of so-called "current status" data, defined as a form of survival data where the information available for each individual is limited to the total length of exposure and the survival status at the time of a survey. Again, appropriate statistical models for the analysis of this type of data, notably binomial regression models, have been developed by statisticians, but are seldom used in demographic analysis. For a discussion of binomial regression in the context of the generalized linear model, see Nelder and Wedderburn (1972) and McCullagh and Nelder (1983). The use of these models for current status survival data is briefly noted in the text by Kalbfleisch and Prentice (1980) and expounded at greater length in the context of accelerated life models by Vanderhoeft (1982) and in the context of proportional hazard models by Diamond, McDonald and Shah (1986).

A limiting case of current status data occurs when the time dimension is degenerate. For example, a woman either initially breastfeeds her infant or she does not. The dependent variable Y_i takes the value 1 if the child is breastfed and 0 otherwise. It is well-known that a linear regression model is inappropriate in this situation because (1) the errors would be heteroscedastic and (2) the predicted values are not constrained to lie between 0 and 1. The most common model employed by demographers to analyze such data is logistic regression, where the probability of breastfeeding is given by

$$E[Y_i] = \pi_i = \exp(x_i'\beta)/[1 + \exp(x_i'\beta)] , \qquad (8\text{-}21)$$

where x_i is the vector of predictor variables for the ith individual and β is the associated parameter vector. In this model, the log-odds (or logit) is a linear function of the predictors: $\log[\pi_i/(1 - \pi_i)] = \text{logit}(\pi_i) = x_i'\beta$. Estimation is easily carried out in the context of generalized linear models (McCullagh and

Nelder, 1983; Baker and Nelder, 1978) by specifying a logit link and binomial error structure. In demographic work, logistic regression has been used not only in this context but also in the analysis of event histories by limiting attention to survival up to a single duration or age (for example, a child either dies or does not die before the fifth birthday). That demographers have not applied binomial regression more generally to current status data is somewhat surprising given both the importance and the ubiquity of this type of information.

The earliest application of current status data in demographic analysis arises in the estimation of (so-called singulate) mean age at first marriage from proportions ever-married at the time of a census of survey, as proposed by Hajnal (1953). In this case the event of interest is marriage, current age determines the length of the observation period, and the simple dichotomy single/ever-married tells us whether the event occurred within the observation period. Hajnal's approach was totally nonparametric, but the same data are amenable to parametric modelling. For a discussion of maximum likelihood estimation of the parameters of the Coale-McNeil (1972) model nuptiality schedule from current status data, see Rodríguez and Trussell (1980).

The recent surge of renewed interest in current status methodology is closely related to the analysis of postpartum infecundity, particularly the duration of breastfeeding, using data of the type collected in the World Fertility Survey (WFS) or its successor, the Demographic and Health Surveys (DHS); see, for example, Lesthaeghe and Page (1980) and Ferry (1981). Both WFS and DHS data on breastfeeding for the most recent births included current status information on whether the child was still breastfeeding at interview or had been weaned, as well as retrospective durations for weaned children. They are thus amenable to conventional life table analysis as well as current status modelling.

There has been considerable debate concerning possible biases of both kinds of data. It is generally accepted that retrospective reports tend to exhibit a marked concentration ("heaping") at multiples of 6 and particularly 12 months. Heaping could, of course, be genuine, reflecting strong normative influences to breastfeed exactly one or two years; but the lack of direct evidence supporting this point led most authors to interpret heaping as an indication of poor data quality. A similar form of heaping occurs with reported durations of amenorrhea, where it is much less likely to be real.

Current status data, on the other hand, are largely free of heaping, but some authors have suggested that they are subject to other biases; see, for example, Bracher and Santow (1981). A careful consideration of sampling frames, however, including a key distinction between samples of women and samples of children, shows that current status measures are unbiased estimates of duration variables for a sample of births occurring in a fixed calendar period (John, Menken, and Trussell, 1988). However, current status measures are subject to larger sampling variances than retrospective data for a sample of the same size. As a result, observed proportions surviving at various durations since start of exposure tend to be highly irregular and often fail to decline monotonically with duration.

To incorporate covariates, Vanderhoeft (1982) has proposed the use of accelerated life models and Diamond, McDonald and Shah (1986) have proposed the use of proportional hazard models, with both papers extending to current status data methods that are well established in survival analysis. In contrast, Smith (1983) has proposed an ad-hoc regression procedure. To our knowledge, none of these procedures have yet been used by researchers other than the original authors. The main difficulty with existing methods is that they require assumptions about the parametric form of the underlying survival distributions (Weibull, log-normal or log-logistic), yet very little is known about the appropriateness of these distributions for breastfeeding or other demographic data.

It is clear that current status estimates have larger variances than retrospective life table estimates based on the same data, although exactly how much precision is lost by using the current status approach has not been addressed. Our very rough preliminary calculations indicate that a totally nonparametric approach leads to variances about 60 times larger for estimates of the survival function at fixed durations, but only about 2.5 times as large for summary statistics such as the mean. A closely related problem concerns the extent to which some of the inevitable loss in precision involved in using current status data may be recovered using parametric models.

At the aggregate level, the preferred approach to the analysis of current status data has been essentially nonparametric. The observation period is grouped in intervals $(\tau_i, \tau_i + 1)$ with $0 = \tau_1 < \tau_2 < ... < \tau_{k+1} = \infty$. Let n_i denote the total number of individuals with observation periods in the ith interval and let Y_i denote the number observed alive. Then $p_i = Y_i/n_i$ is taken as an estimator of $\pi_i = S(t_i)$, the probability of surviving to a duration t_i equal to the mid-point of the ith interval. If the intervals are fine, the p_i will often fail to decline monotonically with duration. The standard solution then is to group the data into coarser intervals. The maximum likelihood solution, however, requires grouping only intervals that fail the monotonicity requirement, in a procedure known as "pooling adjacent violators" algorithm (see Barlow *et al.*, 1972).

We can identify three main approaches for further work on current status data: (1) parametric models of the accelerated life variety, (2) parametric and semi-parametric proportional hazards models, and (3) parametric and semi-parametric proportional odds models, including relational logit models (Bennett, 1983; Lesthaeghe and Page, 1980). We describe briefly the way in which one might proceed.

Let Y_i denote the number of deaths among n_i individuals aged t_i. Following Vanderhoeft (1982) and Diamond, McDonald, and Shah (1986), we will treat Y_i as binomial with parameters n_i and $\pi_i = 1 - S(t_i)$. If the data are not grouped so that $n_i = 1$ and t_i is the exact length of observation for the ith case, then Y_i is exactly binomial. If the data are grouped by combining individuals with slightly different lengths of observation distributed around t_i, then Y_i becomes a sum of independent but not quite identically distributed Bernoulli random variables, and it is therefore only approximately binomial, a point that seems to have eluded earlier workers. The nature of this approximation deserves further investigation.

Suppose now that the underlying distribution of survival times has a known functional form. For concreteness suppose that the distribution is Weibull with parameter λ and index ρ, so that the probability of surviving to time t_i is

$S(t_i) = \exp[-(\lambda t_i)^\rho]$. A simple estimation strategy consists of transforming the survival probabilities to obtain functions that are linear in the parameters. In the present case, the complementary log-log transformation gives $\log[\log(1 - \pi_i)] = \rho\log(\lambda) + \rho\log(t_i)$. The implication of the result is that the parameters of the underlying Weibull distribution may be estimated by treating the current status counts as binomial, using a complementary log-log link and fitting the log of exposure time as a covariate with a linear effect. The slope will then estimate the index ρ and the constant will estimate $\rho\log(\lambda)$, so that exponentiating the ratio of the constant to the slope would yield the MLE of λ.

Covariates are naturally introduced in the context of a Weibull model by keeping the index ρ fixed and letting the parameter λ depend on a vector x_i of characteristics of the individual or subgroup via a log-linear model. The resulting regression model can be shown to be a special case of the family of accelerated life models as well as the family of proportional hazard models, and can be fit to current status counts using essentially the same procedure we have just outlined.

Two other well-known parametric distributions are amenable to the same treatment, although details will be omitted. These are the log-normal and the log-logistic distributions. The appropriate linearizing transformations are the probit and the logit, respectively. In both cases, the logarithm of exposure time enters the model as a covariate with a linear effect. Covariates are naturally introduced in the context of these two models by assuming a constant variance of the underlying normal or logistic distribution of log-survival and letting the mean vary from one observation to another, depending linearly on a vector x_i. The resulting regression models can be shown to belong to the family of accelerated life models but not to the family of proportional hazard models. An interesting feature of the log-logistic regression model is that the hazards for individuals with different values of the covariates tend to become more and more similar as duration increases. This type of phenomenon is often observed and is consistent with the presence of unmeasured heterogeneity.

So far we have had to assume a parametric form for the underlying survival distribution in order to tie together observations at different durations. An alternative semi-parametric approach is found in the application of the proportional hazards model to current status data, a problem considered by Diamond, McDonald, and Shah (1986). For an early application see Page, Lesthaeghe, and Shah (1982). We start from the classic model due to Cox (1972), where the risk at duration t_i for a subject with characteristics x_j has the form $\lambda(t_i, x_j) = \lambda_o(t_i)\exp(x_j'\beta)$, where $\lambda_o(t)$ is a baseline hazard, and $\exp(x_j'\beta)$ is the relative risk associated with characteristics x_j. Under the model, the complementary log-log transformation of the probability of survival to time t_i for a subject with characteristics x_j is

$$\log[-\log(1 - \pi_{ij})] = \log[\Lambda_o(t_i)] + x_j'\beta , \qquad (8\text{-}22)$$

where $\Lambda_o(t)$ denotes the integrated baseline risk. This development gives a generalized linear model on the counts Y_{ij} with binomial error structure and complementary log-log link. Note that the linear structure contains, in addition to the usual linear predictor $x_j'\beta$, a parameter $\alpha_i = \log[\Lambda_o(t_i)]$ for each time point t_i.

The only difficulty with this model is that, since the cumulative risk $\Lambda_o(t)$ is a non-decreasing function of t, the parameters α_i must satisfy certain inequality constraints; specifically, if the t_i are ordered, we require $\alpha_1 \leq \alpha_2 \leq ... \leq \alpha_k$. Fitting generalized linear models under non-linear inequality constraints is not as difficult as it may sound, but it may be extremely tedious without specialized software. The maximum likelihood procedure essentially requires fitting the model (ignoring the constraints) and then pooling adjacent values of t_i each time the constraints are violated. This step is easily accomplished by recoding the factor representing duration. An alternative approach described by Diamond, McDonald, and Shah (1982) involves the use of splines. None of these methods is in wide use, partly because existing accounts make them all look much more complicated than they really are.

We turn next to the issue of unobserved heterogeneity in binomial models. Many of the concepts discussed under extra-Poisson variation in the context of Poisson regression apply also to the presence of extrabinomial variation in regression models for binary data. Let Y_i be a random variable assumed to have a binomial distribution with parameters π_i and n_i and consider for the sake of concreteness a logistic regression model where the logit of the underlying probability of success is a linear function of a vector of covariates x_i, namely $\text{logit}(\pi_i) = x_i'\beta$. Under this model, $E(Y_i) = n_i\pi_i$ and $\text{Var}(Y_i) = n_i\pi_i(1 - \pi_i)$. One way to introduce unobserved heterogeneity in this model is to assume that the π_i are not fixed quantities, but random variables with expectation $\theta_i = 1/[1+\exp(-x_i'\beta)]$. A common assumption is that the π_i have a beta distribution with the required mean. This leads to the so-called beta-binomial model, which may be fitted by maximum likelihood (Williams, 1975; Crowder, 1978).

Recently Williams (1982) has proposed a somewhat more robust approach to the modelling of extra binomial variation. He assumes simply that the π_i are random variables distributed in (0,1) with mean $E(\pi_i) = \theta_i$ as given above and variance $\text{Var}(\pi_i) = \varphi\theta_i(1 - \theta_i)$ for some unknown parameter φ. *Conditional* on π_i the response Y_i is binomial with mean and variance as stated earlier. *Unconditionally*, however, $E(Y_i) = n_i\theta_i$ and $\text{Var}(Y_i) = \sigma_i^2[1+\varphi(n_i-1)]$, where σ_i^2 is the binomial variance. Note that the distribution of Y_i is not fully specified, and maximum likelihood cannot be used. The relationship between the mean and variance, however, is sufficient to specify a quasi-likelihood in the sense of Wedderburn (1974), which leads to estimates of the underlying regression coefficients β. Interestingly, the beta-binomial is one of the distributions that exhibits the mean/variance relationship postulated by Williams, and therefore his quasi-likelihood approach would be appropriate for that parametric model.

A third approach to heterogeneity in binomial models assumes that the logits of the π_i vary about a mean $x_i'\beta$ with constant variance σ^2. One advantage of this specification, investigated by Pierce and Sands (1975), is that heterogeneity is being modelled in the same scale as the effects of the covariates. If the distribution of $\text{logit}(\pi_i)$ is assumed normal, we obtain the logistic-normal model of Aitchinson and Shen (1980). It does not appear to be strictly necessary to assume a distribution for the logits, however, and the model may be fit by quasi-likelihood without additional (and often nearly untestable) assumptions.

So far we have discussed heterogeneity at the level of binomial observations. In the analysis of current status data, however, it may well make more sense to introduce heterogeneity at the level of the underlying survival distribution. Specifically, one might investigate formulations that may be appropriate in a proportional hazards framework, following essentially the work of Heckman and Singer (1982, 1984a, 1984b). Alternatively, one might explore modelling heterogeneity in the context of accelerated life models, where we anticipate that estimates will be more robust because the models make implicit allowances for heterogeneity in the form of an error term.

CONCLUDING REMARKS

In this chapter, we have examined the incorporation of covariates in three statistical models appropriate for the analysis of demographic data. Two models, Poisson regression applied to rates, and binomial regression analysis of current status data, have not been widely used by demographers; hence, these models are more fully developed for observed predictors than is the other model (hazard regression). We also address, for each of the models, the value of incorporating an extra source of variation by adding an unobservable. Unobservables have been extensively explored in the context of hazard regressions used for event history analysis. We review this prior work and explore critically its implications. We conclude that the methods proposed to correct for unobservable heterogeneity deliver less than is commonly assumed, particularly because of an inherent non-identifiability involved when the analyst must rely on observables to assess goodness-of-fit. Resolving the identification problem will in part await progress in defining implicit strata that capture genetic heterogeneity. Genetic epidemiologists are increasingly turning their attention to this task. Successful stratification would drastically improve statistical resolution.

ACKNOWLEDGMENTS

We are indebted to James Heckman, Ken Manton, Burt Singer, and James Vaupel, who have greatly influenced our thinking about unobserved heterogeneity over several years.

REFERENCES

Aitchinson, J., and S. M. Shen, 1980   Logistic-normal distributions: some properties and uses. Biometrika 67: 162-167.
Baker, R. J., and J. A. Nelder, 1978   *The GLIM System, Release 3*. Numerical Algorithms Group, Oxford.
Barlow, R., D. Bartholomew, J. Bremner, and H. Bruuk, 1972   *Statistical Inference under Order Restrictions*. John Wiley, New York.

Bennett, S., 1983   Analysis of survival data by the proportional odds model. Stat. Med. 2: 273-277.

Blumen, I., M. Kogan, and P. J. McCarthy, 1955   The industrial mobility of labor as a probability process. Cornell Studies in Industrial and Labor Relations, #6. Cornell University, Ithaca.

Bracher, M., and G. Santow, 1981   Some methodological considerations in the analysis of current status data. Pop. Studies 35: 425-437.

Breslow, N. E., 1984   Extra-Poisson variation in log-linear models. App. Stat. 33: 38-44.

Brillinger, D. R., 1986   The natural variability of vital rates and associated statistics. Biometrics 42: 693-734.

Brostrom, G., 1985   Practical aspects on the estimation of the parameters in Coale's model for marital fertility. Demography 22: 625 -631.

Cleland, J., and G. Rodríguez, 1988   The effect of parental education on marital fertility in developing countries. Pop. Studies 42: 419-442.

Coale, A. J., and P. Demeny, 1966   *Regional Model Life Tables and Stable Populations*. Princeton University Press, Princeton.

Coale, A. J., 1971   Age patterns of marriage. Pop. Studies 25: 193-214.

Coale, A. J., and D .R. McNeil, 1972   The distribution by age of the frequency of first marriage in a female cohort. J. Am. Stat. Assoc. 67: 743-749.

Coale, A. J., and J. Trussell, 1974   Model fertility schedules: variations in the age structure of childbearing in human populations. Population Index 40: 185-258. See also erratum 41: 572.

Coale, A. J., P. Demeny, and B. Vaughan, 1983   *Regional Model Life Tables and Stable Populations*, Second Edition. Academic Press, New York.

Cohen, J., 1986   An uncertainty principle in demography and the unisex issue. Am. Statistician 40: 32-39.

Cox, D. R., 1972   Regression models and life tables, with discussion. Journal of the Royal Statistical Society, Series B 34: 187-202.

Cox, D. R., 1983   Some remarks on overdispersion. Biometrika 70: 269-274.

Crowder, M. J., 1978   Beta-binomial anova for proportions. Appl. Stat. 27: 34 -37.

Diamond, I., J. McDonald, and I. Shah, 1986   Proportional hazards models for current status data: application to the study of differentials in age at weaning in Pakistan. Demography 23: 607 -620.

Ferry, B., 1981   *Breastfeeding*. World Fertility Survey. Comparative Studies, Number 13. International Statistical Institute, Voorburg.

Frome, E. L., 1983   The analysis of rates using Poisson regression models. Biometrics 39: 665-674.

Hajnal, J., 1953   Age at marriage and proportions marrying. Pop. Studies 7: 111 -136.

Heckman, J. and B. Singer, 1982   Population heterogeneity in demographic models, pp. 567-599 in *Multidimensional Mathematical Demography*, edited by K. Land and A. Rogers. Academic Press, New York.

Heckman, J., and B. Singer, 1984a   A method for minimizing the impact of distributionalassumptions in economic models for duration data. Econometrica 52: 271-320.

Heckman, J., and B. Singer, 1984b   Econometric duration analysis. J. Econometrics 24:63-132.

Heckman, J., and J. Walker, 1987   Using goodness of fit and other criteria to choose among competing duration models: a case study of Hutterite data, pp. 247-307 in *Sociological Methodology 1987*, edited by C. Clogg. American Sociological Association, Washington.

Heckman, J., and J. Walker, 1990   Estimating fecundability from data on waiting times to first conceptions. J. Am. Stat. Assoc., forthcoming.

Henry, L., 1961   Some data on natural fertility. Eugen. Q. 8: 81-91.

Holford, T., 1980   The estimation of age, period, and cohort effects for vital rates. Biometrics 39: 311-324.

Hougaard, P., 1984   Life table methods for heterogeneous populations: distributions describing the heterogeneity. Biometrika 71: 75-83.

John, A. M., J. A. Menken, and J. Trussell, 1988   Estimating the distribution of interval length: current status and retrospective history data. Pop. Studies 42: 115-127.

Judge, G., W. E. Griffiths, R. C. Hill, and T. C. Lee, 1980   *The Theory and Practice of Econometrics*, Chapter 11. John Wiley, New York.

Kalbfleisch, J. D., and R. L. Prentice, 1980   *The Statistical Analysis of Failure Time Data.* John Wiley, New York.

Keyfitz, N., 1985   *Applied Mathematical Demography.* Second Edition. Springer-Verlag, New York.

Lagakos, S. W., 1988   The loss in efficiency from misspecifying covariates in proportional hazards regression models. Biometrika 75: 156-160.

Lagakos, S. W., and D. A. Sshoenfeld, 1984   Properties of proportional hazards score tests under misspecified regression models. Biometrika 40: 1037-1048.

Laird, N., 1978   Nonparametric maximum likelihood estimation of a mixing distribution. J. Am. Stat. Assoc. 73: 805-811.

Lesthaeghe, R. J., and H. J. Page, 1980   The post-partum non-susceptible period: development and application of model schedules. Pop. Studies 34: 143-169.

McCullagh, P., and J. A. Nelder, 1983   *Generalized Linear Models.* Chapman and Hall, London.

Manton, K., B. Singer, and M. Woodbury, 1988   Some issues in the quantitative characterization of heterogeneous populations. Presented at the IUSSP Seminar on Event History Analysis, held in Paris, March 14-17.

Marini, M. M., 1985   Determinants of marital disruption. Presented at the Annual Meetings of the Population Association of America, held in Boston, March 28-30.

Marini, M. M., and B. Singer, 1988   Causality in the social sciences. In *Sociological Methodology 1988*, edited by C. Clogg. American Sociological Association, Washington.

Montgomery, M., and J. Trussell, 1986   Models of marital status and childbearing, pp. 205-270 in *Handbook of Labor Economics*, Volume 1, edited by O. Ashenfelter and R. Layard. Elsevier Science Publishers, New York.

Moore, D. F., 1986   Asymptotic properties of moment estimates for over dispersed counts and proportions. Biometrika 73: 583-588.

Morgan, T. M., 1986   Omitting covariates from the proportional hazards model. Biometrics 42: 993-995.

Nelder, J. A., 1985   Quasi-likelihood and GLIM, pp. 120-127 in *Generalized Linear Models*, edited by R. Gilchrist, B. Francis, and J. Whittaker. Springer-Verlag, New York.

Nelder, J. A., and R. W. M. Wedderburn, 1972   Generalized linear models. J. R. Stat. Soc. A 135: 370-384.

Newman, J., and C. McCullogh, 1984   A hazard rate approach to the timing of births. Econometrica 52: 939-961.

Page, H. J., 1977   Patterns underlying fertility schedules: a decomposition by both age and marriage duration. Pop. Studies 31: 85-106.

Page, H. J., R. J. Lesthaeghe, and I. Shah, 1982   *Illustrative Analysis: Breastfeeding in*

Pakistan. World Fertility Survey, Scientific Reports, Number 37. International Statistical Institute, Voorburg.

Pierce, D. A., and B. R. Sands, 1975   *Extra-Bernoulli Variation in Binary Data*. Technical Report 46, Department of Statistics, Oregon State University.

Rodríguez, G., 1988   Unobserved heterogeneity and extra-Poisson variation in survival models. Presented at the Annual Meetings of the American Statistical Association, held in New Orleans, August.

Rodríguez, G., and J. Cleland, 1988   Modelling marital fertility by age and duration: an empirical appraisal of the Page model. Pop. Studies 42: 241-257.

Rodríguez, G., and J. Trussell, 1980   Maximum likelihood estimation of the parameters of Coale's model nuptiality schedule. World Fertility Survey, Technical Bulletins, Number 7. International Statistical Institute, Voorburg.

Rogers, A., and L. J. Castro, 1981   *Model Migration Schedules*. International Institute for Applied Systems Analysis, Laxenburg, Austria.

Sheps, M. C., and J. A. Menken, 1973   *Mathematical Models of Conception and Birth*. University of Chicago Press, Chicago.

Smith, D. P., 1983   Regression analysis of "current status" life tables on duration of breastfeeding in Sri Lanka. Soc. Biol. 32: 90-101.

Struthers, C. A., and J. D. Kalbfleisch, 1986   Misspecified proportional hazard models. Biometrika 73: 363-369.

Trussell, J., and T. Richards, 1985   Correcting for unmeasured heterogeneity in hazard models using the Heckman-Singer procedure, pp. 242-276 in *Sociological Methodology 1985*, edited by N. Tuma. Jossey-Bass, San Francisco.

Vanderhoeft, C., 1982   Accelerated failure time models: an application to current status breast-feeding data from Pakistan. Genus 38: 135-157.

Vaupel, J. W., K. G. Manton, and E. Stallard, 1979   The impact of heterogeneity in individual frailty on the dynamics of mortality. Demography 16: 439-454.

Vaupel, J., and A. Yashin, 1985a   The deviant dynamics of death in heterogeneous populations, pp. 179-211 in *Sociological Methodology 1985*, edited by N. Tuma. Jossey Bass, San Francisco.

Vaupel, J., and A. Yashin, 1985b   Heterogeneity's ruses: some surprising effects of selection on population dynamics. Am. Statistician 39: 176-185.

Wedderburn, R. W. M., 1974   Quasi-likelihood functions, generalized linear models and the Gauss-Newton method. Biometrika 61: 439-447.

Williams, D. A., 1975   The analysis of binary responses from toxicological experiments involving reproduction and teratogenicity. Biometrics 31: 949-952.

Williams, D. A., 1982   Extra-binomial variation in logistic linear models. Appl. Stat. 31: 144-148.

9

Limitations of a Heterogeneity Technique: Selectivity Issues in Conjugal Union Disruption at Parity Zero in Contemporary Sweden

JAN M. HOEM

In probabilistic models of individual behavior or individual biological or mental development, events like entry into motherhood, recovery from illness, or the completion of an educational degree are seen as moves that people make between distinct statuses in life. They represent the individual's health status, marital status or childbearing parity, employment status or educational level, or some other (possibly multidimensional) status specification. The rate of transition between any two states is modeled as a transition intensity (or "hazard"), which is a function of time and which may depend on fixed or time-varying characteristics of the individual or his environment.

Some of the latter may be endogenous, i.e., their values at any time may be determined by the relevant probability scheme along with the moves between states on which interest primarily focuses. For instance, the disruption intensity for a woman's conjugal union is strongly dependent on the number of children she has, and it may be influenced by her educational level and labor force status. Conversely, any (further) childbearing depends on her civil status, and her employment status changes may affect childbearing as well. As a woman's life unfolds, she continuously updates her life strategy under the influence of a highly dynamic system of resources, experiences, choices, restrictions, and chance outcomes. Her strategy determines her behavior and produces what appears in the probability scheme as mutual causation between her educational and employment careers and her family history. Similarly, the progress of a genetic or other disease may be influenced by the individual's lifestyle (smoking, food and drink, medication, living arrangements, and so on), by his or her biological age or time since onset of initial symptoms, and by the interference of other diseases or other events. Important observable features of this nature would then appear as covariates in the intensities of transitions between the various distinct states of a disease description. Hazard regression is a class of techniques developed to analyze data of this nature.

When a balance is struck between features that are included and others that are left out of a modeling venture, a choice can be forced on the analyst by the lack of relevant information in the data set. (It may be hard to get good data on an individual's personal value position or essential genotype.) Sometimes the effect of omitted variables becomes particularly pressing, and the investigator will look for methods designed to overcome it. In the hazard regression context, omitted variables are subsumed under the notion of *unobserved heterogeneity*, and a technique developed by Heckman and Singer (1982, 1984, and elsewhere) has achieved much attention and prestige. The apparent promise held out by this technique and strongly worded warnings against the potential dangers of disregarding unobserved heterogeneity have led to much pressure toward "always" incorporating it and to much apologetic rhetoric by some who do not.

The usefulness of heterogeneity techniques has been questioned, however, both in terms of the validity of model specifications made, of the feasibility of their implementation, and of how sensible the consequent resource allocation is (Trussell and Richards, 1985; Kiefer, 1988; Trussell and Rodríguez, 1988; Montgomery, 1990; Hoem and Hoem, 1989, 1990). If unobserved heterogeneity is an important ingredient in the behavior investigated, it is bound to show up in the empirical outcome anyway, and it becomes an issue whether it will be interpreted correctly even when less costly statistical methods are applied. It is the purpose of this chapter to contribute to an assessment of such questions by reporting on an attempt to gain improved insight by incorporating persistent unobserved heterogeneity explicitly into a model for a behavior where it is manifestly present and is easily detected by simpler procedures. Most empirical papers will give accounts of successful substantive studies with successful implementations of existing or new methodology. Few reports will present failed experiments. Unconventionally, this paper does. It is published as a warning against undue optimism concerning the usefulness of current heterogeneity techniques. Since unobserved heterogeneity is likely to be present in situations studied in genetics as well as in demography (and in many other fields), there is a message here for geneticist and demographer alike.

THEORIES OF UNION DISRUPTION BEHAVIOR

Our experiment with the Heckman-Singer approach arose out of our interest in patterns of dissolution of marital unions among contemporary Swedes, a population with relatively high disruption risks by international standards. Several studies of marital disruption in modern Western populations have found that a marriage is more dissolution prone if the partners have lived together in a consensual union before marrying than if they have not, even when one controls for other observed factors (Balakrishnan *et al.*, 1987; Bennett, Blanc, and Bloom, 1988; Hoem and Hoem, 1990; Leridon, 1988; Bumpass and Sweet, unpublished results for the United States). This goes against the expectations of a theory that regards nonmarital cohabitation as a premarital sorting ground that ought to "weed out" the less durable unions before marriage, while the weeding out process is otherwise delayed into the early stages of the marriage. According to this *weeding out hypothesis*, women who marry directly should have the higher dissolution risks because they have taken less time to get to know their partner and his behavior in situations that are important for daily family life; the information level and specific investment in the union of those who have the cohabitational experience is simply greater.

The reverse effect, which one typically observes, can be explained as an outcome of differences in norms concerning union *dissolution* that follow from a self-selection in the choice between a consensual union and a marriage as a mode of union *formation*. This *mode-of-entry self-selection hypothesis* posits that by and large, women who enter into a consensual union will have the more liberal and independence-minded attitudes and norms, and that after union formation, it will show up in their union disruption behavior just as it showed up in their choice of mode of entry. So long as individual human experience is cumulative, effects like those predicted by the weeding out hypothesis may be present, but they will be swamped by selection effects.

The self-selection hypothesis is based on the notion that individuals differ from each other in important ways that are not picked up by the observed characteristics one controls for in the empirical analysis, indeed that unobserved heterogeneity dominates certain aspects of behavior. This feature will show up in the present chapter's analysis of first marriage disruption before first successful pregnancy in contemporary Sweden, and it has appeared before in our own investigation of marriage disruption after entry into parenthood in several ways (Hoem and Hoem, 1990). The Swedish population has led the strong increase in prevalence of consensual unions in Western societies over the last two decades. The growing acceptance of nonmarital cohabitation as a possibly long-lasting predecessor of marriage, or even as a replacement for it for many couples, is a reflection of changes in attitudes and norms and possibly even in deeper life values in the general population. In Sweden, such changes have run the gamut from an increasing tolerance of nonconformist modes of family formation to a complete changeover to a situation where to conform implies to behave in a manner that used to be rather deviant. As if by contagion, such conceptual changes have extended to union cohesion and have produced a higher and increasing incidence of dissolution both before and after entry into motherhood for those who marry only after they have started their initial union.

Conversely, over the same period, direct marriage will progressively have become a manifestation of particular religious or other convictions, and the recorded dissolution risks have been lower *and decreasing* for this group (when we control for other variables). This has occurred despite the fact that under the influence of mainstream developments, the family stability even of such individuals is likely to have deteriorated, at least to some extent. When marrying directly becomes rather unusual, as in Sweden, those who do so, progressively come from a select group of people whose conceptual position differs from most people's. The general change in norms will have rubbed off on members of this group as well, but the composition of the group will have shifted sufficiently towards classical family values to make its group level disruption risk decrease

Table 9-1 Women in their first marriage in the Swedish Fertility Survey of 1981, by cohort and by any recorded first subsequent demographic event

		Next Demographic Event	
Cohort Born In	Number of Respondents	Pregnancy	Marriage Dissolution
1936-40	257	226	8
1941-45	434	381	18
1946-50	342	286	15
1951-55	232	151	20
1956-60	62	25	4
Total	1327	1069	65

NOTES:

The next demographic event is recorded only if it occurs before interview.

A successful pregnancy is taken as recorded seven months before entry into motherhood.

in the manner noted. Thus, unobserved heterogeneity manifests itself in risk trends as well as in risk differentials. A good technique for omitted variables ought to be able to handle this case.

DATA FROM THE SWEDISH FERTILITY SURVEY

Our data come from a fertility survey conducted by Statistics Sweden in 1981. It contains usable records for 4,223 women, distributed with 490, 990, 1,014, 1,030, and 699 respondents, respectively, from the five quinquennial cohorts born between 1936-40 and 1956-60. The target sample was selected by simple random sampling from each of these five cohorts, among women born in Sweden, of Swedish nationality, and living in the country as of February, 1981. The uneven size of the five groups of respondents is due mainly to differential sampling fractions in the strata. After some quite extensive cleaning operations (Hoem, 1986), we have usable information about each respondent's family of origin, some of her educational history, her complete cohabitation and marriage history, employment and educational activity (largely month-by-month from September of the year in which she reached age 16 years), and much else. For further general information about the data set, see Lyberg (1984), WFS (1984), and Hoem (1986).

Because the members of our various cohorts had reached different ages at interview, the data contains different amounts of information about them at the life stages relevant for our study. Table 9-1 displays the size of the parts of the data set that directly concern the issue of first marital disruption before first successful pregnancy that is central to this paper. It shows that as little as eight such cases of marital dissolution were recorded in our oldest quinquennial cohort (just four of which occurred during the first eight years of marriage) and only four were recorded in our youngest cohort. Most of our analysis will be based on the middle cohorts, where the number of cases is more adequate.

The case counts in Table 9-1 concern dissolutions that occur before a pregnancy has been recorded. Dissolution risks will be very low for women who

Table 9-2 Dissolution of first union at parity 0, by cohort, civil status, and union duration. Occurrence/exposure rates per 10,000 nonpregnant women per month

Duration of First Union (Months)	Married		Currently Cohabiting
	After Preceding Cohabitation		
	No	Yes	
	Cohorts Born in 1936-50		
0--35	4.3	7.6	51.2
36--71	10.3	9.9	108.9
72--95	19.9	48.4	67.9
	Cohorts Born in 1951-60		
0--35	[a]	30.8	96.5
36--71	[a]	52.6	87.9
72--95	[a]	113.0	175.4

[a] Indicates insufficient group size.

know they are pregnant, and we hedge against the effect of a known pregnancy in our current analysis by censoring all life histories seven months before any entry into motherhood. Note that this practice also eliminates from our current investigation the more highly dissolution prone marriages that have been brought about by an impending first birth, for none of the marriages included occur less than seven months before the birth of the first child.

A first impression of union dissolution levels by cohort, civil status, and union duration can be had from Table 9-2, which displays some of the main features of disruption behavior. We note that for non-pregnant respondents at parity 0, overall, (1) consensual unions were much more fragile than marriages, (2) marriages formed after premarital cohabitation had higher disruption rates than "direct" marriages, (3) union cohesion deteriorated over time and (4) dissolution rates in childless marriages increase at longer durations. Observations (1) to (3) carry over to disruptions after first childbearing (Hoem and Hoem, 1988).

RESULTS WITH NO EXPLICIT INCORPORATION OF UNOBSERVED HETEROGENEITY

Model specification

Some exploratory analysis of the life segment during which a woman lives as nonpregnant at parity 0 in a consensual or (separately) a marital union has led to the intensity models whose fitted relative risks have been listed in Table 9-3. For each transition intensity, we have used a variant of the Cox (1972) hazard regression model of the form

$$h(t) = h_0(t)exp\{\Sigma_{j \geq 1}\Sigma_k b_{jk} z_{jk}(t)\} \, . \tag{9-1}$$

Here, the *baseline intensity* $h_0(\cdot)$ is the hazard for a selected baseline subgroup of respondents. The index $j \geq 1$ represents a set of possibly time-varying covariate factors, each of which has a finite number of factor levels, indexed by k, and $z_{jk}(t)$ is the value at time t (for a given individual) of a binary indicator of level k for factor j. For the baseline group, $z_{jk}(t) = 0$ for all j, k, t.

We have used only categorical factors. The *hazard regression coefficients* $\{b_{jk}\}$ are then probably best interpreted by noting that the "'relative risk'' $exp(b_{jk})$ is the ratio between (1) the transition intensity for a respondent subgroup with level k on factor j and any given level on the other factors, and (2) the corresponding intensity for a subgroup with the baseline level on factor j and the same level as the first group on all other factors. For example, Table 9-3 shows that disruption risks at the life segment in question rose sharply over our birth cohorts. Cohabiting women born in 1936-40 had an estimated 53% of the disruption risks of corresponding women in the baseline cohort born in 1946-50, and the cohort from 1956-60 had 60% higher consensual union dissolution risks than the baseline cohort. This comparison is made for women with the same family background, religious activity level, union starting age, current educational level, and current employment status. We will discuss other features of the estimated covariate effects below.

Several considerations have led to our preference for a categorical representation. Some factors, like "family background,'' are given on a nominal or ordinal scale and naturally are categorical variables. The factor "current educational level'' is based on the number of months of education up to the current month, but we have grouped the covariate, both with a view to overcoming reporting unreliability (Hoem, 1986), and because educational effects essentially come in chunks normally represented by diplomas or degrees achieved, as well as because we have no real empirical or theoretical basis for a more structured representation of its influence on the transition intensities. It is important to avoid projecting nonexisting knowledge about human behavior into the model, and we have taken care not to specify factor linearities or other firm patterns where they are not known to exist. This also explains why we have grouped a "continuous'' covariate like duration of premarital cohabitation. A sensible grouped representation helps prevent functional form misspecification and sometimes makes us discover interesting behavioral patterns that would be hard to detect by other means.

The same consideration leads us to normally prefer a piecewise constant baseline hazard, at least until we have got the time profile of $h_0(t)$ about right, after which one may safely substitute a comparable parametric formula. Otherwise, the specter of baseline hazard misspecification is particularly haunting, for it may produce serious estimation biases in the hazard regression coefficients and it makes them very sensitive to the choice of a heterogeneity distribution. This instability seems to disappear when the baseline hazard is on target. (Documentation has been given by Struthers and Kalbfleisch, 1986, by Montgomery, 1990, in references given by Kiefer, 1988, p. 676, as well as in work by G. Ridder, 1987; L. Engström, K. G. Löfgren, and O. Westerlund, 1988; and R. Aaberge, Ø. Kravdal and T. Wennemo, 1988.)

Table 9-3 Relative risk of union disruption for nonpregnant women at parity 0 in marital and nonmarital first unions, separately. Five quinquennial birth cohorts. No interactions.

Factor/Level	In Consensual Union	Married[a]
Cohort born in		
1936-40[b]	0.53	
1941-45	0.82	1.01
1946-50	1	1
1951-55	1.10	2.77
1956-60	1.60	3.50
Family Background		
Others	1	0.68
Blue Collar, Unskilled	1	0.86
Blue Collar, Skilled + White Collar, Low[c]	1	1
White Collar, Middle + High	1.40	2.08
Religiously Active[d]		
Yes		0.63
No		1
Starting Age[e]		
16-17 years	1.69	
18-19 years	1.27	1.82
20-25 years	1	1
26-35 years	1.23	1.40
Current Educational Level		
Low	1	1.22
Middle	1	1
High	1.17	0.72
Current Employment Status[f]		
Full-time	1	1
Part-time		0.72
Housewife		0.45
Student	1.10	1.24
Other		3.15

NOTES:

A *baseline level* is indicated by a relative risk of 1 (without decimals).

Time variable (grouped):

- For cohabiting women: Months since start of union, censored at entry into month 72.

- For married women: Months since marriage, censored at entry into month 96.

[a] The model for married women also includes duration of premarital cohabitation, grouped into 0 months, 1--9, 10--23, 24--47 months, and 48 months or more.

[b] The respondents born in 1936--40 had too few recorded marriage disruptions for sensible analysis. This cohort has been left out of the analysis for married women, for the behavioral pattern in other cohorts did not extend to it. Despite similarly few recorded disruptions in our youngest cohort, it was not deleted, for it contained correspondingly small exposures.

[c] Daughters of skilled workers and low-grade employees.

[d] Due to the negligible exposures for recorded cohabitants among religiously active respondents, this factor was left out of the analysis of the dissolution of consensual unions.

[e] For cohabiting women: Age at union formation.
For married women: Age at marriage.
The negligible number of respondents who married before age 18 were deleted.

[f] Among nonpregnant respondents in consensual unions, the great majority were students or in full-time employment almost all the time. The student category includes pupils enrolled in all types of educational institutions.

A practical advantage of the piecewise constant hazard is that it makes it easy to include interactions with the time factor, which is not the case with other representations. A non-parametric hazard specification is quite popular and much software has been developed for it, but it has the double disadvantage that (1) it makes it hard to spot the hazard time profile, and (2) it is technically impossible to extend its use to empirical investigations that incorporate an unobservable.

For a mathematical representation of the piecewise constant baseline hazard, we partition the time axis into a finite (and in practice small) number of disjoint time intervals and "assume" that $h_0(t) = exp\{b_{0k}\}$ for all t in interval number k. The $\{b_{0k}\}$ are unspecified constants (parameters), and we can write

$$h_0(t) = exp\{\Sigma_k b_{0k} z_{0k}(t)\} , \tag{9-2}$$

where $z_{0k}(t)$ is an indicator of whether t is in interval number k. This provides an essential symmetry between the effect of the time factor and that of all other recorded factors. As usual, interactions between factors are easily introduced by letting some binary variables $z_{jk}(\cdot)$ be products between other binary variables.

Empirical results

According to the models in Table 9-3, there was an increasing trend over our cohorts in both dissolution risks at parity 0. (Due to differential nonresponse across cohorts, this trend may have been exaggerated for married nullipara.) A significant interaction with union duration (not displayed here) shows that for a first consensual union this increase was concentrated at durations of up to three years, a feature that reflects the changing nature of such unions in the Swedish population. Nonmarital cohabitation has developed into a prevalent institution in its own right, largely replacing the phase when young people used to "go steady" or were engaged as well as the early stages of a marriage. It has lost some of its function as a preamble to a lifelong relationship and has moved towards being merely a practical living arrangement for enjoying many of the trappings of a marriage with much fewer of its bonds. It is to be expected that such unions would become progressively less cohesive over time and would break up more regularly at relatively short durations. (No other interaction had an interesting and reliable pattern.)

Daughters of middle and higher level white collar employees (a group of parents which includes independent academics and members of the educated liberal professions) had consistently higher dissolution risks than others, both at this stage and after entry into motherhood. Presumably, there is something in the bourgeois culture that more easily makes dissolution an acceptable alternative when a union does not function as desired. For marriage disruption risks, a clear gradient over the social groups can be seen. Among married women, those who reported themselves as religiously active had somewhat reduced but still appreciable dissolution risks. (For a discussion of the definition and relevance of this factor, see Section 5.4 in Hoem and Hoem, 1990.)

As we would expect, a woman had an increased union dissolution risk at parity 0 when she entered the union as a teenager, and the risk was higher the younger she was at entry. Unions that started late also had somewhat increased risks (cf. Glick and Norton, 1977 for a similar finding.) Perhaps people become set in their ways and have somewhat unrealistic expectations as to what the partnership can give them as they go beyond the usual starting ages. A selection effect may also be present, in that the most desirable partners may have entered their first union at an age when most young people are teaming up.

Students had much the same risks of experiencing the dissolution of a consensual union as jobholders had. Similarly, women with some education on the university level had much the same risks as others. The scales were slanted slightly against both those who were in school and those who had (some) higher education, however, so a university student ended up having almost a third higher estimated dissolution risk than the baseline category, which is the group of nonstudents with less than university-level education.

For married women, the gradient by educational level was in the opposite direction. Apparently, more highly educated women were more careful in selecting or retaining their marriage partners. We are not sure that this is a real phenomenon, however, because the effect is not significant and it reverses again (once more without being significant) after entry into motherhood (Hoem and Hoem, 1990).

We would expect a married woman with a full-time job to have a higher dissolution risk than a housewife, both because of her better economic position and of her higher general ability to cope with union breakup. She would have more extensive opportunities to meet and get to know men other than her husband through a partly independent social life. Women in part-time jobs may have all of these elements in a reduced "dose". Since we are dealing with life segments prior to childbearing here, being a housewife should constitute a signal of an unusually strong family orientation. These expectations are borne out in Table 9-3. (We will not speculate about the elevated disruption risk of women in "other" employments, for the group is very small at any time.) Note that the current educational level and employment status are time-varying covariates whose values are updated every month of the union, as is its civil status.

The risk of marriage disruption is strongly affected by the mode of entry into marriage. This shows up in the fitted dissolution model for which the last column in Table 9-3 gives the most interesting coefficients. To accommodate so many covariates, we have had to use what turned out to be a rather coarse partition of the premarital duration in that model. To better display the structure of this effect, we prefer to report the relative risks for a finer grouping of this covariate in Figure 9-1, based on a model where all nonsignificant factors have been eliminated and some levels have been combined on the remaining factors. [Beside the time parameter "duration of marriage," that model also contains the factors "cohort" with two levels (1941-50 and 1951-55), "family background" (three levels), "religiously active" (two levels), and "duration of premarital cohabitation". The seven levels on this covariate are indicated in Figure 9-1. The interval grouping was selected so as to minimize the effect of any digital preference in the reporting of the duration of premarital cohabitation, in particular any heaping at six or twelve months.]

As mentioned in the introductory section, much lower risks of marital disruption are seen for women who have married without premarital cohabitation.

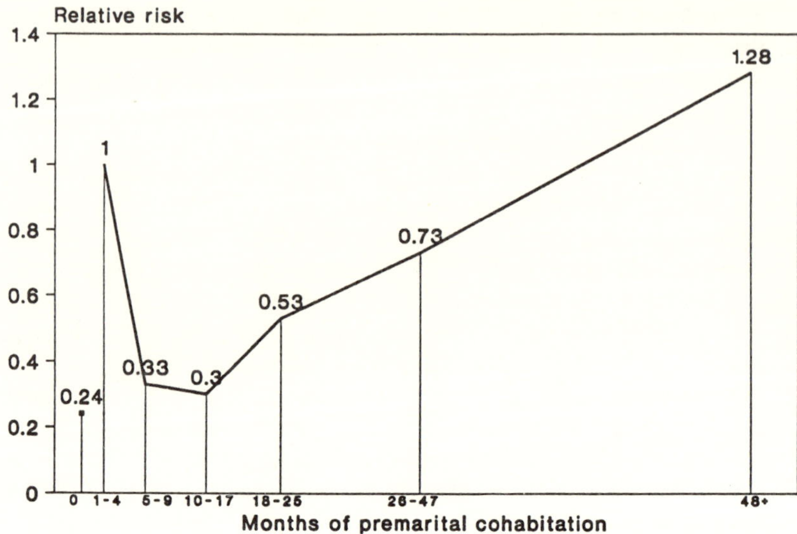

Figure 9-1.   Relative risks of marriage disruption, by duration of any premarital cohabitation. Nonpregnant Swedish women at parity 0, born in 1941-55. Hazard model estimates.

The rest of the risk profile suggests that among other married women, there may be an initial effect along the lines of the weeding out hypothesis, but that the effect wanes for longer durations of premarital cohabitation. Since we are dealing with women at parity 0, such a long premarital duration may indicate a problematic relationship, at least so long as there are no children in the marriage. Similarly, marital dissolution risks increase strongly with marriage duration (not tabulated here) if the marriage remains childless, to the extent that marriages that have stayed at parity 0 for more than four years, have disruption risks that are estimated to be more than three times as high as during the two first years.

Figure 9-1 is evidence of one strong advantage of our practice of using grouped representations of factors that essentially have a continuous basis. It led us to discover the regular but complex effect structure of premarital cohabitation, a structure that could easily have been missed with a less flexible representation with a simple mathematical formula. In particular, this would have been the case if we had used the usual linear formula.

UNOBSERVED HETEROGENEITY

The introduction of unobserved heterogeneity

The findings reported above are outcomes of a separate analysis of each particular type of transition, based on maximum likelihood estimation and likelihood ratio tests. (See Borgan, 1984 for an accessible account of the statistical properties of the procedures we have applied.) In a study of first union disruption, the

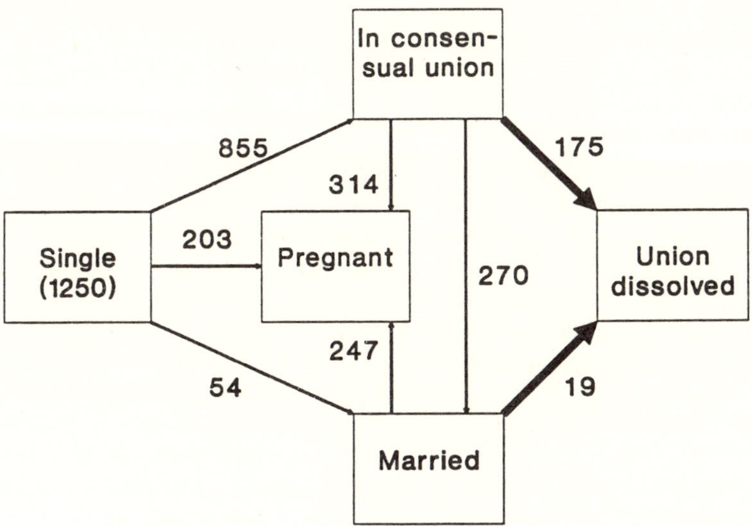

Figure 9-2. Basic state space for union disruption study. Arrows suggest possible transitions. Figures on arrows indicate numbers of transitions in the slimmed data set used for heterogeneity analysis. Cohorts born in 1946-55; 1,250 respondents altogether.

basic state space can be represented as in Figure 9-2, and we have concentrated exclusively on the two rightmost transitions indicated by the thicker arrows, one at a time. During each piece of analysis, other types of transition are ignored or are treated on a par with censoring, as the case may be. Such a procedure is possible because the parameters in ignored intensities are not involved in functional relations with the included intensities, and because unobserved heterogeneity has not been incorporated into the model. The likelihood can then be written as a product of separate factors, one for each intensity in question, and each factor can be maximized on its own.

Generic formulas like (9-1) and (9-2) have been used for the intensity of any selected transition. To indicate that there is a separate set of regression coefficients $\{b_{jk} : j \geq 0\}$ for each type of transition, we now need to add another subscript, say m, and call the parameter b_{jkm}.

The separate analysis of marriage disruption reveals manifest unobserved heterogeneity in the data. To include this feature in a manner that enables us to use current methods of analysis, a further item is added in the formula for each intensity, which then becomes

$$h_m(t; \theta) = exp\{\Sigma_{j \geq 0} \ \Sigma_k b_{jkm} z_{jk}(t) + c_m \theta\} , \qquad (9\text{-}3)$$

for transitions of type m. This is to be understood as follows in our application.

The antilogarithms of the regression coefficients b_{jkm} in (9-3) retain their usual interpretation as relative risks. Formally, all factor binaries $z_{jk}(t)$ are included in all intensities, but we allow for the systematic deletion of some factor(s) from any intensity function by permitting some b_{jkm} to be set to zero.

Beside the set of values on the observed factor level indicators $z_{jk}(t)$ that each respondent has in each month t of observation, she also has an unobserved value θ of the *heterogeneity factor*. To provide any scientific insight, it is important that this factor have a plausible substantive interpretation. It may represent features like lifelong basic personal values, permanent personality traits, or genetic biological traits. In our demographic application, the desired function of the heterogeneity factor is that it should pick up differentials in personal attitudes and values beyond those reflected in our other covariates (cf. Table 9-3). Its value is constant over the investigated life segment and is the same for all transitions involved (a *persistent unobservable*). The values of θ for the individual respondents in a given sample stratum are treated as the outcome of independent random drawings from a stratum-specific mixing distribution. We have no basis for selecting a particular mixing distribution, so we have picked the nonparametric representation introduced by Heckman and Singer (1982, 1984). This means that in the maximum likelihood computations, the mixing distribution is taken to be located on a finite (and usually small) number of points in the unit interval [0, 1], with one point fixed at either end and possibly one or more points at unknown positions inside the interval, and with an unknown distribution over those points.

The impact on transitions of type m of the heterogeneity factor is measured by the *factor loading*, c_m. This is a parameter to estimate on a par with the coefficients b_{jkm} of the observed covariates. Because θ is unobserved, c_m plays a role quite similar to that of a factor loading in factor analysis, just as the b_{jkm} play the same role as regression coefficients do in linear regression analysis. In both cases they have inherited the names of their linear model counterparts. Since the mixing distribution is concentrated on the interval [0,1], we should be able to interpret $exp(c_m)$ as the risk of experiencing a transition of type m for a respondent with the high value 1 on the heterogeneity factor (say, a respondent with very liberal fundamental attitudes to forms of conjugal union formation), relative to the corresponding risk for a completely comparable respondent with the low value 0 on the heterogeneity factor (a respondent with very conservative values concerning conjugal unions) and the same set of levels on all observed factors, in any month in which these individuals are at the risk of transition m. Since we get an equivalent intensity specification with the heterogeneity factor $\theta' = 1 - \theta$, the direction of the factor values is unimportant.

Through its (mixing) distribution, a persistent unobservable provides a kind of cement that binds together the transition intensities it influences and makes it necessary to consider them simultaneously during maximum likelihood estimation. The likelihood is no longer composed of intensity-specific factors that can be maximized separately. Unless there is a simple structure across states or across decrements in the intensities, technical difficulties multiply in situations where there are several states or where it is possible to return to a state that has been left, as when there can be several bouts of unemployment, or sickness relapse, or crime recidivism, or repeated miscarriages or repeated perinatal deaths. The extent of the work and other costs involved may have deterred many analysts from pursuing this route. Unfortunately, leaving out unobserved heterogeneity may lead to a distortion of the impact of recorded factors on individual behavior. In particular, one may expect the gradient of the baseline hazard to be biased downwards in a single decrement situation, for then the heterogeneity mix of "survivors" in the state of origin will be progressively more strongly loaded towards the less decrement prone as time goes on. In the single decrement model, neglecting unobserved heterogeneity or omitting pertinent observed covariates will also lead to an underestimation of covariate effects when the observed covariates are stochastically independent of each other and of the heterogeneity factor. (Available proofs and numerical experiments assume that there are no time-varying covariates and make some mild additional regularity assumptions. Some selected references are Lancaster and Nickell, 1980; Lancaster, 1985; Struthers and Kalbfleisch, 1986; Bretagnolle and Huber-Carol, 1988, as well as unpublished work by G. Ridder and V. Verbakel, 1983.)

As we have noted already, the importance of such misspecification biases and the circumstances in which they really appear is contested. Moreover, unidirectional results like those just mentioned will not hold in cases like ours. We have more complex transitions than those of the single-decrement model, as well as some obvious covariate dependencies. In such a situation, it is not easy to obtain a clear picture of the extent or even the direction of misspecification biases in general. When the heterogeneity factor influences some decrements in one direction and other decrements in the opposite direction, how are we to tell how the mix of "survivors" in the state of origin will develop? Consider, for instance, the status "never in a conjugal union, not pregnant, at parity 0," in which young women are subject to the competing risks of transition into the statuses "pregnant," "in a consensual union," and "married." (See the leftmost state in Figure 9-2. Only the first decremental event counts.) If the heterogeneity factor represents positions on a liberal/conservative personality scale, we would expect the more conservative to have high intensities for the transition into marriage and low intensities for the transition into cohabitation and into pregnancy, while the converse would be true for more liberal individuals. Unless we have some *a priori* knowledge about the strengths of such heterogeneity effects (which we do not), it is hard to tell what the personality mix will be among women who remain nonpregnant, nulliparous, and single. Consequently, we cannot foresee even the direction of any gradient bias in the estimated baseline transition intensities. Similarly complex reflections pertain to corresponding applications in genetics, where a hereditary or other biological trait may weaken an individual's resistance to some illness, say, and may strengthen it against others.

An essential ingredient in our explanation of behavioral trends over the quinquennial birth cohorts that constitute our sample strata is that there has been a drift in values (Hoem and Hoem, 1990). Since we want the heterogeneity factor to represent personal values, it is important to allow the mixing distribution to be cohort-specific and to analyze the records for the cohorts separately to bring out such a feature. Given known difficulties in pinpointing the mixing distribution in empirical work (e.g., Montgomery, 1990) or in recovering it in simulation experiments (Heckman and Singer, 1984), it is probably too much to hope for that insights should arise out of patterns in the mixing distributions fitted to our data, but at least some feature of the empirical outcome ought to reflect value changes. This argument means that the cohort factor helps determine the individual value of θ. So does the family background, while θ is a determinant of the other covariates (with the possible exception of the religious activity level, which may be determined simultaneously). In principle, therefore, the mixing distribution should depend on family background as well as on the cohort. Splitting the data set of each cohort into substrata according to family background would give hopelessly inadequate subgroup sizes, however, so we have tried two different alternative procedures.

First and most extensively, we have worked with a cohort-specific mixing distribution and have let family background appear as a regular observed variable, just like starting age, say. We regard the ensuing mixing distribution as the (cohort-specific) marginal distribution produced by a Bayes argument applied to θ and the family background variable. In a second and less extensive set of experiments, we have disregarded family background completely and have subsumed it under the anonymous set of determinants of the (still cohort-specific) marginal distribution of the heterogeneity factor.

The bulk of the rest of the present account is concerned with the first of these approaches. The outcome of the second approach is quite similar and is reported much more briefly.

Computer implementation

When all observed covariates are categorical and the baseline hazard function is piecewise constant, a matrix of occurrences and a matrix of exposures corresponding to all combinations of factor levels and time intervals constitute a sufficient statistic together. Such an occurrence matrix may have zero marginals, which causes some practical problems with computer programs that do not hedge against this possibility. The algorithm of the best software at our disposal that incorporates unobserved heterogeneity, a program called CTM (K.-M. Yi, B. Honoré, and J. R. Walker, unpublished), is unable to handle zero marginals. The program has not been written with the occurrence-exposure situation specifically in mind, and despite all its other good properties, it also has difficulties converging when the two matrices are sparse.

The question of zero marginals is unique to the grouped representation of factors and baseline hazard. It disappears when the groupings are replaced by more rigid functional forms (linear and quadratic functions, Weibull or Gompertz baselines, and so on). It becomes a practical issue whether the inconvenience caused by any zero marginals is a price worth paying for the added flexibility and the potential for new discovery that it offers. Our investigation appears as a test case, for it is replete with sparseness problems.

There are no marriages at age 17 in our cohorts 1, 3, and 4, no marriages at age 18 in cohort 4, no premarital first pregnancies at the relevant life stage at age 26 in cohort 3, and only one at this age in cohort 4. In cohort 5, there are no recorded marriages (at any age) among daughters of high grade white collar employees and no relevant premarital first pregnancies in the same group nor among the religiously active, and so on. During a process of extensive exploration, we have deleted particularly small or otherwise problematic groups or factors, like the religiously active (too few cases) and the 16- and 17-year-olds (too few marriages). We are particularly disconcerted at having had to remove

Table 9-4 Relative risk of union disruption for nonpregnant, not religiously active women at parity 0. First unions only. Cohorts born in 1946–55, combined.

Factor	In Consensual Union Model: 1	2A	2B	Married Model: 1	2A	2B
Family Background						
Blue Collar, Unskilled	0.81	0.63	0.81	0.67	0.73	0.75
Blue Collar, Skilled + White Collar, Low	1	1	1	1	1	1
White Collar, Mid + High	1.20	1.67	1.20	0.73	0.79	0.80
Age at Union Formation						
-19 years	1.75	2.50	1.75	0.67	0.51	0.49
20-22	1	1	1	1	1	1
23-25	1.22	1.59	1.22	2.27	2.40	2.40
Duration of Premarital Cohabitation						
0 months (= no premarital cohabitation)				0.09	0.00	4.04
1-7 months				1	1	1
8-25				0.24	0.24	0.24
26-47				1.29	1.31	1.32

NOTES:

Time Variable: For cohabiting women: Months since start of union. For married women: Months since marriage. Grouping: 0-23 months, 24-47, 48-71, 72-95 months. All records have been censored at entry into month 96.

The Baseline Level of each factor is indicated by a relative risk of 1 (without decimals).

The factor *duration of premarital cohabitation* does *not* enter into the disruption intensity for women in consensual unions.

unions started before age 18, for surely this removes a select group of women with particular attitudes and personal values and therefore with an interesting influence on the mixing distribution.

We have also deleted records with "other" family backgrounds because the group is heterogeneous and has unpredictable disruption behavior. We have also combined and recombined factor levels, and have settled for those that appear in Table 9-4 below (for union disruptions). The intensities for transitions out of the leftmost state in Figure 9-2 ("single, nonpregnant, at parity 0") use age attained as their basic time variable, grouped into ages 18-19, 20-21, 22-23, and 24-25. Those who remain in the state until age 26 have been censored then. For the states "in consensual union" and "married," the basic time variable is "months since entry into the current state," grouped into four two-year intervals, and the records have been censored at entry into month 96. After some abortive exploration of the data for single cohorts, we have had to abandon our ambition of analyzing each quinquennial cohort separately, and have combined the data for two cohorts to achieve usable results. In fact, we have given up the whole experiment after working on the behavior of the combined cohorts 3 and 4 (born in 1946-55), on which we report below. The reduced data set for these two cohorts has 1,250 respondents. The number of respondents making each transition is indicated in Figure 9-2.

All of our regressors z_{jk} are now independent of t (except of course the time indicators z_{0k}). Although we would dearly have loved to see how the addition of unobserved heterogeneity would influence the estimated impacts

Table 9-5 Fitted likelihoods for models 1 and 2

Model	p_0	Log-Likelihood
1		-11423.07
2, Variant A	0.852	-11418.60
–, Variant B	0.958	-11418.81
–, Variant C	0.849	-11419.31
–, Variant D	0.639	-11418.96

NOTES:

Model 1: Unobserved heterogeneity not incorporated.

Model 2: Two-point heterogeneity distribution with point mass p_0 at 0 and mass $1 - p_0$ at 1.

on union disruption intensities of the time-varying factors "educational level" and "employment status", this has not been possible. These covariates are endogenous processes on a par with the individual-level process depicted in Figure 9-2, and their transition intensities depend on θ just as much as the ones explicitly modeled here. It is hard enough to get sensible results on the basis of a small state space like that of Figure 9-2. To add even a single time-varying covariate with only two levels (say "current employment status" with the levels "employed" and "studying") would mean doubling the state space and more than doubling the number of θ-dependent transition intensities in the model. (Each state in Figure 9-2 would be replaced by two states, one for each of the two employment statuses included.) Because all heterogeneity-dependent transition intensities must be modeled simultaneously, we have seen it as unrealistic from the outset to try to incorporate unobserved heterogeneity into a system of intensities for transitions that include endogenous time-varying covariates.

To keep down the number of regression parameters, we have also avoided any interactions between the factors of the model. Given the outcome of our previous analysis, we would not expect to gain much added insight from interactions in our situation, so this is not such a great loss to us in any case.

Empirical outcome with family background included

After reducing the data set in the manner just described, we have used the program CTM to try to maximize the likelihood. Unfortunately, the likelihood is quite ill conditioned when unobserved heterogeneity is included in the model. It has several local maxima and we have a hard time locating them. We have found four maxima during explorations that include the family background variable, listed as Variants A to D of Model 2 in Table 9-5. Model 2 has a two-point mixing distribution located in the points 0 and 1 and with point mass p_0 in the origin. Note how little difference there is between the values of the likelihood in the four variants of Model 2. Such maxima do not normally merit much attention, but we include a brief discussion of them in order to indicate which features we would be looking for in a meaningful solution.

The natural basis of comparison is a model that does *not* incorporate heterogeneity. We have also fitted such a model (our Model 1), and Table 9-5 contains selected estimated relative risks for the two union disruption intensities in Model 1 as well as in Variants A and B of Model 2. (The coefficients of the other intensities and for Variants C and D do not have sufficient independent interest to be displayed here.) One result for Model 1 bears further testimony to the selectivity of the behavioral processes involved. Note how the dissolution risk of women who married as teenagers was much lower than the corresponding risk for those who married in the next higher age group in Table 9-5, while the relation was the reverse in Table 9-3. This changeover has been produced by the exclusion of respondents who said they were religiously active. (The same empirical feature appears as soon as the records of the religiously active have been deleted in other parts of our investigation.) A plausible explanation is that Swedish women who were not religiously active, and who married as teenagers *without being pregnant* in the cohorts in question, must be a special group with a particularly strong family orientation, for most of those who married at this early age, "had to" do so because they were pregnant. Again, unobserved heterogeneity figures prominently in our explanation of findings based on a model that does not incorporate an unobservable. We would expect a successful heterogeneity model to bring the disruption risk for marrying teenagers into line and to have a similar influence on the corresponding risks for those who marry without premarital cohabitation. We regard it as an indication of the inadequacy of the models as applied to our data that this modest achievement has eluded us.

Except for the coefficients of the two disruption intensities, all relative risks in all four variants of Model 2 are practically the same as in Model 1. We have detected no interesting impact of the introduction of unobserved heterogeneity on the pattern of any baseline intensity. In no variant of Model 2 does the inclusion of θ create any interesting new pattern in the relative risks for any factor beyond the following isolated instance:

In Variant A, the relative disruption risk to married women who did not cohabit before marriage has the nonsensical estimated value of $\exp(-20.5)$. Conversely, Variant B has a most interesting relative disruption risk for this group, namely a value much higher than the relative risk for any group of women who did cohabit before marriage. It shares this feature with Variant D. If either had been the maximum likelihood solution, such a feature could be taken as a vindication of the weeding-out hypothesis, but they are not. Variant C has a value of p_0 that is very close to that of Variant A but its relative risk pattern is much like that of Model 1. Furthermore, none of the local solutions have factor loading patterns conducive to a reasonable interpretation of the heterogeneity factor (not documented here).

It has not seemed worthwhile to search further for the global maximum of the likelihood function for Model 2 for the meager gains we expect, nor has there been any point in extending the heterogeneity model with further mass points, particularly since this led us nowhere in initial experiments (not reported here) with a simpler model for pure decrements from the starting state ("single, nonpregnant, at parity 0"). We suspect that Model 2 is close to being numerically unidentified in our data set, and have worked towards reducing rather than extending the number of parameters in what follows.

Deletion of the family background factor

To reduce the number of model parameters and also to better adjust our approach to the idea that family background helps determine the heterogeneity factor rather

Table 9-6 Factor loadings in our fitted model without the family background factor and with quadratic baseline log-hazards

Transition			
From[a]	To (Status)	Factor Loading	Standard Deviation
Single	- Cohabiting	0.65	1.73
	- Married	−1.85	2.36
	- Pregnant	1.69	11.38
Cohabiting	- Married	−1.71	1.28
	- Union Dissolved	3.16	0.66
	- Pregnant	1.44	0.67
Married	- Union Dissolved	2.45	2.07
	- Pregnant	−40.24	72.10

[a]Women in the sending states are non-pregnant by definition. A pregnancy is recorded seven months before a birth.

than conversely, we have made two further explorative steps. First, we have run the models over again without the family background factor, and subsequently, we have also replaced all piecewise constant baseline log-hazards by quadratic polynomials, which give a fair representation of their time profiles. None of this has led to much improvement. In both cases, the fitted regression coefficients were close to those of the corresponding model without an unobservable. We have found two local maxima for the likelihood when unobserved heterogeneity is incorporated and the baseline log-hazards remain piecewise constant. One of these is very close to having $p_0 = 0$, that is to exclude the unobservable. The only new element is that the sign pattern of the fitted factor loadings becomes consistent with an interpretation of the heterogeneity factor as a measure of personal values when the baseline log-hazards are arcs of parabolas enspace (Table 9-6), with $\theta = 0$ for "conservatives" and $\theta = 1$ for "the liberal minded". (The fitted factor loading for entries into pregnancy among married women should be disregarded, as it is extremely badly determined.)

DISCUSSION

When we have no trouble locating and interpreting effects of unobserved heterogeneity in analyses that do not include this factor explicitly, then why is it so hard to use an instrument devised specifically to enable us to incorporate it? Too few investigations of this nature have appeared to draw any firm conclusions yet, but we suspect that our problems may be inherent in unrealistic demands for data richness made by heterogeneity techniques when applied to the analysis of complex behavior, particularly when the behavior is rather rare (by comparison to, say, union formation or entry into motherhood). We may have compounded the difficulties to some extent by certain choices made in specifying the various elements of our model, in particular when we operationalize our distaste for parametric representations that have insufficient empirical basis. Our final explorations suggest, however, that more general issues are involved.

To have much hope that the numerical algorithm will work, it is highly important that one restricts the number of θ-dependent transition intensities explicitly involved. This has made us adapt our focus by censoring our records as early as on the arrival of the first successful pregnancy and by shedding all endogenous time-varying covariates. We have also cut back on small groups, on fixed covariates, and on their levels. Despite all this, the number of parameters that must be estimated simultaneously remains too large for the data set. In each variant of Model 2 (Table 9-4), there are 73 parameters to estimate altogether, namely 40 ordinary regression coefficients, 24 parameters in the baseline hazards, one factor loading for each intensity involved, and p_0. This is not because each intensity has such a large number of parameters, but rather because there are as many as eight intensities in the model.

Consider for instance, the disruption intensity for married women in Table 9-5, which has more parameters than any other intensity here. It has seven regression coefficients and three parameters in the baseline hazard, plus the factor loading. Deleting the family background variable reduces the number of parameters by two, and switching to a quadratic for the log-baseline intensity removes just one more. Because of the high number of intensities, the parameter count builds up quickly, however, and cutting back three parameters from each intensity (as in our final experiment) reduces the total number by as many as 24. We could delete a few more parameters because our explorations show that a couple of baseline hazards could actually safely be represented as linear or even constant functions of time, but even this would not be enough for a successful application in our case. To delete even further parameters would cut into the whole substantive purpose of the investigation.

At the cost of some programming effort, a more tailor-made algorithm that better accommodates grouped variables could surely reduce the computer time, but the crucial numerical issues would most likely have remained and would impose similarly severe restrictions on the inclusion of behavioral aspects. The model we have ended up with does include unobserved heterogeneity, but otherwise we are left with a bleak and dehydrated picture of the segment of human behavior we are interested in. Along the way, we have spent considerable effort and innumerable computer hours without even getting rid of the effects of mode-of-entry self-selection, which is the minimum that we had expected to achieve. The result of all this toil is that we are no wiser about union disruption patterns in Sweden than we were when before we began our application of the heterogeneity technique.

We emphatically do not question the importance of the notion of persistent unobservables. The attention recently given to unobserved heterogeneity has brought its effects into the limelight as never before. This must inspire people to account for it more in their interpretations of empirical findings, which is all to the good. However, we are quite pessimistic about the possibility that current computer implementations of heterogeneity modeling techniques really open up as many new avenues to substantive insight into human behavior as previously believed. Its practical usefulness for the analysis of a data set of a normal size may be restricted mainly to the simple single or multiple-decrement situation and perhaps to some uncomplicated extensions of it.

ACKNOWLEDGMENTS

The author is grateful to James J. Heckman for making the CTM program available for this investigation and to James R. Walker for many discussions about its usage. Britta Hoem carried out the analysis of marriage disruption at parity 0 reported in part in our Table 9-3 and also provided some further programming assistance. As always, she has been a perceptive discussion partner. Comments from Julian Adams and from two referees helped improve our presentation.

Much of this work was carried out in 1988 during our stay in the Center for Demography and Ecology of the University of Wisconsin at Madison, with financial support from the Swedish Research Council for Social Science and the Humanities, from the U. S. National Institute of Child Health and Human Development Center Grant No. HD 05876, and from the William and Flora Hewlett Foundation. The hospitality of the Center is gratefully acknowledged.

REFERENCES

Aaberge, R., Ø. Kravdal, and T. Wennemo, 1988   Unobserved heterogeneity in models of marriage dissolution. Central Bureau of Statistics of Norway, Discussion Paper No. 42.

Balakrishnan, T. R.,   K. Vaninadha Rao,   E. Lapierre-Adamcyk, and K. J. Krotki, 1987   A hazard model analysis of the covariates of marriage dissolution in Canada. Demography 24: 395-406.

Bennett, N. G., A. K. Blanc, and D. E. Bloom, 1988   Commitment and the modern union: Assessing the link between premarital cohabitation and subsequent marital stability. Am. Sociol. Rev. 53: 127-138.

Borgan, Ø., 1984   Maximum likelihood estimation in parametric counting process models, with applications to censored failure time data. Scand. J. Stat. 11: 1-16.

Bretagnolle, J., and C. Huber-Carol, 1988   Effects of omitting covariates in Cox's model for survival data. Scand. J. of Stat. 15: 125-138.

Cox, D. R., 1972   Regression models and life tables (with discussion). J. R. Stat. Soc. B: 187-202.

Engström, L., K.-G. Löfgren, and O. Westerlund, 1988   Intensified employment services, unemployment durations, and unemployment risks. University of Umeå: Umeå Economic Studies No. 186.

Glick, P., and A. J. Norton, 1977   Marrying, divorcing and living together in the U. S. today. Pop. Bull. 32 (5).

Heckman, J. J., and B. Singer, 1982   Population heterogeneity in demographic models, pp. 567-599   in *Multidimensional Mathematical Demography,*   edited by K C. Land and A. Rogers. Academic Press, New York.

Heckman, J. J., and B. Singer, 1984   Econometric duration analysis. J. of Econometrics 24: 63-132.

Hoem, B., and J. M. Hoem, 1989   The impact of women's employment on second and third births in modern Sweden. Pop. Studies 43: 47-67.

Hoem, B., and J. M. Hoem, 1990   Union dissolution in contemporary Sweden. To be published in *Demographic Applications of Event History Analysis*, edited by J. Trussell, R. Hankinson, and J. Tilton. Clarendon Press, Oxford, to appear.

Hoem, J. M., 1986   The impact of education on modern family-union initiation. Eur. J. of Pop. 2: 113-133.

Kiefer, N. M., 1988   Economic duration data and hazard functions. J. of Econ. Lit. 26: 646-679.

Lancaster, T., and S. J. Nickell, 1980   The analysis of re-employment probabilities for the unemployed (with discussion). J. R. Statis. Soc. A 143: 141-165.

Lancaster, T., 1985   Generalized residuals and heterogeneous duration models: With applications to the Weibull model. J. Econometrics 28: 155-169.

Leridon, H., 1988 *Analyse des Biographies Matrimoniales dans l'Enqûete sur les Situations Familiales*. Dossiers et recherches 19. Institut National d'Études Démographiques, Paris.

Lyberg, I., 1984   *Att Frâga om barn: Teknisk Beskrivning av Under-sökningen 'Kvinnor och Barn'*. Bakgrundsmaterial från Prognosinstitutet 1984:4. Statistics Sweden, Stockholm.

Marini, M. M., 1985   Determinants of marital disruption. Paper presented at the annual meetings of the Population Association of America, Boston, MA.

Montgomery, M. R., 1988   Household formation and home ownership in France, in *Demographic Applications of Event History Analysis*, edited by J. Trussell, R. Hankinson, and J. Tilton. Clarendon Press, to appear.

Ridder, G., 1987   The sensitivity of duration models to misspecified unobserved heterogeneity and duration dependence. University of Amsterdam: Department of Actuarial Science and Econometrics.

Ridder, G., and W. Verbakel, 1983   On the estimation of the propotional hazards model in the presence of unobserved heterogeneity. University of Amsterdam: Department of Actuarial Science and Econometrics.

Struthers, C. A., and J. D. Kalbfleisch, 1986   Misspecified proportional hazard models. Biometrika, 73: 363-369.

Trussell, J., and T. Richards, 1985   Correcting for unmeasured heterogeneity in hazard models using the Heckman-Singer procedure, pp. 242-276 in *Sociological Methodology*, edited by N. Tuma. Jossey-Bass, San Francisco.

World Fertility Survey, 1984   *Fertility in Sweden, 1981; A Summary of Findings*. World Fertility Survey, Voorburg, The Netherlands.

10

Kindred Lifetimes: Frailty Models in Population Genetics

JAMES W. VAUPEL

How in a heterogeneous population do individual life-history traits that are theoretically important but largely unobservable, affect observed population dynamics? How can inferences be drawn about the underlying traits from the population patterns? As illustrated by Trussell and Rodríguez, Hoem, and Weiss (Chapters 8, 9 and 12), this pair of questions is of convergent interest to both geneticists and demographers; this chapter presents a new method for addressing some aspects of it.

Consider, for example, a life-history trait—lifetime—that is of fundamental interest to both geneticists and demographers. In studies of the duration of life, the data typically consist of a distribution of individual lifetimes and a corresponding age-specific survival curve. Geneticists ask: to what extent is the variation in lifetimes due to genetic versus environmental variation among individuals? Demographers ask: what is the shape of the survival curve for individuals and how does this trajectory differ across individuals? If individuals are classified by genotype (and if the effects of common environment are unimportant), then the demographers' question is the same as the geneticists', because variation across individuals is then genotypic variation and the variation in lifetimes implied by differing survival curves is the residual environmental variation.

Even with recent advances in mapping genes, genotypes are still largely unobserved and many of the details of environmental variation are also unobserved. Some environmental covariates, such as year of birth, caloric intake, and temperature may be measured in some empirical studies, but it is not practicable to measure all environmental and behavioral perturbations. Thus, as suggested by Trussell and Rodríguez in this volume, a metaphor for the geneticists' and demographers' question is the decomposition of a known quantity C into the sum of two unknown quantities A and B. Unless some further information is added, the equation $C = A + B$ has infinitely many solutions.

The tack taken by geneticists is to combine (1) theories of how genes are transmitted and (2) theories and assumptions about how genes interact with other genes and with environmental factors with (3) empirical data on related individuals, such as twins, siblings, or parents and children. The observed population variation then can be uniquely decomposed into the two components of genetic and environmental variation. This classic and widely used method unfortunately has numerous limitations (e.g., Feldman and Lewontin, 1975; Falconer, 1981).

Demographers, in contrast, proceed by specifying functional forms for their A (the shape of survival curves for individuals) and B (the variation in survival curves across individuals) and for the relationship between A and B. This approach can also be severely criticized, especially when there is little ancillary evidence concerning the true nature of these functional forms [as discussed by Trussell and Rodríguez (Chapter 8) and by Hoem (Chapter 9) in this volume]. Following the geneticists' lead, this chapter develops a method that is grounded in genetic theory and that exploits data on related individuals.

Following the demographers' lead, the method is based on modern ideas of survival analysis and frailty modeling (rather than on the decomposition of variance). The hybrid method has some limitations and some advantages compared with existing methods. In some life-history applications, it may provide geneticists and demographers with a useful, convergent supplement to their current, disparate approaches.

For expository simplicity, the chapter focuses the analysis of lifetimes. As discussed in two companion articles (Vaupel, 1990; Larsen and Vaupel, 1989), the method developed for analyzing mortality is directly applicable to a variety of other life-history characteristics, including fertility, migration, marriage, and morbidity, and to data sets on repeated events as well as related individuals: instead of ages at death for relatives, the data might consist of times to successive conceptions. The illustrative examples used in the chapter pertain to human twins and to adopted children and their biological and adoptive parents, but applications to other sets of relatives and to other species can be developed. To facilitate extensions to various life-history traits, kin groupings, and species, mathematical results are presented in quite general terms. The mathematics, however, is not difficult and the only results presented are those of direct interest to geneticists and demographers who are analyzing survival or duration data.

A FRAILTY MODEL FOR GENOTYPES

Consider first a population of individuals who are classified into groups with identical genotypes. The data may pertain, for example, to monozygotic twins or to a set of inbred lines and the offspring derived from crosses between them (F_1 crosses). Suppose these data are of the kind typically studied in the branch of statistics known as survival analysis (Kalbfleisch and Prentice, 1980; Cox and Oakes, 1984). In particular, suppose the data include the lifetime X_{ij} of individual i in genetic group j, an indicator δ_{ij} equal to 1 if X_{ij} is a death time and 0 if X_{ij} is the oldest age when the individual was known to be alive prior to being censored (i.e., lost to further observation) in an uninformative way, and, perhaps, a vector of covariates v_{ij} that may vary with age or time.

Frailty models for analyzing survival data focus on estimating the age trajectory of the force of mortality (i.e., hazard of death). The force of mortality $\mu(x)$ at age x is related to the survival curve $s(x)$, which gives the probability of surviving to age x, by

$$\mu(x) = \frac{-ds(x)/dx}{s(x)} \tag{10-1}$$

and

$$s(x) = e^{-H(x)} , \tag{10-2}$$

where the cumulative hazard $H(x)$ is given by

$$H(x) = \int_0^x \mu(t)dt . \tag{10-3}$$

The probability density function of age at death (i.e., lifetime) is given by

$$f(x) = \mu(x)s(x) . \tag{10-4}$$

In frailty models it is assumed that the force of mortality for an individual can be separated into two multiplicative components called frailty and the baseline force of mortality (Vaupel, Manton, and Stallard, 1979). This may not be entirely true in particular applications, but no model is a perfect representation of reality. The operative question is whether the model is useful: is it simple enough to be tractable and understandable but sophisticated enough to shed some new light on reality? The simplicity of the multiplicative frailty approach is analogous to the simplicity of linear regression; in a variety of theoretical and empirical applications, frailty models have provided useful insights (e.g., Vaupel, Manton, and Stallard, 1979; Manton, Stallard, and Vaupel, 1981 and 1986; Heckman and Singer, 1984; Vaupel and Yashin, 1985a,b; Hougaard, 1986a; Aalen, 1987, 1988).

For data classified by genotypic groups, the force of mortality for individual i in group j would be $z_j \mu_{ij}(x)$, where z_j denotes the frailty (or relative risk) of each of the individuals in group j and $\mu_{ij}(x)$ gives the baseline force of mortality. The value of z_j is not known; it is described by a probability density function $g_j(z)$. In many applications the same g will hold for all genotypic groups and this g may be interpreted as the distribution of genotypic frailty in the population. The key idea is that genotype determines frailty rather than the phenotypic trait (lifetime) *per se*. In genetics, the concept of liability is sometimes used, the notion being that an individual is susceptible to, say, some cause of death only if the individual's liability exceeds some threshold (Falconer, 1981). Frailty is fundamentally different from this kind of liability: frailty is a relative risk such that the greater an individual's frailty with regard to some cause of death (or death in general) the greater the individual's susceptibility to the cause of death. (See Vaupel, 1988 for further discussion of frailty with regard to overall mortality and see Weiss, Chapter 12, for some innovative ideas concerning frailty with respect to specific diseases).

The baseline force of mortality $\mu_{ij}(x)$ is a function of the individual's age x and any covariates v_{ij}. Often the log-linear form

$$\mu_{ij}(x) = e^{c'v_{ij}} \mu^o(x) \tag{10-5}$$

is used, where c is a vector of parameter values. The function $\mu^o(x)$, which describes the underlying age pattern of the force of mortality, is frequently represented by the Gompertz function ae^{bx} or the Weibull function ax^b. Because a wide variety of other representations of $\mu_{ij}(x)$ may, however, be more reasonable in genetic and demographic research, throughout this chapter the general notation $\mu_{ij}(x)$ will be used to describe the baseline hazard faced by an individual with a given set of covariates. In the simplest case, no covariates are observed and $\mu_{ij}(x)$ is given by the same $\mu(x)$ for every individual. In another simple case, the only covariate is year of birth and the subscripts ij on μ merely indicate that different birth cohorts may suffer different levels and patterns of mortality.

In empirical applications of frailty models, the observed survival data are used to estimate the parameters of the distribution of frailty $g(z)$ and the baseline force of mortality $\mu_{ij}(x)$. Usually the parameters are estimated so as to maximize the likelihood of the observed data. The likelihood L of the survival data on a set of genotypes is the product of the likelihood L_j for each genotypic group:

$$L = \prod_{j=1}^{J} L_j \;. \tag{10-6}$$

A key mathematical result of this chapter is to derive a formula for the genotypic likelihoods. The formula is based on the theory of survival analysis as explained in such standard texts as Kalbfleisch and Prentice (1980) and Cox and Oakes (1984); it is related to a stream of biostatistical research, reviewed by Hougaard (1987), on so-called multivariate survival analysis. Multivariate, in this context, refers not to multiple covariates but to groupings of survival times; the terms "kindred-survival analysis" and "kindred-frailty models" are used in this chapter.

The required formula can be expressed as:

$$L_j = e^{h_j} g^{\dagger}(M_j, m_j) \;. \tag{10-7}$$

The formula involves three statistics that summarize the data. The first, h_j, is the total log hazard at observed death times:

$$h_j = \sum_{i=1}^{I_j} \delta_{ij} \, \log \mu_{ij}(X_{ij}) \;, \tag{10-8}$$

I_j being the number of individuals in genotypic group j. The second summary statistic, M_j, is the total cumulative hazard:

$$M_j = \sum_{i=1}^{I_j} H_{ij}(X_{ij}) \;, \tag{10-9}$$

with

$$H_{ij}(X_{ij}) = \int_0^{X_{ij}} \mu_{ij}(x)dx \;. \tag{10-10}$$

Finally, m_j is the number of deaths,

$$m_j = \sum_{i=1}^{I_j} \delta_{ij} \;. \tag{10-11}$$

The integral transform $g^\dagger$ is given by

$$g^\dagger(M,m) = \int_0^\infty z^m e^{-zM} g(z)dz \; . \tag{10-12}$$

Proof of (10-7) is straightforward. It follows from standard methods of survival analysis that the probability of the survival data for a genotypic group given the value of z is

$$L_{jz} = \prod_{i=1}^{I_j} \left[z\mu_{ij}(X_{ij})\right]^{\delta_{ij}} e^{-zH_{ij}(X_{ij})} \; . \tag{10-13}$$

Furthermore,

$$L_j = \int_0^\infty L_{jz} g(z)dz \; . \tag{10-14}$$

Rearranging terms yields (10-7).

The frailty transform $g^\dagger(M,m)$ has some mathematically interesting properties. Furthermore, it appears frequently in the probability distributions and likelihoods used in kindred-frailty analysis, as illustrated by several formulas in this chapter, and it serves as a bridge between ordinary survival analysis and frailty modeling (on unrelated individuals and events) and kindred-frailty analysis. These properties and features are discussed in Vaupel (1990). For the purposes of statistical estimation, what is most important is that closed-form expressions can be derived for the transform for a variety of frailty distributions $g(z)$.

Suppose, for instance, that the distribution of frailty follows a gamma distribution, as assumed by Beard (1963); Clayton (1978); Vaupel, Manton and Stallard (1979); Oakes (1982); Wild (1983); Clayton and Cuzick (1985); and others:

$$g(z|\lambda,\kappa) = \lambda^\kappa z^{\kappa-1} e^{-\lambda z}/\Gamma(\kappa) \; . \tag{10-15}$$

Then it is readily shown that the frailty transform is

$$g^\dagger(M,m) = \frac{\Gamma(\kappa + m)}{\Gamma(\kappa)} \cdot \frac{\lambda^\kappa}{(\lambda + M)^{\kappa+m}} \; . \tag{10-16}$$

Alternatively, suppose that frailty follows a two-point distribution, such that individuals are either frail or robust (or either movers or stayers), as assumed in analyses of hidden heterogeneity by Blumen, Kogan, and McCarthy (1955), Shepard and Zeckhauser (1980), Keyfitz and Littman (1980), Trussell and Richards (1985), Vaupel and Yashin (1985b), and others. For this simple discrete distribution,

$$g(z_1) = p_1 \; , \qquad 0 < p_1 < 1 \; , \tag{10-17}$$

and

$$g(z_2) = 1 - p_1 = p_2 \; . \tag{10-18}$$

The likelihood formula (10-7) straightforwardly generalizes to discrete distributions, with a summation replacing the integrals and a probability mass function replacing the probability density function in the frailty transform. In particular, for the two-point distribution,

$$g^\dagger(M,m) = p_1 z_1^m e^{-z_1 M} + p_2 z_2^m e^{-z_2 M} \; . \tag{10-19}$$

This result immediately generalizes to N-point distributions. Expressions for other distributional forms of $g(z)$ are given by Vaupel (1990).

APPLICATION TO DANISH MONOZYGOTIC TWINS

In studies of the genetic and early environmental components of the longevity of monozygotic (MZ) twins, the frailty z_j of a twin pair might be defined as the relative risk the two twins share (Hougaard, 1986b, Vaupel, 1988). Data are available on the day, month, and year of birth and death of Danish twins born from 1870 through 1930 (Hauge *et al.*, 1968, Holm, 1983) and a proposal to computerize and analyze these data has been prepared by Vaupel, Holm and others. To explore the estimability of frailty models applied to the Danish twin data, 5 mortality data sets were generated that might resemble the actual data set for Danish male MZ twins. It was assumed that for twin pairs unbroken at age 35, mortality rates were given by the Gompertz trajectory

$$\mu(x) = ae^{-ry+bx} , \tag{10-20}$$

where x is age, y is the birth cohort (varying from zero in 1870 to 60 in 1930), a determines the level of mortality, r is the rate of progress in reducing this level, and b determines how quickly mortality rates increase with age. In the simulation, a was 0.0002, b was 0.1 and r was 0.01. Frailty was assumed to be Gamma distributed with a mean of 1 and a variance of .25; in the simulation the inverse of the variance, k, was used and set equal to 4. The data set generated consisted of 15 twin pairs in the 1870 cohort, gradually increasing to 45 twin pairs in the 1930 cohort. The last year of observation was 1991 (the final year of the proposed data updating and computerizing); all survivors were censored at this time.

Parameter values were then estimated from the simulated data using the likelihood function in (10-7). The results are given in Table 10-1. Reassuringly, the parameter estimates are close to actual values, with no evidence of important bias, and the estimated standard deviations are consistent with the standard deviations of the estimates.

INTERPRETATION OF THE PARAMETERS OF A FRAILTY MODEL

In a frailty model like the one described above, the parameters of a hazard function and a frailty distribution are estimated. Given an observed distribution of lifetimes, the parameters are linked in the following way. As the variance of the distribution of frailty increases, the variance of the distribution of lifetimes for each frailty group decreases. Equivalently, as the distribution of frailty spreads out, the baseline hazard function becomes steeper. In the limit, as the variance in frailty approaches infinity, the hazard function becomes vertical, implying that the level of frailty precisely determines age at death. At the other extreme, when the variance in frailty is zero, the population is homogeneous and the hazard function for the various, equivalent genotypic groups is also the hazard function for the entire population.

Table 10-1 Comparison of actual and estimated parameters of five simulated data sets generated by the frailty model described in the text

| | Parameters | | | | |
	a	b	ρ	k	σ^2
Actual Values:	.0002	.1	.01	4.	.25
Estimated Values and (S.D.'s):					
Data Set 1	.00019 (.00002)	.101 (.002)	.0107 (.0015)	3.87 (.49)	.26
2	.00019 (.00002)	.098 (.002)	.0081 (.0015)	4.70 (.69)	.21
3	.00023 (.00003)	.102 (.002)	.0130 (.0015)	4.10 (.57)	.24
4	.00022 (.00003)	.102 (.002)	.0125 (.0016)	3.50 (.43)	.29
5	.00019 (.00002)	.099 (.002)	.0090 (.0015)	4.00 (.53)	.25

Table 10-2 Comparison of variances

Variance in Lifetimes for Entire Population	Variance in Lifetimes for Subpopulation with Frailty Equal to One	Variance in Frailty
161.4	161.4	0.
161.4	142.6	0.25
161.4	114.1	1.
161.4	52.3	10.
161.4	17.1	100.
161.4	0	∞

A simple numerical example provides an illustration. Suppose that the observed distribution of lifetimes is that implied by the Gompertz hazard function ae^{bx}. Further suppose that frailty is Gamma distributed with mean 1 and variance c. Finally, suppose that there is a common baseline hazard $\mu(x)$. Following Vaupel, Manton, and Stallard (1979) it can then be shown that this hazard function has the form

$$\mu(x) = ae^{bx+(ac/b)(e^{bx}-1)} . \tag{10-21}$$

Specifically, suppose a is .00005 and b is .1; using numerical methods it can be calculated that the population life expectancy (mean lifetime) is 70.3 with a variance of 161.4. If various values are specified for c, then numerical methods can be used to calculate the variance in lifetimes for "standard" individuals with frailty 1. Some results are shown in Table 10-2. Vaupel (1988) presents some additional results on the relationship between variance in lifetimes and variance in frailty.

COMPETING RISKS

The analysis of genetic factors in various causes of death is an active research frontier in genetics and epidemiology; Weiss' chapter (Chapter 12) in this volume provides a stimulating example. To extend frailty modeling to this area of research, suppose the force of mortality for individual i in genotypic group j from cause k is $z_j^k \mu_{ij}^k$. If the same value of z governs frailty with regard to two or more causes, these causes can be collapsed in a frailty model into a single, combined cause. Suppose, on the other hand, that different, independent values of z determine frailty with regard to different causes. If the causes of death are observed, then if follows from the standard methods of survival analysis and from (10-7) that the likelihood of the data for a genotypic group j can be expressed as:

$$L_j = \prod_{k=1}^{K} L_j^k \, , \tag{10-22}$$

where

$$L_j^k = e^{h_j^k} \cdot g_k^{\dagger}(M_j^k, m_j^k) \, . \tag{10-23}$$

The transform and three summary statistics in this formula are analogous to those used earlier. The total log hazard is given by:

$$h_j^k = \sum_{i=1}^{I_j} \delta_{ij}^k \, \log \, \mu_{ij}^k(X_{ij}) \, , \tag{10-24}$$

where the indicator δ_{ij}^k is one when the individual is known to have died of cause k and zero otherwise. The total cumulative hazard from cause k is given by:

$$M_j^k = \sum_{i=1}^{I_j} \int_0^{X_{ij}} \mu_{ij}^k(x) dx \, , \tag{10-25}$$

and the number of deaths from cause k is

$$m_j^k = \sum_{i=1}^{I_j} \delta_{ij}^k \, . \tag{10-26}$$

The frailty transform is taken with respect to the probability density function of z^k:

$$g_k^{\dagger}(M, m) = \int_0^{\infty} z^m e^{-zM} g^k(z) dz \, . \tag{10-27}$$

The parameters pertaining to cause k, that is the parameters of the distribution of z^k and of the hazard function $\mu_{ij}^k(x)$, that maximize the likelihood of the data can, in the case of independent causes of death with independent frailties, be estimated by maximizing

$$L^k = \prod_{j=1}^{J} L_j^k \, . \tag{10-28}$$

This convenient result implies that as in the case of cause-of-death data on unrelated individuals, data on independent causes of death for genotypic groups can be analyzed separately without reference to other, competing causes.

HIDDEN COMPETING RISKS

In intermediate cases where cause-specific frailty values are neither perfectly dependent nor independent, more structure is required. One approach is to assume that there are generalized frailty factors that affect two or more causes of death as well as specific factors that affect a single cause. In the simple case of two causes of death (perhaps the cause of interest and a group of other causes), the model might be that a genotype's hazard from cause 1 is

$$(z_j^0 + z_j^1)\mu_{ij}^1(x) \,, \tag{10-29}$$

and from cause 2 is

$$(z_j^0 + z_j^2)\mu_{ij}^2(x) \,. \tag{10-30}$$

This blending of risks can be viewed, at least mathematically, in the usual context of competing risks. However, if an individual dies, say, from cause 1, it is not known whether the operative hazard was the $z^0\mu^1(x)$ or the $z^1\mu^1(x)$ component of the risk. Thus the problem can be interpreted as one of hidden competing risks.

It turns out that methods for analyzing kindred-survival data with hidden competing risks are useful in several other applications of convergent interest to geneticists and demographers. Several specific examples are given subsequently; they involve changes in the impact of genotypic frailty over age, the analysis of premature vs. senescent death, and the analysis of survival data on relatives other than MZ twins.

Suppose K causes of death are known to exist but are not observed. Assume that each frailty group j consists of I_j individuals who share independent frailties $z_j^1, \ldots, z_j^K$ with respect to these causes. Let k_{ij} denote the unobserved cause of death for the ith individual in the jth group; if the individual is lost to follow-up let k_{ij} be zero. Let the vector $(k_{1j}, \ldots, k_{Ij})$ represent a possible set of causes of death and let $L_j^{(k_1,\ldots,k_I)}$ denote the likelihood of this set. This likelihood can be calculated by (10-22) with δ_{ij}^k equal to 1 when k equals k_{ij} and zero otherwise. The situation here is exactly the same as with observed causes of death because the possible set of causes is assumed to be the actual set.

The likelihood of the actual data on the genotypic group is simply the sum of these cause-specific likelihoods over all possible sets of causes of death:

$$
\begin{aligned}
L_j &= \sum_1^K \cdots \sum_1^K L_j^{(k_1,\ldots,k_I)} \\
&= \sum_1^K \cdots \sum_1^K e^{h_j^{(k_1,\ldots,k_I)}} g_1^\dagger(M_j^1, m_j^1) \ldots g_1^\dagger(M_j^K, m_j^K) \,,
\end{aligned}
\tag{10-31}
$$

where M_j^k and m_i^k are given by (10-25) and (10-26) as before, and where the summations are taken over all individuals whose age at death is known. For censored individuals, lost to follow-up, the value of k is zero.

As a simple example of this formula, consider the case of two hidden causes of death and two MZ twins with known lifetimes X_{1j} and X_{2j}. Then

$$L_j = L_j^{(1,1)} + L_j^{(1,2)} + L_j^{(2,1)} + L_j^{(2,2)} \,, \tag{10-32}$$

where, e.g.,

$$L_j^{(1,1)} = \mu_{1j}^1(X_{1j})\mu_{2j}^1(X_{2j})g_1^\dagger(M_1,2)g_2^\dagger(M_2,0) \qquad (10\text{-}33)$$

and

$$L_j^{(1,2)} = \mu_{1j}^1(X_{1j})\mu_{2j}^2(X_{2j})g_1^\dagger(M_1,1)g_2^\dagger(M_2,1) \ . \qquad (10\text{-}34)$$

Proof of (10-31) is straightforward. It follows from standard methods of survival analysis that

$$L_j = \int_0^\infty \cdots \int_0^\infty \prod_{i=1}^{I_j}\left\{\sum_{k=1}^{K} z_j^k \mu_{ij}^k(X_{ij})\right\}^{\delta_{ij}} \qquad (10\text{-}35)$$

$$\cdot\, e^{-\int_0^{X_{ij}}\sum_{k=1}^{K} z_j^k \mu_{ij}^k(x)dx} dz^1 \ldots dz^K \ .$$

Rearranging terms and substituting h, M, and m, yields (10-31).

CHANGING FRAILTY AND THE GERONTOLOGICAL PARADIGM

In some analyses it may be appropriate to assume that an individual's frailty changes with age (e.g., Yashin, Manton, and Vaupel, 1985 and Vaupel, Yashin, and Manton, 1988). In the case of the frailty shared by MZ twins, for instance, it may be plausible that the twins' shared frailty, due to common genotype and early environment, becomes less significant as the twins age and cumulatively experience different environmental influences. That is, it might be hypothesized that as they become older, twins become more like unrelated individuals.

This hypothesis could be modeled by setting the force of mortality for twins equal to

$$\{w(x)z_j + [1 - w(x)]\}\mu_{ij}(x) \ , \qquad (10\text{-}36)$$

with $w(x)$ being a weighting function between zero and one that starts off at one and declines with age. For instance $w(x)$ might be given by e^{-bx}. The parameters of this model can be estimated using the method given above for hidden competing risks, with μ_{ij}^1 equal to $w(x)\mu_{ij}(x)$ and $\mu_{ij}^2(x)$ equal to $[1 - w(x)]\mu_{ij}(x)$ and with z_j^1 equivalent to z_j and z_j^2 equal to one for all genotypes. Because the population is homogeneous with regard to the second "cause of death," the frailty transform with respect to cause 2 reduces to $\exp(-M_j^2)$.

A second example of changing frailty (and of hidden competing risks) is provided by the hypothesis that there are two theoretically-important kinds of death at older ages that may be impossible to distinguish in practice: premature death due to some disease or mishap and genetically-predetermined death due to senescence. This hypothesis, which many gerontologists believe to be correct, was popularized by Fries and Crapo (1981). Two MZ twins might share two frailties with regard to these two broad categories of death and their force of mortality could be modeled by

$$z_j^1 \mu_{ij}^1(x) + z_j^2 \mu_{ij}^2(x) \ . \qquad (10\text{-}37)$$

The force of mortality from senescent death is thought by many gerontologists to be an inexorable consequence of aging and essentially independent of environmental influences, including personal behavior and medical interventions. Hence, the model might be reduced to

$$z_j^1 \mu_{ij}^1(x) + z_j^2 \mu^2(x) \, , \tag{10-38}$$

with $\mu^2(x)$ being close to zero until old age and then rising precipitously so that by age 85 or so it becomes the dominant cause of death. As noted earlier, if $\mu^2(x)$ rises sharply, then the distribution of z^2 has a very large variance. Also note that the model implies that the twins' overall frailty starts off at z_j^1 and then moves toward z_j^2 at older ages, with z^2 being more important in determining age at death than z^1 (because it has a large variance and its associated hazard function is much steeper). Thus, in contrast to the previous model, this gerontological model postulates that genotypic factors become more important at advanced ages in determining mortality.

MONZYGOTIC VS. DIZYGOTIC TWINS

As another example of the use of (10-31), consider the analysis of survival data on MZ and dizygotic (DZ) twins. Such data is sometimes used by human geneticists to try to separate the observed variance in, say, lifetimes into the three components of genetic variance, variance due to common environment, and variance due to other environmental factors (Falconer, 1981). In a corresponding frailty model, the force of mortality for MZ twins might be assumed to be given by

$$w z_j^1 \mu_{ij}^1(x) + (1 - w) z_j^2 \mu_{ij}^2(x) \, , \tag{10-39}$$

whereas the force of mortality for DZ twins might be given by

$$\frac{1}{2} w [z_j^1 + z_{ij}^1] \, \mu_{ij}^1(x) + (1 - w) z_j^2 \mu_{ij}^2(x) \, , \tag{10-40}$$

with w being a weight between zero and one.

The model can be interpreted as implying that there are two causes of death for MZ twins, due to common genotype and common early environment, but that there are three kinds of death for DZ twins. The first kind is due to common genes and the third kind to different genes, the weight of $\frac{1}{2}$ reflecting the fact that DZ twins share half their genes. The assumption here is that the genetic determinants of frailty are additive; if there are important dominance or interaction effects, or if the effects are additive on some other scale, such as log frailty, then the model needs more structure. Also note that the two twins have different frailties for the third cause of death: under the usual assumption of random mating, these two z's can be assumed to be independently drawn from the population distribution of genotypic frailty.

The likelihood function for the MZ twins is identical to (10-32): there are two causes of death and two individuals per genotypic group. The likelihood function for the DZ twins is somewhat more complicated because there are three causes of death and because the twins differ from each other in their frailty with respect to the third cause. That there are three causes of death implies that

$$\begin{aligned} L_j = {} & L_j^{(1,1)} + L_j^{(1,2)} + L_j^{(1,3)} + L_j^{(2,1)} + L_j^{(2,2)} + L_j^{(2,3)} \\ & + L_j^{(3,1)} + L_j^{(3,2)} + L_j^{(3,3)} \, . \end{aligned} \tag{10-41}$$

Deriving the formula for each of these eight terms requires a slight digression to understand the likelihood of survival data on unrelated individuals.

Consider a specific cause of death and suppose the survival data pertain to individuals with different frailties. Then there are really two groups, with only one member each, instead of one group with two members. The likelihood of the combined data is the product of the likelihoods for each of these single-individual groups:

$$L_j = \{\mu_{1j}(X_{1j})^{\delta_{1j}}g^\dagger(H_{1j},\delta_{1j})\}\{\mu_{2j}(X_{2j})^{\delta_{2j}}g^\dagger(H_{2j},\delta_{2j})\}$$
$$= e^{h_j}g^\dagger(M_{1j},m_{1j})g^\dagger(M_{2j},m_{2j}) , \tag{10-42}$$

where in this instance M_{ij} and m_{ij} are equivalent to H_{ij} and δ_{ij}. So instead of a single transform, it is necessary to use the product of two transforms. More generally, whenever individuals share a common frailty, their survival data should be included within the same frailty transform, whereas if they have different frailties, the data should be separated into different transforms.

Returning to the problem of DZ twins, it can now be seen that, e.g.,

$$L_j^{(1,1)} = e^{h_j^{(1,1)}}g^\dagger(M_j^1,2)g^\dagger(M_j^2,0)g^\dagger(H_{1j}^1,0)g^\dagger(H_{2j}^1,0) \tag{10-43}$$

and

$$L_j^{(2,3)} = e^{h_j^{(2,3)}}g^\dagger(M_j^1,0)g^\dagger(M_j^2,1)g^\dagger(H_{1j}^1,0)g^\dagger(H_{2j}^1,1) . \tag{10-44}$$

The model for MZ twins can be written as $z\mu(x)$, with z equal to $wz^1 + (1-w)z^2$.

A geneticist might ask: what is the variance of the population distribution of z and how does this variance compare with the genotypic variance of z^1 vs. the common-environment variance of z^2. Since

$$\text{Var}(z) = \alpha^2\text{Var}(z^1) + (1-\alpha)^2\text{Var}(z^2) , \tag{10-45}$$

the proportion of the variance in overall frailty due to genotypic vs. common-environment variance is

$$\frac{\alpha^2\text{Var}(z^1)}{\alpha^2\text{Var}(z^1) + (1-\alpha)^2\text{Var}(z^2)} . \tag{10-46}$$

As indicated earlier, the variance in lifetimes is a function of the variance in frailty and the variance in lifetimes among individuals with specific levels of frailty; Vaupel (1990) discusses this. The formulas are, in general, messy, but can be evaluated by numerical methods. Thus, the customary decomposition of variance can be retrieved from a frailty analysis. As discussed, however, by Vaupel (1990), frailty models provide a much richer description than that provided by a decomposition of variance, and this complexity can be used to gain a deeper, multifaceted understanding of the nature of genetic and environmental influences and their interaction.

ADOPTED CHILDREN

As a final example of the use of frailty models with hidden competing risks, consider data on the lifetimes of adopted children and their adoptive and biological parents (Soerensen *et al.*, 1988). A simple, first-cut approach to frailty modeling of such data would be to use the data to construct three separate data sets, one on biological fathers and children, one on biological mothers and children, and the third on adoptive parents and children.

For the first two of these data sets, the hazard function for each parent/child pair might be

$$\frac{1}{2}z_j\mu_{ij}(x) + \frac{1}{2}z_{ij}\mu_{ij}(x) ,\tag{10-47}$$

because parents and children, like DZ twins, share half their genes. Then the likelihood would be

$$L_j = e^{h_j}g^\dagger(M_j,m_j)g^\dagger(H_{1j},\delta_{1j})g^\dagger(H_{2j},\delta_{2j})\tag{10-48}$$

For the trios of adoptive parents and children, the hazard function and corresponding likelihood might simply be

$$z_j^*\mu_{ij}^*(x)\tag{10-49}$$

and

$$L_j = e^{h_j}g^\dagger(M_j,m_j) ,\tag{10-50}$$

where z^* would now be interpreted not as genetic frailty but as frailty due to common environment.

By comparing the distributions of z and z^* and the shapes of μ and μ^*, some insights might be gained into the interaction among nature, nurture, and subsequent environment in influencing the longevity of adopted children.

DISCUSSION

The examples in this chapter have concerned survival data on MZ and DZ twins and on adopted children and their adoptive and biological parents. Many other data sets on the lifetimes of related individuals exist for humans (e.g., the Utah genealogical data base described by Bean, Chapter 15) and various other species. Furthermore, as discussed in two companion articles (Vaupel, 1990 and Larsen and Vaupel, 1989), the frailty models developed in this chapter can be extended to other life-history traits and to data on such related events as an individual's waiting times to successive conceptions. Thus there are broad possibilities for research by geneticists and demographers in developing and applying appropriate frailty models to analyze various kinds of data on related individuals and events.

Geneticists have developed a large body of knowledge about how genes are transmitted and about how genes interact with each other and with environmental influences to produce phenotypic outcomes. Evolutionary theory places strong constraints on genetic properties and recent advances in mapping genes and in understanding the effects of specific genes are leading to detailed knowledge of the nature and influence of genetic factors. The theories and empirical findings of geneticists are crucial in constructing frailty models, both in the determining the general form of such models (as illustrated in this chapter) and in determining the functional forms to be used for frailty distributions and hazard functions. In this chapter, frailty was assumed to be, say, Gamma distributed and the force of mortality was assumed to follow, say, a Gompertz trajectory.

There is some evidence that such assumptions are reasonable for some kinds of analyses, but in other cases, as discussed by Trussell and Rodríguez in this volume, it is mathematical convenience more than biological reality that dictates the assumptions made. Consequently, as suggested by Weiss (Chapter 12), an important convergent area of research for geneticists and demographers is the study of the biological underpinnings of and constraints on the functional forms used in frailty models. An example of such research is a study in progress by J. W. Curtsinger and the author of the shape of the force of mortality function for several Drosophila genotypes: survival data are being gathered on four inbred lines and their six F_1 crosses, each population consisting of 5,000 individuals raised under similar conditions.

In addition to research on developing more powerful and appropriate methods and models for frailty analysis, geneticists and demographers can engage in research in applying the frailty methods that have been developed. Although existing methods have major weaknesses and shortcomings, they may lead to some different and perhaps deeper insights. In particular, the methods adumbrated in this chapter, which represent a hybridization of two very different approaches currently used by geneticists and demographers, may help researchers in both disciplines as well as providing a basis for productive cross-disciplinary research.

For geneticists, the methods of kindred-frailty modeling provide an alternative, in the analysis of life-history traits, to customary methods of decomposition of variance. A key strength of the frailty approach is that it provides a rich description of reality as summarized by hazard functions and frailty distributions. Furthermore, frailty models highlight the interaction between nature and nurture, because genotypic and environmental influences are fundamentally intertwined in the multiplicative relationship between frailty and the baseline hazard function.

For demographers, the methods of kindred-frailty modeling provide a means for taking advantage of life-history data on related individuals and events: nearly all demographic analyses to date have treated such data as if the individuals and events were unrelated. Furthermore, as indicated above, kindred-frailty models can help demographers clarify what is meant by "frailty." In kindred-frailty models, as illustrated in this chapter, the meaning of frailty is clear. In models of frailty for unrelated individuals, it is sometimes difficult to sort out the conceptual basis for dividing the heterogeneity among individuals into a frailty component and a residual component given by the distribution of lifetimes of individuals with the same frailty. Finally, from kindred-frailty models grounded in genetic theory and findings, demographers can gain a deeper understanding of the biological constraints on the distribution of frailty and on the shape of hazard trajectories.

ACKNOWLEDGMENTS

The author thanks Joanna H. Shih for her extensive statistical and mathematical assistance, and Julian Adams, Donald A. Berry, Morris L. Eaton, Philip Hougaard, Niels Kieding, David A. Lane, Ulla M. Larsen, Thomas A. Louis, James Trussell, Kenneth Weiss, John Wilmoth and Anatoli A. Yashin for their comments.

REFERENCES

Aalen, O. O., 1987   Two examples of modelling heterogeneity in survival analysis. Scand. J. Stat. 14: 19-25.

Aalen, O. O., 1988   Heterogeneity in survival analysis. Stat. Med. 7: 1121-1137.

Beard, R. E., 1963   A theory of mortality based on actuarial, biological, and medical consideration. In *Proceedings of the International Population Conference.* International Union for the Scientific Study of Population, London, UK.

Blumen, I., M. Kogan, and P. J. McCarthy, 1955   *The Industrial Mobility of Labor as a Probability Process.* Cornell University Press, Ithaca, NY.

Clayton, D., 1978   A model for association in bivariate life tables and its application in epidemiological studies of familial tendency in chronic disease incidence. Biometrika 65: 141-151.

Clayton, D., and J. Cuzick, 1985   Multivariate generalisations of the proportional hazards model (with discussion). J. R. Stat. Soc. 148: 82-117.

Cox, D. R., and D. Oakes, 1984   *Analysis of Survival Data.* Chapman Hall, London, UK.

Falconer, D. S., 1981   *Introduction to Quantitative Genetics.* Second Edition. Longman, New York.

Feldman, M. W., and R. C. Lewontin, 1975   The heritability hang-up. Science 190: 1163-1168.

Fries, J. F., and L. M. Crapo, 1981   *Vitality and Aging: Implications of the Rectangular Curve.* W. H. Freeman, San Francisco.

Heckman, J., and B. Singer, 1984.   Econometric duration analysis. J. Econometrics 24: 63-132.

Hauge, M., B. Harvald, M. Fischer, K. Gotlieb Jensen, N. Juel-Nielsen, J. Raebild, R. Shapiro, and T. Videbech, 1968   The Danish twin register. Acta Genet. Med. Gemellol. 17: 315-332.

Holm, N. V., 1983   Tvillingstudiers anvendelse til belysning af årsagsforholdene for sygdomme af kompleks aetiologi med cancer som eksempel (The use of twin studies to investigate causes of diseases with complex etiology, with a focus on cancer). Ph.D. Thesis (in Danish), 236 pages. Odense University, Denmark.

Hougaard, P., 1986a   Survival models for heterogeneous populations derived from stable distributions. Biometrika 73: 387-396.

Hougaard, P., 1986b   A class of multivariate failure time distributions. Biometrika 73: 671-678.

Hougaard, P., 1987   Modelling multivariate survival. Scand. J. Stat. 14: 291-304.

Kalbfleisch, J. D., and R. L. Prentice, 1980   *The Statistical Analysis of Failure Time Data.* John Wiley, New York.

Keyfitz, N., and G. Littman, 1980   Mortality in a heterogeneous population. Pop. Studies 33: 333-343.

Larsen, U. M., and J. W. Vaupel, 1989   Frailty models of Hutterite fertility. Working paper WP-89-09-1, Center for Population Analysis and Policy, University of Minnesota.

Manton, K. G., E. Stallard, and J. W. Vaupel, 1981   Methods for comparing the mortality experience of heterogeneous populations. Demography 18: 389-410.

Manton, K. G., E. Stallard, and J. W. Vaupel, 1986   Alternative models for the heterogeneity of mortality risks among the aged, J. Am. Stat. Assoc. 81(395): 635-644.

Oakes, D., 1982   A model for association in bivariate survival data. J. R. Statis. Soc. B44: 414-422.

Shepard, D. S., and P. J. Zeckhauser, 1980   Long term effects of intervention to improve survival in mixed populations. J.Chron. Dis. 33: 413-433.

Soerensen, T. I. A., G. G. Nielsen, P. K. Andersen, and T. W. Teasdale, 1988   Genetic and environmental influences on premature death in adult adoptees. N. Eng. J. Med. 315: 727-732.

Trussell, J., and T. Richards, 1985   Correcting for unmeasured heterogeneity in hazard models using the Heckman-Singer procedure. pp. 242-276 in *Sociological Methodology*, edited by N. B. Tuma. Jossey-Bass, London, UK.

Vaupel, J. W., 1988   Inherited frailty and longevity. Demography 25(2): 227-287.

Vaupel, J. W., 1990   Relatives' risks: Frailty models of life history data. Theor. Pop. Biol. In press.

Vaupel, J. W., K. G. Manton, and E. Stallard, 1979   The impact of heterogeneity in individual frailty on the dynamics of mortality. Demography 16: 439-454.

Vaupel, J. W., and A. Yashin, 1985a   The deviant dynamics of death in heterogeneous populations. pp. 179-211 in *Sociological Methodology*, edited by N. B. Tuma. Jossey-Bass, London, UK.

Vaupel, J. W., and A. Yashin, 1985b   Heterogeneity's ruses: Some surprising effects of selection on population dynamics. Am. Statist. 39: 176-185.

Vaupel, J. W., A. I. Yashin, and K. G. Manton, 1988   Debilitation's aftermath: Stochastic process models of mortality. Math. Pop. Studies 1(1): 21-48.

Wild, C. J., 1983   Failure time models with matched data. Biometrika 70: 633-641.

Yashin, A. I., K. G. Manton, and J. W. Vaupel, 1985   Mortality and aging in a heterogeneous population: a stochastic process model with observed and unobserved variables. Theor. Pop. Biol. 27: 154-175.

11

Heterogeneity in Fecundability:
The Effect of Fetal Loss

JAMES W. WOOD AND
MAXINE WEINSTEIN

During the 1980s, heterogeneity among individuals in the risk of vital events has emerged as a central theoretical concern in demography (see, for example, Keyfitz and Littman, 1980; Heckman and Singer, 1982; Vaupel and Yashin, 1985; Manton and Stallard, 1988). Such heterogeneity is also central to population genetics. Natural selection, for example, requires at a minimum that variation in vital rates exists; the rate of random genetic drift is also affected by demographic heterogeneity insofar as such heterogeneity reduces effective population size (Pollak, 1980). In demography, uncontrolled heterogeneity is typically viewed as a nuisance, one which introduces biases and renders parameter identification impossible. Both quantitative and population geneticists, in contrast, view heterogeneity as an opportunity: without phenotypic heterogeneity, whatever its source, there is nothing to analyze.

In this paper we adopt a population geneticist's viewpoint while considering a demographic problem: how heterogeneity in the risk of fetal loss—that is, the spontaneous death *in utero* of either embryos or fetuses (WHO, 1978)—affects the distribution of birth interval components. Birth intervals are of fundamental demographic importance because they determine a woman's lifetime reproduction, as well as fertility at the population level; in addition, heterogeneity in birth intervals can lead not only to fertility differentials, but also to differential infant and early childhood mortality (see, for example, Hobcraft, McDonald, and Rutstein, 1983; Trussell and Pebley, 1984; Palloni and Tienda, 1986). It has long been known that fetal loss lengthens birth intervals, but the extent to which it does so has not been fully appreciated until quite recently (Wood, 1989). It now appears that the risk of fetal loss which characterizes a woman is a major determinant of both her fecundity (that is, her biological capacity for reproduction) and her fertility (the actual number of live births she will produce). Any heterogeneity in risk among women of a given population, or among populations, must translate directly into differences in fertility.

Demographic and epidemiological evidence indicates that heterogeneity in the risk of fetal loss probably does exist and may in fact be substantial (Leridon, 1976; Wilcox and Gladen, 1982), while genetic studies suggest that at least some of this heterogeneity may be associated with genetic variation(Beer *et al.*, 1983).

BACKGROUND

Heterogeneity in fetal loss affects birth intervals via its influence on *effective fecundability*. Gini (1924) defined fecundability in general as the probability that a couple conceives during a month of exposure to unprotected intercourse, given that both partners are biologically able to conceive. For many purposes, it is useful to classify fecundability into three categories: total fecundability, which is defined for all pregnancy outcomes, but which is unobservable owing to our inability to detect the earliest stages of gestation; apparent fecundability, which is conditioned on survival of the conceptus to a diagnosis of pregnancy; and effective fecundability, which we focus on in this paper, limited to *fertile conceptions*, that is, conceptions that survive to term. Thus, effective fecundability is the monthly probability of a conception that terminates in a live birth. Clearly, both apparent and effective fecundability are influenced to some degree by the level of fetal loss.

Demographers are concerned with fecundability because it determines a substantial portion of each birth interval, the fecund waiting time to a subsequent conception. In a homogeneous population, the waiting time to a next conception has a simple relation to fecundability: the waiting time is a geometric random variable with parameter equal to the value of fecundability, and the mean waiting time is simply the inverse of fecundability. In a population with heterogeneous fecundability, the relation between waiting times and fecundability is more complex. When couple-to-couple variation in fecundability (or month-to-month variation for any given couple) is purely random, the relation is unaltered except that the expected waiting time to conception is the inverse of the mean fecundability rather than some constant value (Sheps, 1964). When, however, there exist systematic and persistent differences in fecundability among couples, the entire distribution of waiting times is altered (Figure 11-1). The distribution becomes more right-skewed, and the mean waiting time to conception may be increased substantially because of a selection process that renders the sample of couples less fecund at higher conception waits (Mode, 1985). The mean waiting time in the heterogeneous case is equal to the inverse of the harmonic mean fecundability (Sheps, 1964). Since, in a heterogeneous population, the harmonic mean is necessarily smaller than the arithmetic mean, the expected waiting time is correspondingly larger. Thus, heterogeneity in fecundability, by itself, increases the average birth interval and thereby lowers fertility.

Several investigators have analyzed data on waiting times to a recognizable conception under the assumption that heterogeneity in fecundability among couples exists and can be modeled (following Henry, 1964) using a beta distribution.

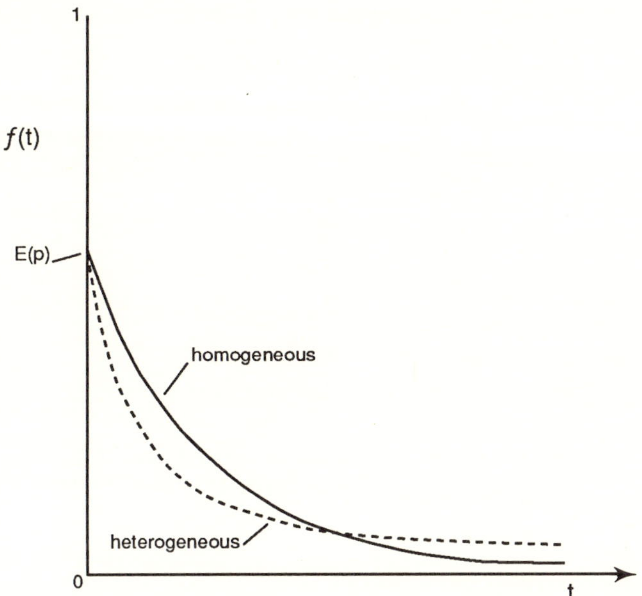

Figure 11-1. Schematic representation of the distributions of fecund waiting times to conception when fecundability is homogeneous (solid line) and heterogeneous (broken line). For simplicity, each distribution is drawn as if it were continuous.

In these studies, the expected waiting time to conception and the variance of the waiting time have been found to be substantially higher in the heterogeneous case than in the corresponding homogeneous case (Table 11-1). These estimates suggest that mean waiting times when heterogeneity is present in the population may be 30%-70% greater than when the population is homogeneous, while the variance in waiting times may be 4 to 72 times greater. Empirically, then, it appears that couple-to-couple heterogeneity in fecundability often has a large effect on birth interval distributions.

What are the important sources of heterogeneity in fecundability? Somewhat surprisingly, the observed levels of variation in coital rates within populations are *not* a likely source of the degree of heterogeneity that beta models imply for waiting times (Weinstein, Wood, and Chang, 1990). Using observed levels of heterogeneity in coital rates to model apparent fecundability induces at most a 15% increase in the waiting time to conception relative to the homogeneous case—only a fraction of the increase reported in most empirical studies. Indeed, increases in waiting times of only 25% were induced by *doubling* the observed variation in coital rates for a population with a relatively low mean coital rate, while in a population with a high mean coital rate, variation in coital rate had to be increased by 50% over the observed level to increase the waiting time by just under one-fourth.

In contrast, sensitivity analyses suggest that effective fecundability may be strongly influenced by the level of fetal loss (Wood and Weinstein, 1988). In this paper, therefore, we explore the effects of variation in fetal loss on couple-

Table 11-1 Estimates of apparent fecundability, mean waiting time to first recognizable conception, and related parameters

Population	Mean Fecundability	Variance In Fecundability	Mean Wait[a]	Variance In Waits	E(T) Ratio[b]	Var(T) Ratio[b]
USA	0.14	0.006	10.0	287.5	1.4	7.0
Taiwan	0.16	0.006	8.2	139.3	1.3	4.4
Peru[c]	0.17	0.009	9.6	513.1	1.6	16.9
Crulai	0.18	0.009	8.1	208.0	1.4	8.2
Brazil	0.19	0.011	7.9	250.7	1.5	11.6
Tourouvre	0.21	0.014	7.7	423.1	1.6	23.6
Mexico[c]	0.21	0.014	7.4	310.8	1.6	18.1
Historical France	0.23	0.011	5.5	56.6	1.3	4.0
Tunis	0.25	0.020	6.7	864.3	1.7	72.0
Hutterites	0.27	0.014	4.8	44.9	1.3	4.5
Historical Canada	0.31	0.027	5.2	317.7	1.6	44.3

Sources: Henripin, 1954 (reanalyzed by Bongaarts, 1975); Vincent, 1956 (reanalyzed by Leridon, 1977); Gautier and Henry, 1958 (reanalyzed by Bongaarts, 1975); Ganiage, 1960; Henry, 1964; Potter and Parker, 1964; Berquo *et al.*, 1968 (reanalyzed by Leridon, 1977); Jain, 1969; Charbonneaux, 1970; Majumder and Sheps, 1970; Balakrishnan, 1969.

[a] In months.

[b] E(T) Ratio and Var(T) Ratio are ratios of the mean and the variance in fecund waiting times, respectively, in the heterogeneous case versus the homogeneous case in which $f = \mathrm{E}(f)$.

[c] Women age 25+ only.

to-couple heterogeneity in effective fecundability and on the waiting time to next fertile conception.

Each death *in utero* increases the waiting time to next fertile conception in two ways: first, there is a nonsusceptible period associated with the pregnancy that ends in death plus subsequent "postpartum" infecundability and, second, there is an additional fecund waiting time to the next conception following death. The number of fetal deaths that occur in each birth interval is then determined by the overall probability of loss per conception (Wood, 1989). While, in theory, any of these aspects of fetal loss may vary among women, we concentrate here on variation in the overall probability of loss.

Earlier studies of clinically diagnosed pregnancies have revealed a typical J-shaped pattern of fetal loss with maternal age (for a review, see Alberman, 1987). The probability of loss is known to increase after age 25, a pattern which apparently has been stable over at least the past two generations (Wilcox, Treloar, and Sandler, 1981). Some studies have suggested that the probability of fetal loss is also elevated in teenage pregnancies (Harlap, Shiono, and Ramcharan, 1980). Little attention, however, has been given to the question of whether there exists significant variation in the risk of loss among women net of age. That such variation probably does exist is suggested by two lines of evidence: epidemiological evidence suggesting that fetal loss is affected by a variety of environmental and maternal risk factors, including but not restricted to maternal age (Shapiro and Bross, 1980), and genetic evidence suggesting that the risk of loss may be affected by the parents' genotypes at the major histocompatibility complex or by their degree of biological relatedness (Beer *et al.*, 1983). Leridon (1976) is one of few demographers to have addressed this question directly; in an analysis of data from a French obstetrics clinic, he estimated that the variance among women in the probability of loss at each age is about 0.007.

Is this estimated variance high or low? What sort of effect is it likely to have on the distribution of waiting times to next fertile conception? Is such variation in fetal loss likely to be an important source of heterogeneity in effective fecundability? We explore these questions by performing sensitivity analyses on a probability model of effective fecundability that we have recently developed.

A MODEL OF FECUNDABILITY

As a full description of our model appears elsewhere (Wood and Weinstein, 1988), we summarize it briefly here.

Let c be the event that conception occurs in one ovarian cycle of exposure to unprotected intercourse. Then

$$Pr(c) = \sum_{n=0}^{\infty} Pr(c \mid N = n)Pr(N = n) , \qquad (11\text{-}1)$$

where N is the number of acts of unprotected intercourse per cycle. This equation partitions the probability of conception into two independent components: a "behavioral" component, $Pr(N = n)$; and a "physiological" component, $Pr(c \mid N = n)$. The physiological component can be further partitioned as

$$Pr(c \mid N = n) = \sum_{i=0}^{n} Pr(c \mid I = i)Pr(I = i \mid o, N = n)Pr(o) , \qquad (11\text{-}2)$$

where o is the event that a cycle is ovulatory and I is the number of times that intercourse occurs during one fertile period, that is, the periovulatory portion of the cycle during which insemination has any nonzero probability of resulting in conception. If acts of intercourse are independent, then

$$Pr(I = i \mid o, N = n) = \binom{n}{i} \left(\frac{F}{C}\right)^{i} \left(1 - \frac{F}{C}\right)^{n-i} , i = 0,\ldots,n , \qquad (11\text{-}3)$$

where F is the length of the fertile period and C is total cycle length (both in days).

The probability of conception can now be found as follows. Let $p_r(i)$ be the conditional probability that exactly r days with at least one act of intercourse occur in the fertile period given that $I = i$. Then

$$p_r(i) = \binom{F}{r}\binom{i-1}{r-1} \bigg/ \binom{F+i-1}{i} , r = 1,\ldots,F . \qquad (11\text{-}4)$$

If α is the probability that conception occurs from a single day of intercourse during the fertile period, then

$$Pr(c \mid I = i) = \sum_{r=1}^{F} p_r(i)[1 - (1 - \alpha)^r] , i > 0 , \qquad (11\text{-}5)$$

and $Pr(c \mid I = 0) = 0$.

We specify the behavioral component of Eq. (11-1), first, by treating the number of acts of intercourse per cycle as a Poisson random variable, such that

$$Pr(N = n) = [exp(-mC) \cdot (mC)^n]/n! \, , n = 0, 1, 2, \ldots , \qquad (11\text{-}6)$$

where m is the mean daily probability of intercourse. This specification implies that acts of intercourse are randomly and independently distributed across the menstrual cycle, consistent with equation 11-3, and that the probability of intercourse does not vary by cycle day. While neither assumption is likely to be true, exploratory analyses indicate that more complex alternative specifications make very little difference to the final outcome.

Next, we allow m to decline as a function of marital duration, as has been observed in several empirical studies (Udry, 1980; James, 1983). Thus, at maternal age x,

$$m = \frac{\int_0^x h(x - y)g(y)dy}{\int_0^x g(y)dy} , \quad y < x , \qquad (11\text{-}7)$$

where $g(y)$ is the probability density of women marrying at age y, and $h(\cdot)$ is an empirically-derived function describing the decline in daily coital rate with marital duration (for details, see Wood and Weinstein, 1988). Under this model, coital frequency declines with maternal age, not because of any effect of age *per se*, but rather because age and marital duration are highly correlated; that is, the average coital rate among women of a given age reflects the distribution of marital durations at that age, and these durations generally become longer as women grow older. The function $g(y)$ in equation 11-7is calculated from the Coale-McNeil marriage model with a mean and variance in age at marriage of 23 and 43, respectively, close to the European standard (Coale and McNeil, 1972).

The probability of conception per cycle estimated by combining the behavioral and physiological components can be converted into the probability of conception per month, that is, total fecundability, as shown by Wood and Weinstein (1988). Assuming homogeneity, effective fecundability (f_e) is then related to total fecundability (f) as:

$$f_e = f(1 - p)/(1 + fpt) , \qquad (11\text{-}8)$$

where p is the total probability of loss per conception and t is the mean duration of the nonsusceptible period (gestation plus any residual "postpartum" infecundability) associated with each loss. From a reanalysis of data reported by French and Bierman (1962), we estimate t to be approximately 1.8 months (see Wood, 1989).

In order to model the effects of heterogeneity in the risk of fetal loss, we assume that the probability of loss follows a beta distribution among women at each maternal age. This distribution, which was used for similar purposes by Leridon (1976) and by Wilcox and Gladen (1982), was chosen simply because it is flexible and because it constrains the value of p to lie between zero and one, which as a probability it should. (We know too little about how the risk of loss varies to make a stronger, empirical justification for the beta distribution.) The mean or expected probability of loss at each maternal age follows the curve shown in Figure 11-2, which we have estimated from data on nine populations in which induced abortion is not a competing risk (Wood and We-

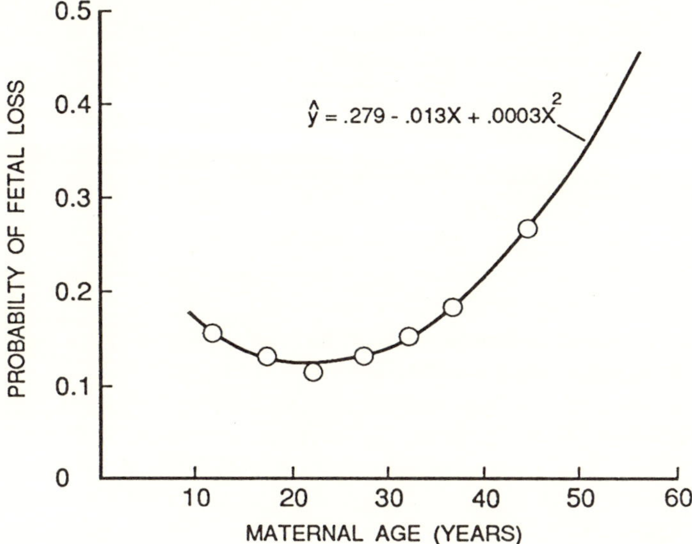

Figure 11-2. Age pattern of fetal loss in women from several populations in which induced abortion is not a competing risk. Open circles represent pooled data from nine populations, each adjusted to yield an overall loss across all ages of 150 per 1,000 conceptions. Solid curve is a quadratic equation fit to these data by least squares ($r^2 = 0.998$). (Redrawn from Wood and Weinstein, 1988).

instein, 1988). The level of this curve corresponding to maternal age 23 has been adjusted upward to 0.3, consistent with recent estimates based on assays of urinary chorionic gonadotropin (Wilcox *et al.*, 1988), and the rest of the curve is then allowed to vary in parallel with the curve in Figure 11-2. Because an elevation in mean risk of loss among adolescents is not a universal finding of population-based studies, we restrict attention to maternal ages 20 and above.

In the sensitivity analyses reported here, it is the variance of the beta distribution which we allow to vary and whose impact on fecundability and waiting times we attempt to assess. In particular, we consider two scenarios: one in which the variance is constant with maternal age, and the other in which the variance is itself a positive function of the mean risk of loss and therefore increases with maternal age (again ignoring teenage pregnancies). Based on current evidence, it is uncertain which of these two scenarios is biologically more realistic.

Within each scenario we consider four cases, ranked from low to high variance as shown in Table 11-2. For midreproductive age women, the variance estimated by Leridon (1976) from the French data (which to our knowledge is the only available empirical estimate of this variance) falls near the middle of this range. We consider it likely, therefore, that our sensitivity analyses bracket the "true" variance in the risk of loss.

Since equation 11-8 assumes that fecundability is homogeneous, we compute effective fecundability separately for each of a series of classes, within which the probability of fetal loss is fixed; the frequency of these classes is then determined by the beta distribution for variation in the risk of loss. The means

Table 11-2 Variances in the risk of fetal loss at each age used in the sensitivity analyses, where p is the mean probability of fetal loss at a given age

Type of Variance	Value of Variance			
	(1)	(2)	(3)	(4)
Constant	0.001	0.005	0.01	0.02
Increasing	$p^4/10$	$p^4/2$	p^4	$2p^4$

and variances in both effective fecundability and the waiting time to next fertile conception reported below are thus weighted averages.

Finally, the other components of the model, that is, those not pertaining to fetal loss, follow the age patterns that have been observed in normal, healthy western women (for computational details, see Wood and Weinstein, 1988). For the age pattern of variation in cycle length and in the onset of menarche and menopause, we use the data of Treloar *et al.* (1967) and Treloar (1974); estimates of the probability that a cycle at each age is ovulatory are taken from Vollman (1977). Although these data sets are comparatively old, they remain the largest ones available, and more recent work has confirmed their reliability (see Lenton, Landgren, and Sexton, 1984; Lenton *et al.*, 1984). We do not allow either the length of the fertile period or the conditional probability of conception given a day of intercourse within the fertile period to vary with age simply because no one has yet convincingly documented such variation. In accordance with Royston's (1982) results, which have recently been confirmed by a WHO study (WHO, 1983), the length of the fertile period is set at four days and the probability of conception from an insemination in the fertile period is set at 0.6, the average daily probability of conception over the whole fertile period, adjusted to the mean age of Royston's sample. For none of the components unrelated to fetal loss do we allow any variation apart from maternal age. All heterogeneity net of age in the following analyses is attributable solely to variation in the risk of fetal loss.

In generating waiting time distributions, we face a censoring problem which remains refractory to solution. Under the beta distribution, some women will have an absolute risk of fetal loss equal to one and therefore an infinite waiting time to next fertile conception. Fortunately the density of the beta distribution at exact value one is infinitesimal. But there are still some women at very high risk (though less than one) who will have extremely long waiting times, longer in fact than the age intervals used in our analysis. Some fraction of waiting times at each age will thus be censored. We are currently developing a life table analysis to correct this problem, but have not used this correction here. Consequently, our waiting time distributions are biased downward to some degree. Our sense at present is that this bias is small.

RESULTS

Figure 11-3 and Figure 11-4 show some of the probability density functions (PDFs) of effective fecundability at various maternal ages implied by the beta distribution of fetal loss. In each figure the two extremes, low and high variance in loss, are shown in the top and bottom panels, respectively. Figure 11-3 represents the case in which the variance is constant across age, while Figure 11-4 represents the case in which the variance increases with age.

In all instances, the PDFs change systematically with maternal age, as might be expected, but the changes are far more dramatic when the variance increases with age. At extremely high variances, such as those occurring at high maternal ages when the variance changes with age, effective fecundability takes on a curious U-shaped distribution that does not span the full interval from zero to one. This distribution occurs for two reasons: (1) At high variances the beta distribution of fetal loss itself becomes U-shaped, with greatest density near zero and one (Johnson and Kotz, 1970), and (2) even when the risk of loss is set equal to zero, effective fecundability is not one at higher maternal ages because of the fecundity-reducing effects of long and/or anovulatory cycles. Since it is unlikely that such U-shaped distributions are found in reality, their occurrence here may suggest either that the beta distribution is an inappropriate specification for the PDF of fetal loss, or that the highest variances investigated here are too high to be biologically realistic.

From a statistical point of view, it is interesting that none of the PDFs of effective fecundability in Figures 11-3 and 11-4 can be well modeled with a beta distribution, despite the fact that this distribution is the one most widely used in estimating heterogeneous fecundability (see Table 11-1). All the PDFs are too leptokurtic, and the U-shaped ones do not vary continuously from zero to one. Thus, if the beta distribution describes variation in fetal loss reasonably well, it cannot at the same time be used to model variation in apparent or effective fecundability. One conclusion of our sensitivity analysis, therefore, is that the estimates of the mean and variance of fecundability in Table 11-1 may be seriously biased.

Figure 11-5 and Figure 11-6 give the mean and variance in effective fecundability at all ages from 20 to 44 predicted by our model for all levels of variation in fetal loss considered, including the purely homogeneous case. (In the homogeneous case, the variance in effective fecundability is, of course, equal to zero at all ages.) There is only a small effect of variation in fetal loss on mean effective fecundability, except at higher ages when the risk of loss increases with age. The effect of variation in fetal loss on the *variance* in effective fecundability, on the other hand, is substantial. This is especially true when the variance in loss increases at later maternal ages. (Note the change in scale of the Y-axis between the panels of Figure 11-6.) Variances in fetal loss on the order of 0.01 and greater generate predicted variances in fecundability which are large fractions of the empirically-estimated variances in Table 11-1.

What are the implications of these variances for the waiting time to next fertile conception? Figure 11-7 shows the ratio of the mean waiting time in the case of heterogeneous fetal loss to that expected in the purely homogeneous case—comparable to the $E(T)$ ratios in Table 11-1—for several levels of variation in fetal loss. At younger maternal ages, heterogeneity in fetal loss has only a modest effect on the expected waiting time; before age 32, even the highest levels of heterogeneity considered here increase the waiting time by less than 10%. At later ages, however, the effect becomes more marked, especially when the variance in fetal loss increases with age. (Again note the change in scale between the two panels.) By the midforties, approaching the end of the female

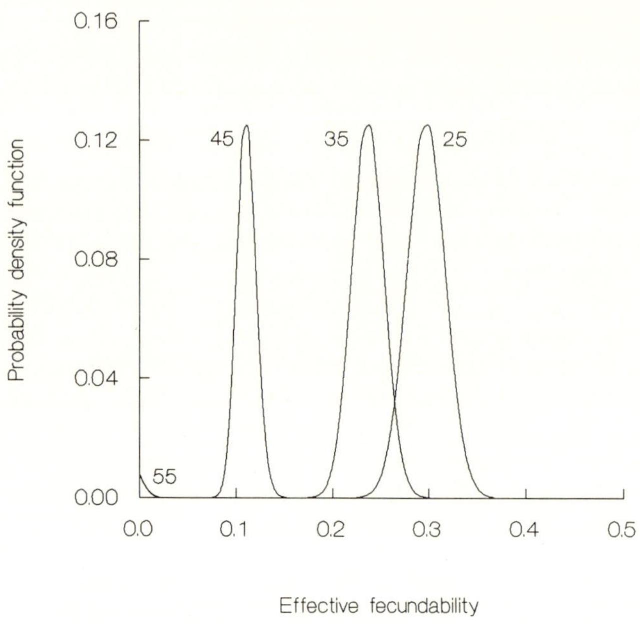

Low variance in fetal loss of 0.001.

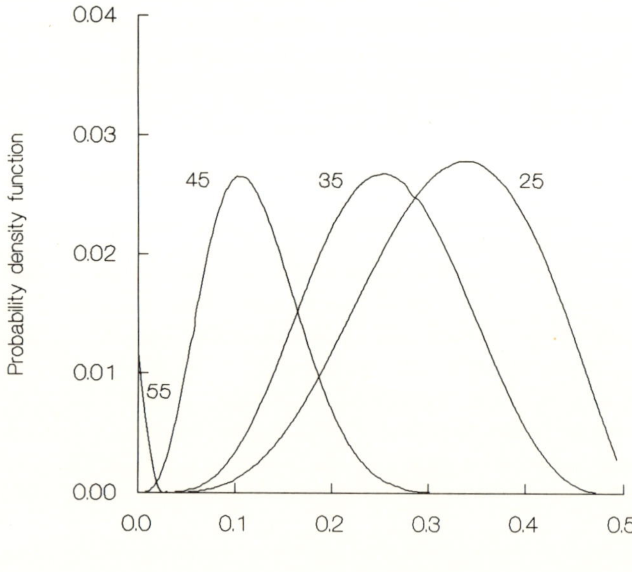

High variance in fetal loss of 0.02.

Figure 11-3. Model-predicted probability density functions for effective fecundability when the variance in the risk of fetal loss is constant across ages. Four maternal ages, 25, 35, 45, and 55 years, are shown.

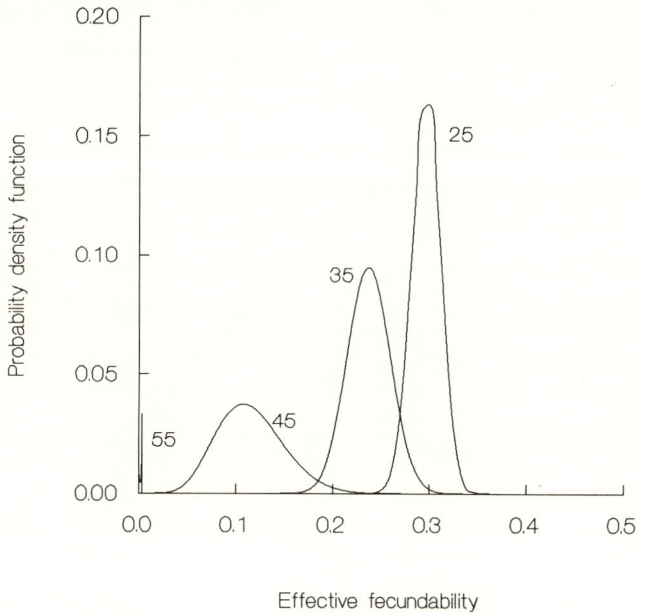

Low variance in fetal loss of $p^4/10$.

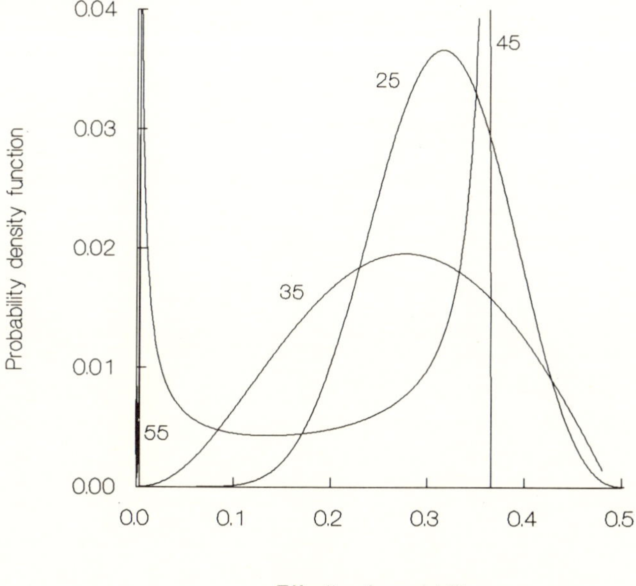

High variance in fetal loss of $2p^4$, where p is the mean probability of fetal loss at each age.

Figure 11-4.   Model-predicted probability density functions for effective fecundability when the variance in the risk of fetal loss increases with age. Four maternal ages, 25, 35, 45, and 55 years, are shown.

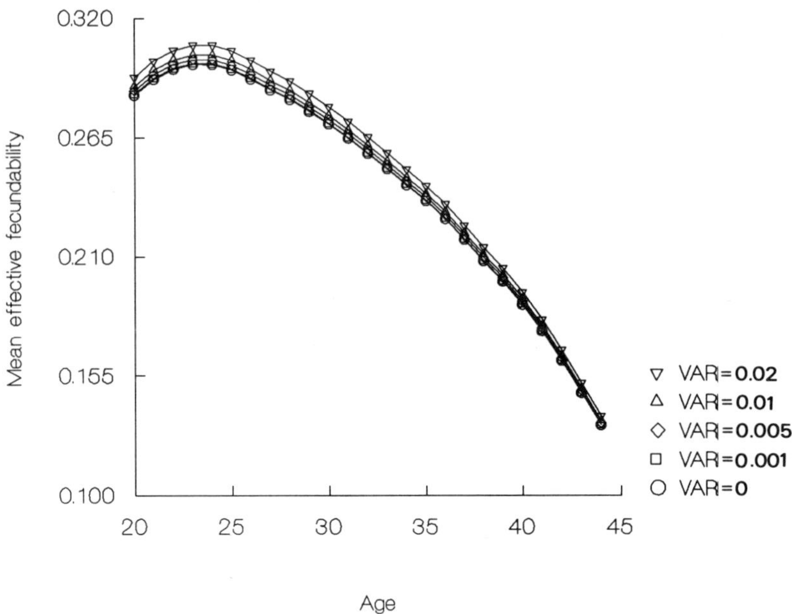

Variance in fetal loss constant across ages.

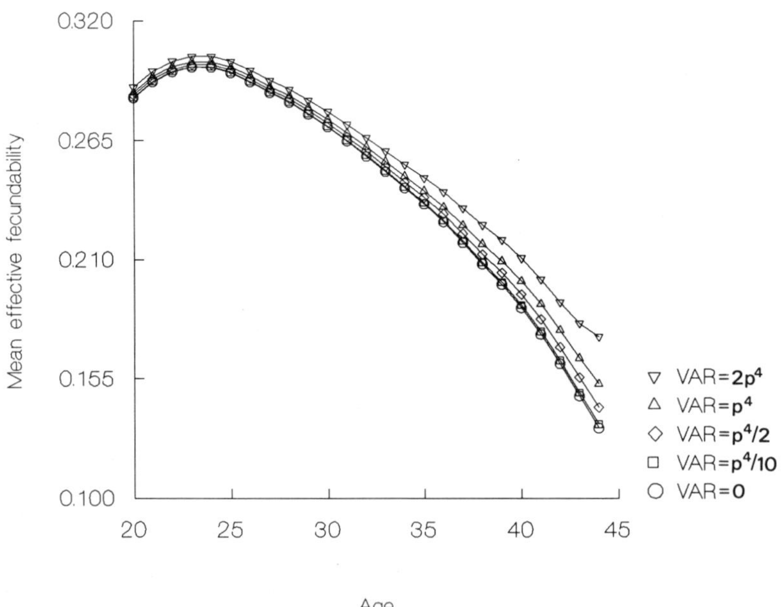

Variance in fetal loss increasing with age.

Figure 11-5. Mean effective fecundability by maternal age for several variances in fetal loss.

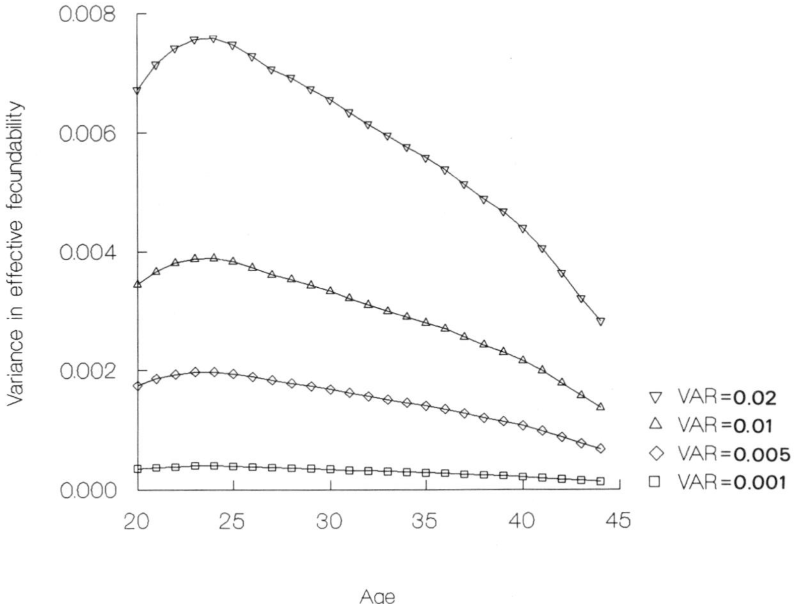

Variance in fetal loss constant across ages.

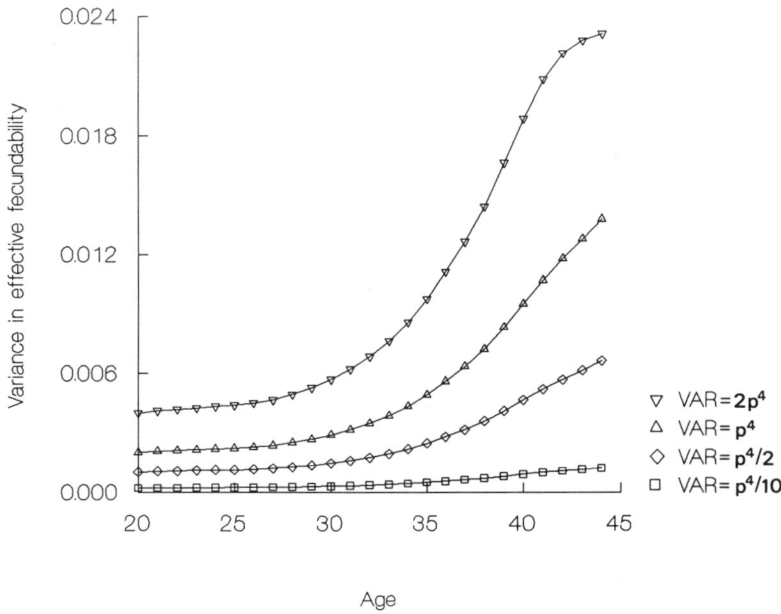

Variance in fetal loss increasing with age.

Figure 11-6. The variance in effective fecundability by maternal age for several variances in fetal loss.

reproductive span, waiting times may be anywhere from 16 to 700% longer on average than in the homogeneous case. Since birth intervals are already growing longer at these ages because of senescent changes in ovarian function, an increase in the mean risk of fetal loss and, for most couples, declining coital frequency (Wood, 1989), heterogeneity in fetal loss is likely to induce extremely long birth intervals, long enough to dominate the reproductive patterns of those older women who are not already practicing contraception or induced abortion.

DISCUSSION

Our analyses suggest that variation among women in the risk of fetal loss may be a powerful source of heterogeneity in effective fecundability. At the same time, analyses reported elsewhere indicate that effective fecundability is probably not sensitive to observed levels of couple-to-couple variation in coital frequency (Weinstein, Wood, and Chang, 1990). These conclusions may come as a surprise to many demographers, who tend to think that fertility differentials originate overwhelmingly from behavioral causes. At least in the present case, however, this appears to be untrue: we believe that the age pattern of fecundability and heterogeneity in fecundability are both attributable principally to physiological factors, especially fetal loss (see also Wood, 1989).

Our conclusions about the importance of fetal loss are partly conditioned on assumptions about how the variance in risk of loss changes with maternal age. As our analyses clearly indicate, the impact of fetal loss is considerably greater if heterogeneity in the risk of loss increases with age. At present we do not know whether this is the case or not; it is nearly impossible to measure the degree of heterogeneity at any one age, let alone at all ages simultaneously. However, what we know about the etiology of fetal loss leads us to suspect that the variance in risk of loss may indeed grow larger toward the end of the reproductive life span.

It is well established that a large fraction of spontaneous abortions are associated with chromosomal aberrations, especially autosomal trisomies, as well as other major genetic defects (Boué, Boué and Lazar, 1975; Boué et al., 1976; Creasey, Crolla, and Alberman, 1976). At least in the case of autosomal trisomies, incidence increases with maternal age approximately as a Gompertz function (Hassold et al., 1984), which is linear on a log-log scale. The intercept of this relation is constrained so that all women experience fairly low risks of trisomy at ages 20-24. Consequently, any variation among women in the slope parameter must translate directly into increasing heterogeneity among women at later ages. We do not know if this slope does in fact vary among women since the Gompertz function is always fitted to aggregate data, but it seems not unreasonable that any variation among women in chromosomal repair mechanisms, hormone levels, or exposure to mutagens might induce such variation in slope. Moreover, if Stein et al. (1980) are correct in believing that senescent changes in the endometrium are responsible for a fraction of losses, especially among euploid abortions, it is possible that rates of senescence differ among women, thus leading to increasing heterogeneity of fetal loss at later ages.

At any rate, our analyses point to the importance of fetal loss, both as an overall determinant of fecundability and as a potential source of heterogeneity

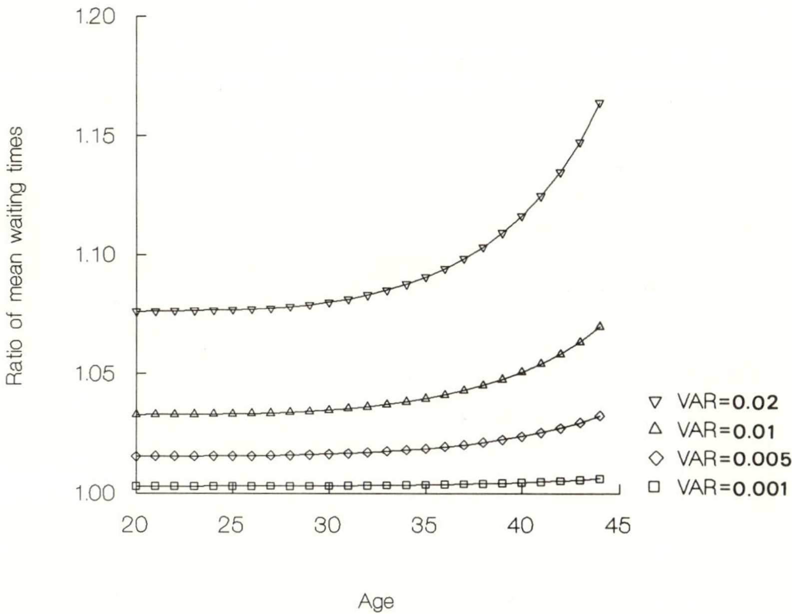

Variance in fetal loss constant across ages.

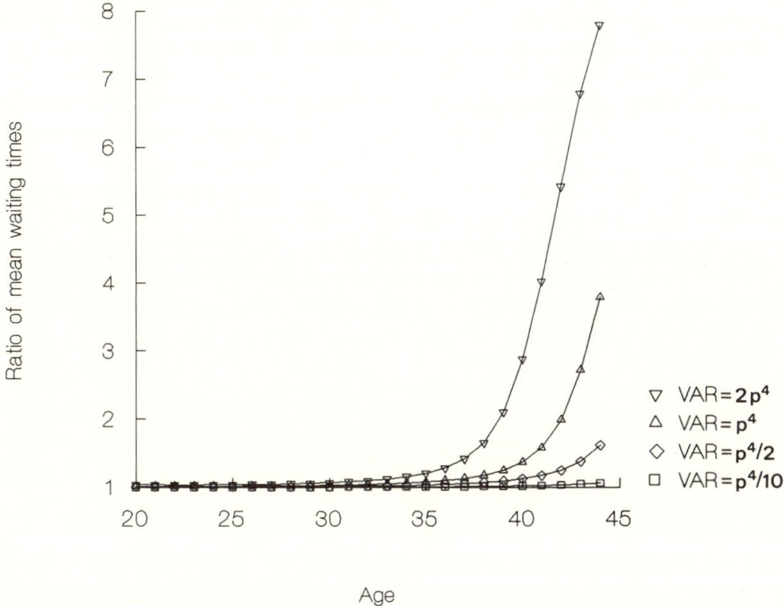

Variance in fetal loss increasing with age.

Figure 11-7. The ratio of the mean waiting time to next fertile conception when the risk of fetal loss varies among women versus that in the purely homogeneous case, by maternal age and by several levels of variation in fetal loss.

in fecundability. Although there remains much to be learned about the processes involved, especially with respect to age patterns of heterogeneity, we suggest that the lengthening birth intervals observed at later maternal ages in natural fertility populations may be caused primarily by an increase in both the mean and variance of the risk of fetal loss.

ACKNOWLEDGMENTS

We are grateful to Freddy Christiansen, Patricia Johnson and Ann Riley for comments and suggestions. We also thank Anne Buchanan and Daniel Greenfield for help in programming the analyses and in plotting the figures. This research is supported by NIH grant 5-RO1-HD20989-02.

REFERENCES

Alberman, E., 1987   Maternal age and spontaneous abortion, pp.77-89 in *Spontaneous and Recurrent Abortion*, edited by M. J. Bennett and D. K. Eemonds. Blackwell Scientific, Oxford.

Balakrishnan, T. R., 1969   Probability of conception, conception delay, and estimates of fecundability in rural and semiurban areas of certain Latin American countries. Soc. Biol. 26: 226-231.

Berquo, E. S., R. M. Marques, M. L. Milanesi, J. S. Martins, E. Pinho and I. Simon, 1968   Levels and variations in fertility in São Paulo. Milbank Memorial Fund Quarterly 46: 167-185.

Beer, A. E., J. F. Quebbeman, A. E. Semprini, P. E. Smouse, and R. F. Haines, 1983   Recurrent abortion: analysis of the roles of parental sharing of histocompatibility antigens and maternal immunological responses to paternal antigens, pp. 185-195 in *Reproductive Immunology*, edited by S. Isojima and W. D. Billington. Elsevier Science Publishers, Amsterdam.

Bongaarts, J., 1975   A method for the estimation of fecundability. Demography 12: 645-660.

Boué, J., A. Boué, and P. Lazar, 1975   Retrospective and prospective epidemiological studies of 1500 karyotyped spontaneous human abortions. Teratology 12: 11-26.

Boué, J., E. Phillippe, A. Giroud, and A. Boué, 1976   Phenotypic expression of lethal chromosomal anomalies in human abortuses. Teratology 14: 3-20.

Charbonneaux, H., 1970   *Tourouvre-au-Perche au XVIIème et XVIIIème Siècle: Étude de Démographie Historique*. Presses Universitaires de France, Paris.

Coale, A. J., and D. R. McNeil, 1972   The distribution by age at first marriage in a female cohort. J. Am. Statis. Assoc. 67: 743-749.

Creasey, M. R., J. A. Crolla and E. D. Alberman, 1976   A cytogenetic study of human spontaneous abortions using banding techniques. Hum. Genet. 31: 177-184.

French, F. E., and J. E. Bierman, 1962   Probabilities of fetal mortality. Publ. Health Rep. 77: 835-847.

Ganiage, J., 1960   *La Population Européene de Tunis au Milieu du XIXème Siècle*. Presses Universitaires de France, Paris.

Gautier, E., and L. Henry, 1958   *La Population de Crulai, Paroisse Normande: Étude Historique*. Presses Universitaires de France, Paris.

Gini, C., 1924   Premières recherches sur la fécondabilité de la femme. Proceedings of the International Mathematical Congress 2: 889-892.

Harlap, S., P. H. Shiono, and S. Ramcharan, 1980   A life table of spontaneous abortions and the effects of age, parity, and other variables, pp. 145-158 in *Human*

Embryonic and Fetal Death, edited by I. H. Porter and E. B. Hook. Academic Press, New York.

Hassold, T., P. Jacobs, J. Kline, Z. Stein, and D. Warburton, 1984   Relationship of maternal age and trisomy among trisomicspontaneous abortions. Am. J. Hum. Genet. 36: 1349-1356.

Heckman, J. J., and B. Singer, 1982 Population heterogeneity in demographic models, pp. 567-599 in *Multidimensional Mathematical Demography*, edited by K. Land and A. Rogers. Academic Press, New York.

Henripin, J., 1954   *La Population Canadienne au Début du XVIIIème Siècle*. Presses Universitaires de France, Paris.

Henry, L., 1964   Mortalité intra-utérine et fécondabilité. Population 19: 899-940.

Hobcraft, J., J. W. McDonald and S. Rutstein, 1983   Child-spacing effects on infant and early child mortality. Pop. Studies 49: 585-618.

Jain, A. K., 1969   Fecundability and its relation to age in a sample of Taiwanese women. Pop. Studies 23: 69-85.

James, W. H., 1983   Decline in coital rates with spouses' ages and duration of marriage. J. Biosoc. Sci. 15: 83-87.

Johnson, N. L., and S. Kotz, 1970 *Distributions in Statistics: Continuous Univariate Distributions*. John Wiley and Sons, New York.

Keyfitz, N., and G. Littman, 1980   Mortality in a heterogeneous population. Pop. Studies 33: 333-343.

Lenton, E. A., B.-M. Landgren, and L. Sexton, 1984   Normal variation in the length of the luteal phase of the menstrual cycle: identification of the short luteal phase. Br. J. Obstet. Gynaecol. 91: 685-689.

Lenton, E. A., B.-M. Landgren, L. Sexton, and R. Harper, 1984   Normal variation in the length of the follicular phase of the menstrual cycle: effect of chronological age. Br. J. Obstet. Gynaecol. 91: 681-684.

Leridon, H., 1976   Facts and artifacts in the study of intra-uterine mortality: a reconsideration from pregnancy histories. Pop. Studies 30: 319-336.

Leridon, H., 1977 *Human Fertility: The Basic Components*. University of Chicago Press, Chicago.

Majumder, H., and M. C. Sheps, 1970   Estimators of a type I geometric distribution from observations on conception times. Demography 7: 349-360.

Manton, K. G., and E. Stallard, 1988   *Chronic Disease Modelling*. Charles Griffin, London.

Mode, C. J., 1985   *Stochastic Processes in Demography and Their Computer Implementation*. Springer-Verlag, Berlin.

Palloni, A., and M. Tienda, 1986   The effects of breastfeeding and pace of childbearing on mortality at early ages. Demography 23: 31-52.

Pollak, E., 1980   Effective population numbers and mean times to extinction in dioecious populations with overlapping generations. Math. Biosci. 52: 1-25.

Potter, R. G., and M. P. Parker, 1964   Predicting the time required to conceive. Pop. Studies 18: 99-116.

Royston, J. P., 1982   Basal body temperature, ovulation, and the risk of conception, with special reference to the lifetimes of sperm and egg. Biometrics 38: 397-406.

Shapiro, S., and D. Bross, 1980   Risk factors for fetal death in studies of vital statistics data: inference and limitations, pp. 89-105 in *Human Embryonic and Fetal Death*, edited by I. H. Porter and E. B. Hook. Academic Press, New York.

Sheps, M. C., 1964   On the time required for conception. Pop. Studies 18: 85-97.

Stein, Z., J. Kline, E. Susser, P. Shrout, D. Warburton, and M. Susser, 1980   Maternal age and spontaneous abortion, pp. 107-127 in *Human Embryonic and Fetal Death*, edited by I. H. Porter and E. B. Hook. Academic Press, New York.

Treloar, A. E., 1974   Menarche, menopause, and intervening fecundability. Hum.

Biol. 46: 89-107.

Treloar, A. E., R. E. Boynton, B. G. Behn, and B. W. Brown, 1967   Variation of the human menstrual cycle through reproductive life. Int. J. Fertil. 12: 77-126.

Trussell, J., and A. R. Pebley, 1984   The potential impact of changes in fertility on infant, child, and maternal mortality. Studies in Family Planning 15: 267-280.

Udry, J. R., 1980   Changes in the frequency of marital intercourse from panel data. Arch. of Sexual Behav. 9: 319-325.

Vaupel, J. W., and A. I. Yashin, 1985   Heterogeneity's ruses: some surprising effects of selection on population dynamics. Amer. Statistician 39: 176-185.

Vincent, P., 1956   Données biométriques sur la conception et la grossesse. Population 11: 59-82.

Vollman, R. F., 1977   *The Menstrual Cycle*. W. B. Saunders, Philadelphia.

Weinstein, M., J. W. Wood, and M.-C. Chang, 1990   Age patterns of fecundability, in *Biomedical and Demographic Determinants of Human Reproduction*, edited by R. Gray, H. Leridon, A. Spira and M. John. Oxford University Press, Oxford (to appear).

WHO, 1978   *International Classification of Disease, Ninth Revision*. World Health Organization, Geneva.

WHO, 1983   A prospective multicentre trial of the ovulation method of natural family planning. III. Characteristics of the menstrual cycle and of the fertile period. Fertil. Steril. 40: 773-778.

Wilcox, A. J., and B. C. Gladen, 1982   Spontaneous abortion: the role of heterogeneous risk and selective fertility. Early Hum. Dev. 7: 165-178.

Wilcox, A. J., A. E. Treloar, and D. P. Sandler, 1981   Spontaneous abortion over time: Comparing occurrence in two cohorts of women a generation apart. Am. J. Epidemiol. 114: 548-553.

Wilcox, A. J., C. R. Weinberg, J. F. O'Connor, D. D. Baird, J. P. Schlatterer, R. E. Canfield, E. G. Armstrong, and B. C. Nisula, 1988   Incidence of early loss of pregnancy. N. Engl. J. Med. 319: 189-194.

Wood, J. W., 1989   Fecundity and natural fertility in humans, pp. 61-109 in *Oxford Reviews of Reproductive Biology, Volume 11*, edited by S. Milligan. Oxford University Press, Oxford.

Wood, J. W., and M. Weinstein, 1988   A model of age-specific fecundability. Pop. Studies 42: 85-113.

12

Biology, Homology, and Epidemiology

KENNETH M. WEISS

Our evolutionary history is written clearly in the construction of our bodies, in the pattern of our development, and in the DNA sequences of our chromosomes. The biological similarity of organisms is inversely related to the time since they last shared a common ancestor (Nei, 1987). This is one of the bases for using animal models in biomedical research.

Comparative genetic studies are a molecular extension of comparative anatomy. Genes in related mammals typically have similar functions and even similar chromosomal positions (e.g., Nadeau, 1989). Genetic studies in one mammal species can often be directly transferred to humans since the genes are *homologous* in the evolutionary sense: they are descended from an ancestral gene shared in common by the two species. Homologies of this nature can reach striking proportions. Many of the basic processes of cellular differentiation and tissue development are essentially similar between humans and mice, arrangements unchanged in nearly 100 million years since the species diverged. Indeed, at least some of the genes involved in mammal tissue differentiation share striking similarities with genes having similar function in insects and even in yeast.

The distinction between *analogy* and homology is an important one for this chapter. Analogy refers to evolutionarily independent similarity of function, but homology refers only to traits whose similarity is due to common evolutionary ancestry. Diabetes mellitus illustrates these points. Instances of insulin-dependent (type I) diabetes mellitus (IDDM) in both man and mouse involve autoimmune attacks on the insulin-producing "islet" cells of the pancreas, in both cases related to the same defect in evolutionarily homologous genes of the histocompatibility system. Diabetes can also be manifest in different ways *within* a single species as well as between species. In some instances, the resultant diseases are only analogous, or phenotypically but not genotypically similar. Diabetes produced by auto-antibodies is analogous, but not homologous, to diabetes resulting from pancreatic cancer (which destroys the islet cells), or to type II diabetes, whose cause is not known but in which the islet cells function normally.

Homologous disease can also occur within a single species. The human hemoglobin molecule is made up of two components, known as α and β chains, each coded for by a separate gene. These genes are located on separate chromosomes, but their nucleotide (DNA) sequences clearly show that they evolved by gene-duplication many millions of years ago, with subsequent evolutionary diversification (Lewin, 1987). In fact, there are two major clusters of related globin genes, on different chromosomes, and the history of the gene duplications producing them from a single common ancestral gene can be inferred (e.g., Lewin, 1987). Several of these genes are only expressed at certain times in the human life cycle. Today, if either an α or a β chain is disrupted by mutation, the individual manifests one of several forms of anemia known as thalassemia (Antonarakis, Kazazian, and Orkin, 1985). These can be classed as *homologous diseases.*

There are many known examples of gene-duplications of this kind, involving essentially all physiological systems, from the simplest metabolic defects to complex behavioral phenotypes. Indeed, the entire human genome can be viewed as a complex tree of genes, with related genes on each branch having arisen by gene duplication, and having subsequently diverged in function and detailed structure by mutation, natural selection, and other processes. These are important evolutionary processes generally (Ohta, 1980, 1988).

This chapter explores some of these issues for a specific class of diseases, human cancers, in terms of their possible relevance to the age-specific incidence/mortality functions for cancer (known in demography as the "hazard function"). Cancer is a genetic problem at the cellular level, resulting from processes which usually require decades to complete. The biological basis of these processes is reflected in the hazard functions for cancer within a given population, in interesting ways which reflect the underlying biological processes.

AGE-DEPENDENT HAZARD FUNCTIONS FOR HUMAN CANCER

Age-specific mortality rates for most human cancers increase in an accelerating way with age. The nature of this age pattern, along with knowledge of the mutational nature of cancer, has led many investigators to develop multistage (stochastic) statistical models for the hazard functions for cancer (e.g., Whittemore and Keller, 1978; Moolgavkar, 1989). These models describe cancer as the result of a series of "hits" or stages which must occur in a given cell, and have been found to fit age-specific cancer incidence rates quite well. For example, the logarithm of the hazard function increases linearly with the logarithm of age for many cancer types:

$$\log[h(t)] \approx \alpha + \beta\log(t) . \tag{12-1}$$

The closeness of fit of models like this Weibull function to cancer data has led many authors to accept the general reality of multistage models, not only as good statistical descriptors of the information in the cancer hazard function, but also as literal evidence that the modeled multistage processes are in fact responsible for carcinogenesis.

Many different multistage models fit cancer data about equally well. Here, results from a particular simple multistage model are presented (Chakraborty and Weiss, 1989), with no claim that it is biologically better than other models in the literature (e.g., Whittemore and Keller, 1978; Moolgavkar, 1989). This model formally hypothesizes b necessary events, each occurring stochastically with fixed probability a per unit time, treated as independent Poisson processes. The probability, $P(t)$ of disease by age t is

$$P(t) = \left(1 - e^{-at}\right)^b . \tag{12-2}$$

The definition of the hazard function, given the distribution function, is

$$h(t) = P'(t)/[1 - P(t)] \tag{12-3}$$

or, when using (12-2)

$$h(t) = abe^{-at}\left(1 - e^{-at}\right)^{b-1}/\left[1 - \left(1 - e^{-at}\right)^b\right] . \tag{12-4}$$

A log-log-linear form like (12-1) can be derived for (12-4) if we introduce the power series approximation

$$e^{-at} \approx 1 - at , \tag{12-5}$$

in (12-2) and apply (12-3) to obtain $h(t)$. This has been a common approximation (as well as end point) of many different multistage models.

In 1969, Cook, Doll, and Fellingham (1969) showed that (12-1) closely fitted a wide range of international cancer hazard functions, and found that the slope on the log-log plot was approximately $4 - 6$ for many tumors; this has been interpreted as an estimate of the number of etiological stages required for cancer. The similarity of slopes across populations implies that hazard functions for different exposure groups differ approximately by a multiplicative constant. In epidemiological studies this translates to a constant relative risk following exposure to carcinogenic agents. To a reasonable approximation, this is what smoking does in a dose-dependent way to the risk of lung cancer (Doll and Peto, 1978), or exposure to ionizing radiation to the risks of many kinds of adult cancer (UNSCEAR, 1988).

In this chapter the results of fitting (12-4) to a wide variety of cancer age-specific mortality data are presented, drawing from work done originally in 1980 by R. H. Ward, R. Chakraborty, and myself, belatedly published as Chakraborty and Weiss (1989). We assumed that the number of stages (b) should be constant across populations for a given tumor type, to within statistical error, and that variation in a would reflect environmental exposures. Our intent was to study the genetic basis of different tumor types with this model.

It quickly became apparent that using (12-5) to derive a log-log-linear approximation (equation 12-1) ignores important curvature in the cancer data, a point also noted by Moolgavkar (1977). Multistage parameters estimated from (12-1) are substantially different from those estimated directly from (12-4), to the extent that if the approximation is *not* used, estimates of b are biologically unrealistic. The latter estimates seem too high, relative to what is known about

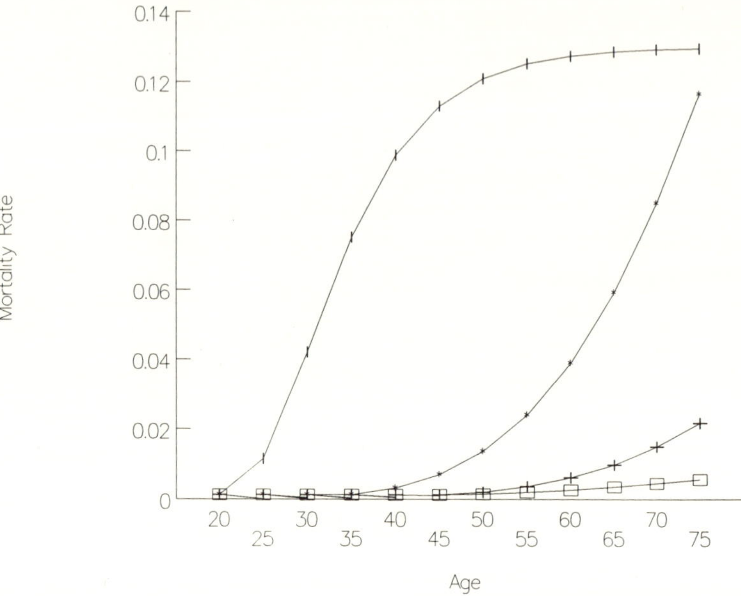

Figure 12-1 Colon cancer age-specific mortality rates, smoothed by maximum-likelihood fitting of equation (12-4). Key: open square = Thailand, plus = Mexico, star = USA, vertical line = FAP patients. For display on the same graph, all rates are ×100 except for FAP. Source: Chakraborty and Weiss, 1989.

the number of mutations needed to cause cancer; in addition, the estimated b's differ so much among populations as to imply that a different number of mutations needed to produce the same cancer in different populations (or even sexes). This is not a valid way to think of cancer in different populations. For this and other reasons, the parameters of these models cannot be interpreted literally in terms of mutational stages and their transition rates (e.g., Chakraborty and Weiss, 1989; Weiss and Chakraborty, 1984, 1990).

Even so, with certain exceptions and caveats (Chakraborty and Weiss, 1989), equation (12-4) is quite useful as a good descriptive fit to much of the available cancer data. Here, its use is strictly for that purpose.

PARAMETRIC MODELS OF CANCER INCIDENCE

Figure 12-1 illustrates cancer hazard functions with data for colon cancer, covering the worldwide range of national population data, plus the experience of victims of Familial Adenomatous Polyposis (*FAP*), a single-gene defect strongly predisposing to colon cancer. Except for the genetic tumor, to which we will refer later, these differences are presumably due varying environmental exposures across populations. The curves shown are smoothed by (12-4), which closely fits each data set; despite looking quite different, these schedules have a similar mathematical shape.

Using data of this kind from the World Health Organization for most major tumor sites, for 1970, we fitted equation (12-4), obtaining maximum likelihood estimates of a and b. Data were grouped in 5-year age classes from 35-39 through 70-74, sexes fitted separately. Thus, for tumor site i in population j, we obtained the points $(\widehat{a}_{i,j}, \widehat{b}_{i,j})$. Figure 12-2 provides a scatter plot of these points for colon cancer (excluding FAP, see below), and for leukemia as well as stomach, lung, and prostate cancer.

These figures reveal a high correlation between the parameters a and b across populations for a given tumor site, so that the points on the 'a-b' scale lie tightly along a straight line. This is true for all of the cancer types. In addition, Figure 12-3 shows the scatter plots for panels of Figure 12-2 plotted together. In this figure it is essentially impossible to see the individual points, because the linear patterns for these *different* tumor sites are also very similar, a result qualitatively like the observation that cancer hazard functions are approximate multiples of each other across risk groups. Table 12-1 provides data on additional sites. Some exceptions, for example prostate cancer, will be discussed below. One implication of this result is that given appropriate environmental exposure levels, essentially the same entire age-specific mortality schedule can arise for a variety of different tumor sites.

Differences in cancer risk appear to be largely due to differences in environmental exposures, such as to dietary fat, sunlight, or cigarette smoke (Doll and Peto, 1981). In our hands, the points a, b are correlated with age-standardized crude incidence rates. Environmental exposures seem to shift the entire complex age-onset pattern in a linear fashion *on the regressed a-b scale*. This is not the same as a constant relative risk, but rather is a re-scaling within a restricted subset of a family of mathematical forms.

Although some curvature in the hazard function can be caused by heterogeneity (e.g., Manton and Stallard, 1988), our results are not an artifact of heterogeneity, nor of the use of mortality rather than incidence data, to secular trends in risk, or the statistical fitting methods used. Chakraborty and I (1989) speculated that the mathematical similarity of the hazard functions is due to similar or even homologous constraints placed on the carcinogenic process by the genetic mechanisms involved, even for different tissue sites. Absolute risk differences would reflect variation in environmental factors as well as tissue-specific differences in the numbers of at-risk cells, cell turnover rates, and the like. Is this speculation consistent with the biological basis of cancer?

RELEVANT ASPECTS OF THE BIOLOGY OF CANCER

It used to be debated whether it was appropriate to use a single term, "cancer", to classify the tumors which arose in different organs or whether these different neoplastic diseases shared only their ability to mestasize, that is, to shed cells which could colonize other organs. This debate is over: cancers are clearly related diseases. Essentially, a cancer is a *clonal* genetic disease. A tumor and its metastases are comprised of cells all of which are descendents of a single original "transformed" cell in the host individual. That progenitor cell contained mutations in critical genes responsible for tissue-specific growth and development. All but one of these mutations were inherited by that cell either from the body cells which gave rise to it, or by the individual from his/her parents. By definition, the cell in which the final mutation occurs is the tumor progenitor. (This may be somewhat of an overstatement: cancer may arise when

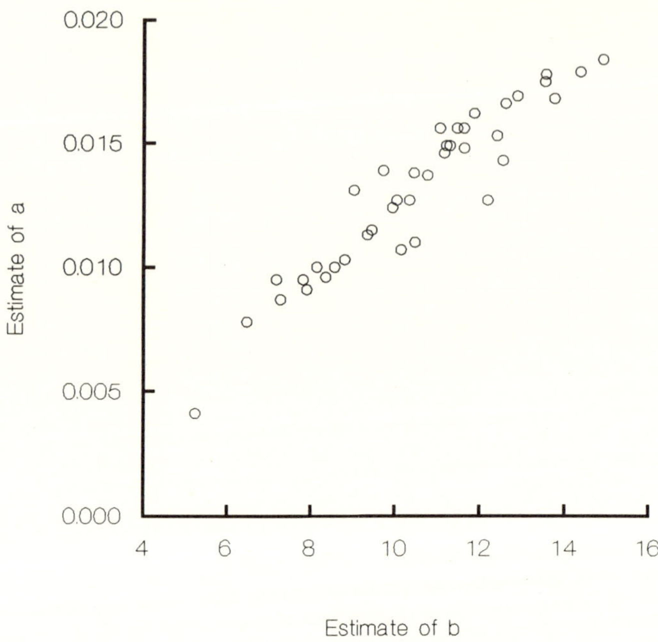

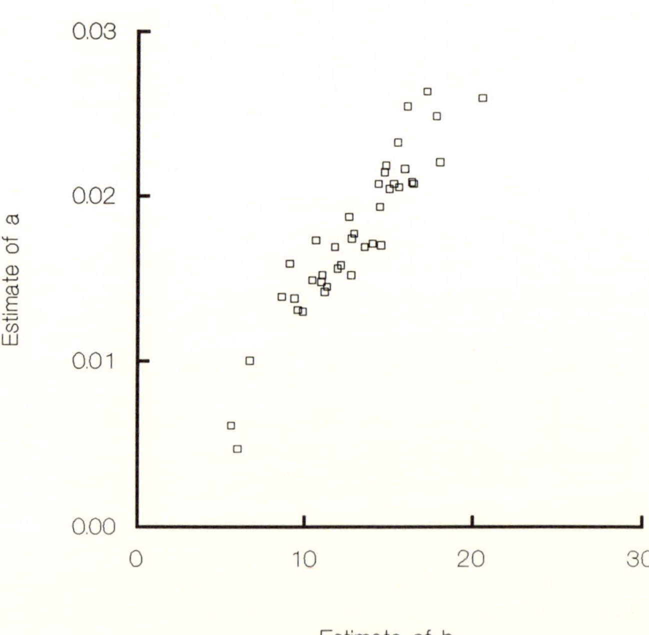

Figure 12-2.   Parameter estimates from a worldwide selection of populations for site-specific age patterns of mortality fitted to equation (12-4). Source: Chakraborty and Weiss, 1989. (a) Colon; (b) Stomach; (c) Leukemia; (d) Lung; (e) Prostate.

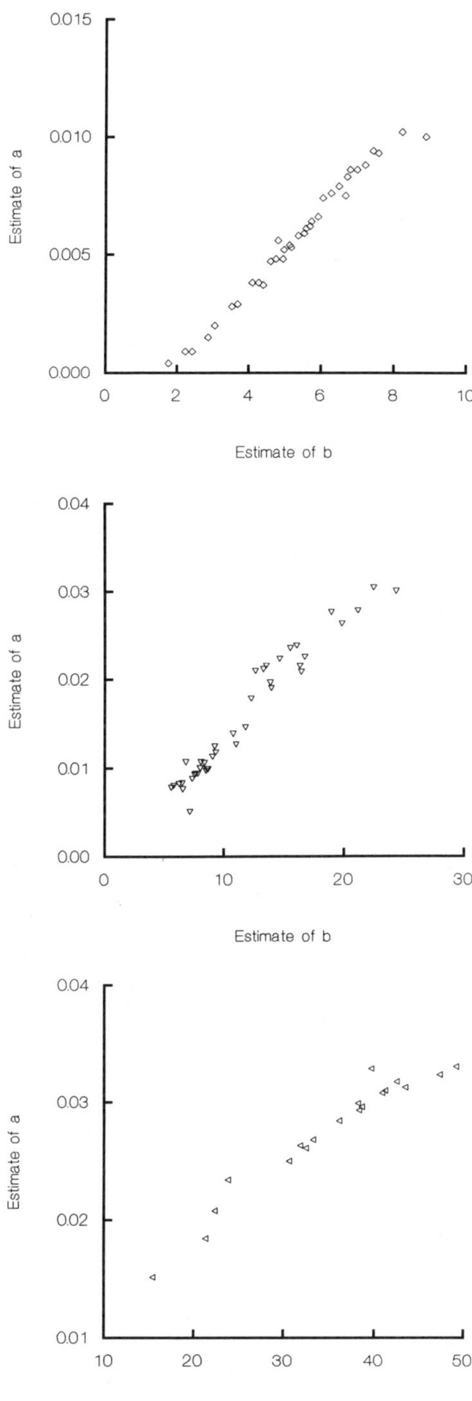

Figure 12-2. Continued.

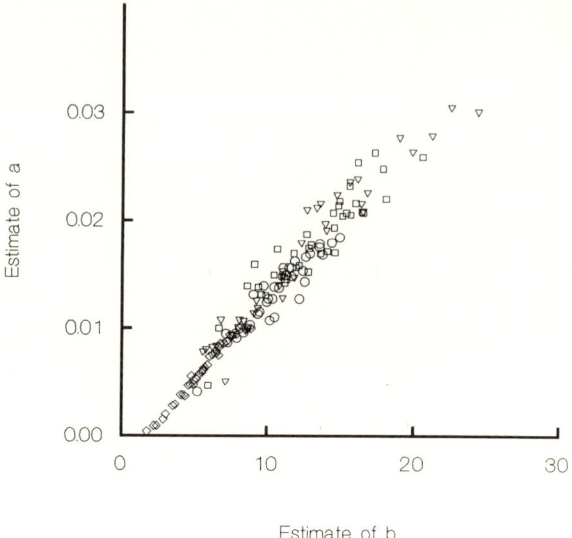

Figure 12-3. Plot of parameter estimates from Figure 12-2 for cancer of colon, lung, stomach and leukemia, using symbols as in Figure 12-2.

Table 12-1 Summary of hazard function fitting for various cancer sites

| Site | *a-b* Regression Parameters | | Correlation |
	Slope	Intercept	
Esophagus	0.00084	0.00251	0.82
Stomach	0.00136	-0.00021	0.96
Colon	0.00128	-0.00138	0.94
Lung	0.00148	-0.00141	0.98
Bone	0.00117	-0.00210	0.90
Skin	0.00111	-0.00150	0.96
Breast*	0.00194	-0.00282	0.96
Cervix*	0.00161	-0.00130	0.90
Uterus*	0.00121	-0.00112	0.94
Prostate	0.00049	0.01024	0.97
Leukemias	0.00158	-0.00269	0.99
Lymphomas	0.00150	-0.00209	0.96

Source: From Chakraborty and Weiss, 1989, based on 1970 World Health Organization Mortality data; see text and original paper for details.

*Fitted for ages 25-44 only.

a mutated cell is stimulated by growth factors in a quantitative, rather than just a qualitative, way).

This mutational process appears generally to occur one step at a time, randomly among cells having the capacity to divide, and, within those cells, randomly among the six billion or so constituent elements of the DNA there. Mutations occur and/or proliferate at cell division; therefore, hormones, irritants, or other factors which stimulate cell growth may assist the process, even though not directly causing mutational changes. Mutagens are sometimes called ''initiators'' and growth agents ''promotors''; some agents may have both features.

This mutational process would seem to be ideally suited to multistage statistical modeling if only we knew enough about the number of stages, cells, cell divisions, mutation rates and the like. The process is so similar in many tumor types, that cancers are also good candidates for investigating homologous diseases in man.

The molecular biology of cancer is complex. Genes known to play a role in cell transformation involve many, if not all, of the biochemical steps regulating cell growth. Cancer results from disrupted, aberrant, or histologically ectopic expression of these genetically-controlled pathways. These stages in cell growth include the initial stimulation of growth by hormones or growth factors such as those which circulate in the blood; the detection of such growth factors by cell-surface receptor molecules; the activation of chemical reactions on the inside of cell surfaces by activated receptors; the transport of growth-directing information to the nucleus of a triggered cell; and the stimulation or regulation of transcription of specific growth-related genes by chemical reactions within the nucleus. Table 12-2 briefly summarizes the functional role of many cancer-related genes. Details are available in many recent reviews (e.g., Kahn and Graf, 1986; Teich, 1986).

As Table 12-2 shows, a number of these genes are involved in more than one kind of tumor. A tumor may result from abnormal expression of a gene in its tissues in which it is normally expressed, or by its expression in an unusual tissue, it may be similar enough to a homologous gene product normally produced there to mislead the cell. It appears that some, perhaps most, tumors require mutations in more than one of these kinds of genes (e.g., Vogelstein *et al.*, 1988). However, it is not clear to what extent the process requires a mandatory set, or sequence, of mutations, or whether any of a number of oncogenes can produce the same effect. The current dogma has been that an enumerable set of genes are responsible, but reports like Vogelstein *et al.* (1989) suggest that many alternate genetic pathways may lead to similar ends, much as quantitative traits in genetics are modelled.

There are many homologies among cancer-related genes (e.g., Hanks *et al.*, 1988), and the *logic* of the message-transduction systems is common to many, if not most, cell types. These relationships may be extraordinarily deep in evolutionary time; for example, the *ras* family of cancer-related genes involves intracellular information-transduction proteins, and appears to have homologous relationship even to genes (also known as *ras* genes) in yeast. Systems laid down for major physiological functions, such as growth regulation and cell differentiation, are often very highly conserved (e.g., Whitman and Melton, 1989).

It is now so commonly accepted that different tumors involve the same, or homologous, genes that once such genes are discovered in one type of tumor, tests for their activity are used in the search for the genetic basis of other tumors. The *ras* genes are one of many examples.

Inherited cancers

There are two basic modes of action a mutant gene may have in a cell. If a mutant allele in only one of the two chromosomes carrying a given gene is suffi-

Table 12-2 Summary of cancer gene classification

Normal Function and Tissue	Oncogene Name	Chromosomal Location	Tumor Association
1. Plasma Membrane (GTP binding signal detectors/transducers)			
	Ha-ras	11p	Erythroleukemia, Sarcoma
	Ki-ras	6	Erythroleukemia, Sarcoma, Lung, Colon, and Bladder Carcinomas
	N-ras	1	Neuroblastoma, Breast cancer
2. Growth Factor Related			
	sis	22q	Sarcomas
3. Plasma Membrane (tyrosine kinase activity signal transducers)			
	yes	18	Sarcomas
	abl	9q	Leukemias, Sarcoma
	src	2q	Sarcomas
	fps,fes[a]	15q	Sarcomas
	abl	9q	Sarcomas
		12p	
4. Plasma Membrane Growth-Factor Receptor			
	erbB	7	Erythroblastosis, Vulval Breast and Ovarian Carcinomas
	fms	5q	Sarcomas
5. Cytoplasm			
	mos[b]	8?	Sarcomas, Erythroleukemia
	erbA	17	?
	raf[b]	?	Sarcomas
	mil[b]	?	Same as myc
6. Nucleus			
	ets	11q	Erythro- and Myeloblastosis
	myb	6q	Erythro- and Myeloblastosis Colon Cancer
	fos	14q	Osteosarcoma
	c-myc	8q	Myelocytoma, Granulocytic Leukemia, Sarcoma, Colon and Small-cell, Lung Carcinoma
	n-myc		Neuroblastoma, Retinoblastoma
	ski	1q	Squamous Carcinoma

Source: Modified from Teich, 1986; Kahn and Graf, 1986.

[a] Different names in chicken, cat.

[b] Also has serine/threonine kinase activity. Of the kinases, *mos* can be activated by overexpression, whereas *erbB, fms, abl, src* activated by mutation or deletion. Mutant *ras* genes have been associated with human tumors of bladder, lung, mammary gland, colon, neuroblasts, melanomas, fibrosarcomas, leukemias, and teratocarinoma.

cient to express its deleterious effect, that effect is cellularly "dominant". Some leukemias may involve cellularly dominant genes; for example, the "Philadelphia" chromosome associated with a chronic adult leukemia is produced when a growth-related gene is aberrantly transferred from its normal chromosomal location to a different chromosome, near an active antibody-producing gene in a white blood cell. A mutation is said to be cellularly "recessive," on the other hand, if the defect it causes can only be expressed when both chromosomes carry the mutation. A variety of embryonal tumors involve such recessive mutations (Friend, Dryja, and Weinberg, 1988).

Despite all this cellular genetics, by and large, cancer seems *not* to be an inherited disease. There are few if any known inherited cancers which are due solely to a cellularly dominant mutation. Such a mutation might be incompatible with successful fetal development. However, a number of heritable cancers are known which involve recessive genes, and two of these in particular have provided much of what is known about the molecular biology of cancer. The first is *retinoblastoma (RB)*, an embryonal cancer of the retina. The other is colon cancer occurring in an inherited condition known as FAP (see above); this leads to the formation of numerous small growths, or polyps, in the intestine, one or more of which eventually becomes cancerous.

Retinoblastoma usually develops in infancy. Most cases are not hereditary but are sporadic, and almost invariably arise as a single primary tumor in one eye. The remaining 10-15% of cases are hereditary. In these, multiple family members may be involved. The disease typically involves independent tumors arising in both eyes and/or multiple primary tumors in a given eye.

Retinal cells lose their ability to divide after they have initially differentiated, in late fetal or early infant life. The hazard function for retinoblastoma is thus constrained to young ages in genetic as well as sporadic cases. However, the number of primary tumors and their laterality suggested a genetic explanation, and in 1971 A. G. Knudson hypothesized (Knudson, 1971) that, in the familial form of RB, a germinal mutation is inherited from one of the parents, and the other, originally normal chromosome by chance experiences a similar *somatic* mutation during retinal development. Cells in which this occurs become progenitors of primary retinal tumors. Sporadic (nonfamilial) RB is rare because the probability of two separate mutations occurring in the same retinal cell due to environmental events is very small. However, the number of retinal cells is enough to essentially ensure that if all the cells have the first stage by inheritance, at least one of them will experience the second.

This hypothesis, one of the most prescient in the history of cancer research, for the most part has proven to be correct. The cellularly recessive *rb* gene is on human chromosome 13. Its normal function is to repress cell division. A routine mutation in *rb* can serve as the first event. The second event often appears to involve a process known as mitotic recombination, in which the normal process of cell division produces one daughter cell with two copies of the defective gene. A cell which contains one deleted and one normal chromosome 13 divides to produce two daughter cells but, unlike normal cell division in which each daughter is a copy of the original cell, one daughter contains two copies of the normal chromosome, the other two copies of the deleted chromosome. The latter is the tumor progenitor.

Once its basic genetics had been worked out, aberrant *rb* expression was quickly found to be involved in a number of other cancers including cancers of bone, breast, and lung.

The same cellularly recessive mechanism for a gene on chromosome 5 known as *fap* was found to be involved in colon cancer, first in FAP family members, where it appears to be the basis of the polyposis characterizing the syndrome, but also in at least some sporadic tumors and polyps (Bodmer *et al.*, 1987). This *fap* gene appears to be sufficient to cause the growth of the initial polyps which line the intestine. However, when—and perhaps only when—this gene is made homozygous in a polyp cell by somatic recombination and/or some other mutation, the cell becomes transformed (e.g., Bodmer *et al.*, 1987; Vogelstein *et al.*, 1988). In an FAP victim there are so many dividing colonic cells, all having inherited the first mutation, that (if not treated prophylactically) the subsequent transforming events are nearly inevitable. These events also often involve somatic recombination.

If RB is confined to infancy, one would expect that the large bowel, with many more cells at risk, would be as well. However, while most *fap* carriers do develop colorectal cancer, this process requires up to six decades to complete. Polyps do not begin to arise in large numbers and size until early adulthood. Even so, the gradual onset of cancer thereafter suggests that transforming events other than *fap* are required. In fact, recent studies have indeed implicated mutation at a number of loci in colon cancer, including a frequent involvement of specific regions of chromosomes 17 and 18 as well as in the *ras* genes (Baker *et al.*, 1989; Vogelstein *et al.*, 1988; 1989).

The cancer hazard function for FAP, relative to sporadic colorectal cancer, may be informative of the nature of this process. The hazard in FAP is very similar in Europe, the US, and Japan, despite very different baseline colorectal cancer risk levels. As the latter variation is presumably of environmental origin, it may be that the final transformation processes in FAP are strictly endogenous, not rate-limited by exposure to exogenous agents.

Sporadic colon cancer has a classic multistage hazard function. This is also true for FAP-associated colon cancer, which is fitted well by equation 12-4. The FAP data are based on very small samples, and it is difficult to obtain data on the total cumulative incidence. However, it appears that virtually all persons with the gene develop cancer by age 60 or so (Ashley, 1969, Lipkin, Sherlock, and DeCosse, 1980; Utsonomiya, Murata, and Tanimura, 1980). Parameters fitted to these FAP data are about an order of magnitude higher than the values for any national population. The best-fitting values are in the region $\hat{a} \approx 0.10 - 0.12$ and $\hat{b} \approx 25 - 50$ (Weiss and Chakraborty, 1984, 1990). However, in this range the fitting surface is quite flat, especially in respect to b. If the parameters are constrained to lie on the it a-b line for colon cancer, the best estimate of b is in the range $95 - 105$ for the available data sets; there is also a local unconstrained minimum in this range.

Unfortunately, this finding is difficult to analyze in a more precise way, because the FAP data are based on small samples, and precise onset information is critical for parameter estimation given the extremely rapid acceleration of risk. Currently, the condition is treated prophylactically, so we are unlikely to obtain better data. However, at the least it seems reasonable to say that the hazard function, even in this extreme population, is constrained to have shape very similar to the hazard functions in ''normal'' populations.

A developing picture may be emerging from a number of diverse research efforts. Genetic studies comparing tumor tissue from normal tissue of the same individual have shown that somatic mutation to *fap* homozygosity occurs even in some sporadic polyps and tumors (Solomon *et al.*, 1987; Vogelstein *et al.*, 1988), and a variety of studies suggest that most, perhaps even all, colorectal cancers develop from adenomatous (secretory-like) polyps, a high fraction of which may be familial (Cannon-Albright *et al.*, 1988; Sugarbaker, Gunderson, and Wittes, 1985). However, it is not clear what fraction of sporadic colon cancer specifically involves the *fap* gene. Nor is it known to what extent the other identified mutations reflect tumor progression and clinical details rather than primary transforming events (e.g., Slamon *et al.*, 1989; Vogelstein *et al.*, 1988, 1989), or even reflect mutations at some currently undetected ''mutator'' locus.

HOMOLOGOUS MECHANISMS?

These uncertainties notwithstanding, the reason that we know as much as we do know about colon (and other) cancers is that mechanisms and genes found active in one tumor type have transferred rather directly to others. The action of the same and/or homologous genes is involved in the growth and differentiation of many tissues. The numbers of cells, their turnover rates, environmental exposures, and other details may differ from one tissue to the next, but the general genetic groundplan of their differentiation, bears deep evolutionary homologies. That the hazard functions should be constrained in some way by these similarities is not unexpected.

Until the development of models which map specific statistical parameters directly to genetic and cellular mechanisms, it will be difficult to interpret the regularities in the cancer hazard functions. However, there are essentially only three basic explanations for those regularities: homology, analogy, and coincidence. The latter would include the kind of robustness by which a diversity of processes lead to similar outcomes (like the central limit theorem in statistics). However, the fact that prostate cancer deviates from the common *a-b* line, and that there are other life-history based exceptions (below) shows that coincidence is not likely to be explanation.

It is difficult to see how simple analogy could explain the patterns observed. It would seem to be highly unlikely that totally different biological processes would lead to such a high degree of regularity in the life-long expression of a diversity of diseases. Even if the same *kind* of mathematical function, such as (12-4), fit the data on many tumors, there is no inherent mathematical or statistical reason why the *values* of the parameters should be so highly constrained.

To invoke homology in this instance, even when its specifics cannot be identified, does not seem to be carrying speculation beyond reason. There is nothing speculative about saying that genetic homologies are involved in different cancer sites. The only connection which has not yet been clearly made is with the hazard function. In principle, the one-parameter nature of variation shown within tumor sites may also suggest that environmental exposures affect only one, or one aspect, of the carcinogenic process. The same could apply between sites. Indeed, it has long been a tacit form of recognition of the unity of cancer that the same mathematical models are applied to cancer at many sites.

DISCUSSION

The purpose of this chapter has been primarily to describe the highly regular pattern found across a diverse range of human cancer hazard functions, to show

how biological factors can operate strongly to constrain the expression of human disease, that is, to provide a clearly convergent issue in genetics and demography. In addition, I have speculated about possible mechanisms underlying those constraints in light of what is known about the biology of cancer.

The population data fitted here are mixtures of individuals with very different genetic or environmentally-caused risk. Mixing of this kind generates convexity in the hazard function, because at older ages individuals in higher-risk subsets differentially drop out of the surviving population; this can distort model-fitting results so that, for example, a mixture of subgroups with Weibull hazard function [Equation (12-1)] may have a convex hazard function (e.g., Manton and Stallard, 1988). However, Chakraborty and I (1989) found by simulation that the parameters estimated from (12-4) for whole populations were approximate mixture-weighted averages of the parameters of the population subsets, if those subset parameters lay along the lines shown in Figure 12-3 and Table 12-1. In similar ways, secular trends can affect the hazard function when using cross-sectional data; indeed, if individuals themselves change exposures during their lifetimes, as often happens, no cancer hazard data may be from a 'pure' risk process. But I think the *general* regularity we find is robust to these effects.

Tissue dynamics are known to be affected by a variety of growth-related factors such as menarche, the adolescent growth spurt, and cascades of tissue proliferation activity stimulated by injury or infection. The restriction of retinoblastoma to very early childhood, the concentration of certain bone cancers to the adolescent long-bone growth plates, the relationship of risk of female reproductive tumor to hormonal status, and many experimental tumors show this to be so. There are many examples of such life-history dependencies. Generally, they can be accommodated to the patterns described in this paper; for instance, female breast cancer fits equation 12-4 up to the age of menopause (Chakraborty and Weiss, 1989), even if for the whole life-course, more complex models would be needed (e.g., Moolgavkar, 1986). Prostate cancer also has a unique hazard function. The very slow-growing, frequently occult, nature of prostatic tumors may make the hazard function, for which the data are based on clinical diagnosis, an unreliable reflection of the disease process.

Another important exception are the cancers due to infection by exogenous viruses. Epstein-Barr virus involvement in nasopharyngeal cancer, certain leukemias, and cervical cancer are examples and they have characteristic, and somewhat atypical, hazard functions which clearly are related to exposure. Even in these cases, the ultimate mechanism of carcinogenesis is mutation, and involves similar genetic mechanisms. For example, cervical cancer can be caused when papilloma virus genes are integrated into the host DNA in an unlucky gene-location; it is not clear if this acts in a cellularly dominant way, or if other mutations are required (Meanwell, 1988). Clearly, hazard functions for tumors of this type must be analyzed cautiously.

Moolgavkar (1986, 1989) has developed a two-stage model of carcinogenesis in which the product of both mutation rates appears as a single term $\mu_1\mu_2$ outside an integral which specifies advantageous growth for clones of partially transformed cells. He has even recently tried to relate his model to the kinds of genetic mechanisms discussed here to explain the effect of smoking on cancer (Moolgavkar, Dewanji, and Luebeck, in press). The model, which fits much of the available cancer data, can behave one-dimensionally across environmental exposures and thus has a certain appeal. At present, however, there is inconclusive evidence that intermediate-stage cells have a natural growth advantage. For example, adenomatous polyps represent one of the best understood cancer precursor lesions. In the normal colon dividing cells are located in structures called ''crypts''; the number of crypts may change slowly over time (Potten, Chwalinski, and Khokar, 1982), and there is a suggestion that only some 'master' crypts have the potential thus to divide and produce crypts, and these replacements may increase in polyposis (Cheng *et al.*, 1986). The polyps themselves do represent an abnormal growth advantage for the partially-transformed polyp precursor cell, but they do not all expand without limit, and it is not clear whether their growth advantage is sufficient to generate the extremely high hazard rates in Moolgavkar's model, or is sufficiently general to apply to other tumors.

Natural experiments may exist for testing some of these ideas. Ionizing radiation is a carcinogen for which the pathogenic mechanism is reasonably well known, and dose-response data can be collected from natural or even laboratory experiments. Retinoblastoma gene carriers are at greatly elevated risk of bone cancer, and this involves the *rb* gene, as it also does in many sporadic bone cancers. Children with heritable RB who are irradiated to the bony orbit to treat retinoblastoma, experience a high subsequent risk of bone cancer in the radiation field, with a generally accelerating hazard persisting for 40 years to date, and a cumulative risk by age 20 of about 20% (summarized in UNSCEAR, 1988). This does not occur in children irradiated for sporadic RB. Elevated risk following radiation is expected, but every exposed cells in the heritable RB patients carries the *rb* gene. Why should radiation effects take decades to arise, and yet not be carcinogenic in sporadic patients? Clearly, radiation is not immediately transforming. Are genes other than *rb* involved?

Some years ago, Peto *et al.* (1975, 1976) wondered why mice should have age of onset patterns of cancer which are similar to those of humans, scaled in proportion to the mouse lifespan. With orders of magnitude fewer cells, cell divisions, and times of exposure to exogenous mutagens, mice should have much lower cancer risks than humans. In fact, their hazard functions are much like ours re-scaled. The fact that we are homologous to mice may lead us to a generic explanation, at the least.

The age of onset patterns of cancer have been the object of study for many reasons. Among these are an assessment of the dose-response patterns of cancer following exposure to carcinogens. Peto (1976) stressed the importance of examining the effects on the hazard function for understanding the underlying epithelial processes, results Chakraborty and I (1989) would extend to some nonepithelial tissues. This kind of risk projection is needed in setting limits on environmental and occupational exposure to carcinogenic chemicals and agents such as ionizing radiation. Such efforts can only be helped by a better understanding of the hazard function and how it varies.

Geneticists traditionally have been much more aware of heterogeneity in the hazard function, known in genetics as the penetrance function, than have demographers (Weiss, 1990), but neither group has, in my view, made sufficient effort to incorporate the facts of biology into their models for these processes. The views expressed in this paper are an attempt to show that opportunities exist to make the concept of heterogeneity in the hazard function epidemiologically useful and biologically meaningful.

REFERENCES

Ashley, D. J. B., 1969   Colonic cancer arising in polyposis coli. J. Med. Genet. 6: 376-378.

Antonarakis, S. E., H. H. Kazazian, and S. H. Orkin, 1985   DNA polymorphism and molecular pathology of the human globin gene clusters. Hum. Genet. 69: 1-14.

Baker, S. J., E. R. Fearon, J. M. Nigro, S. R. Hamilton, A. C. Preisinger, J. M. Jessup, P. VanTuinen, D. H. Ledbetter, D. F. Barker, Y. Nakamura, R. White, and B. Vogelstein, 1989   Chromosome 17 deletions and p53 gene mutations in colorectal carcinomas. Science 244: 217-221.

Bodmer, W. F., C. J. Bailey, J. Bodmer, H. J. R. Bussey, A. Ellis, P. Gorman, F. C. Lucibello, V. A. Murday, S. H. Rider, P. Scambler, S. Sheer, E. Solomon, and N. K. Spurr, 1987   Localization of the gene for familial adenomatous polyposis on chromosome 5. Nature 328: 614-616.

Cannon-Albright, L. A., M. H. Skolnick, T. Bishop, R. G. Lee, and R. W. Burt, 1988   Common inheritance of susceptibility to colonic adenomatous polyps and associated colorectal cancers. N. Engl. J. Med. 319: 533-537.

Chakraborty, R., and K. M. Weiss, 1989   Age-specific risks for cancer as determined by multi-stage models of carcinogenesis, pp. 64-91 in *Statistics in Medicine*, edited by T. Krishnan. Himalaya Publishing House, Bombay. (Reprint available from authors).

Cheng, H., M. Bjerknes, J. Amar, and G. Gardiner, 1986   Crypt production in normal and diseased human colonic epithelium. Anat. Rec. 216: 44-48.

Cook, P., R. Doll, S. Fellingham, 1969   A mathematical model for the age distribution of cancer in man. Int. J. Cancer 4: 93-112.

Doll, R., and R. Peto, 1978   Cigarette smoking and bronchial carcinoma: dose and time relationships among regular and lifelong non-smokers. J. Epidemiol. Comm. Health 32: 303-313.

Doll, R., and R. Peto, 1981   *The Causes of Cancer*. Oxford University Press, Oxford.

Friend, S. H., T. P. Dryja, and R. A. Weinberg, 1988   Oncogenes and tumor-suppressing genes. N. Engl. J. Med. 318: 618-622.

Hanks, S. K., A. M. Quinn, and T. Hunter, 1988   The protein kinase family: conserved features and deduced phylogeny of the catalytic domains. Science 241: 42-52.

Kahn, P., and T. Graf, editors, 1986   *Oncogenes and Growth Control*. Springer-Verlag, Berlin.

Knudson, A. G., 1971   Mutation and cancer: statistical study of retinoblastoma. Proc. Natl. Acad. Sci. USA 68: 820-834.

Lewin, R. *Genes*, Third Edition, 1987   John Wiley, New York.

Lipkin, M., P. Sherlock, and J. J. DeCosse, 1980   Risk factors and preventive measures in the control of cancer of the large intestine. Current Problems in Cancer, IV, Number 10.

Manton, K. G., and E. Stallard, 1988   *Chronic Disease Modelling*. Charles Griffin, Oxford.

Meanwell, C. A., 1988   The epidemiology of human papillomavirus infection in relation to cervical cancer. Cancer Surv. 7: 481-497.

Moolgavkar, S. H., 1977   The multistage theory of carcinogenesis. Int. J. Cancer 19: 730-731.

Moolgavkar, S. H., 1986   Hormones and multistage carcinogenesis. Cancer Surv. 5: 635-648.

Moolgavkar, S. H., 1989   Stochastic models of carcinogenesis, in *Handbook of Statistics*, Vol. 8, edited by C. R. Rao and R. Chakraborty, in press.

Moolgavkar, S. H., A. Dewanji, and G. Luebeck, 1989   Cigarette smoking and lung cancer: reanalysis of the British doctors' data. J. Natl. Cancer Inst. 81: 415-420.

Nadeau, J. H., 1989   Maps of linkage and synteny homologies between mouse and man. Trends Genet. 5: 83-87.

Nei, M., 1987   *Molecular Evolutionary Genetics*. Columbia University Press, New York.

Ohta, T., 1980   *Evolution and Variation of Multigene Families*. Lecture Notes in Biomathematics, Springer-Verlag, New York.

Ohta, T., 1988   Evolution by gene duplication and compensatory advantageous mutations. Genetics 120: 841-847.

Peto, R., 1976   Epidemiology, multistage models, and short-term mutagenicity tests, pp. 1403-1428 in *Origins of Human Cancer*, Vol. C, edited by H. H. Hiatt, J. D. Watson, and J. A. Winsten, Cold Spring Harbor Laboratory, Cold Spring Harbor, New York.

Peto, R., F. J. C. Roe, P. N. Lee, L. Levy, and J. Clack, 1975   Cancer and ageing in mice and men. Br. J. Cancer 32: 411-426.

Potten, C. S., S. Chwalinski, and M. T. Khokar, 1982   Spatial inter-relationships in surface epithelia: their significance in proliferation control, pp. 285-300 in *The Functional Integration of Cells in Animal Tissues*, edited by J. D. Pitts and M. E. Finbow, Cambridge University Press, Cambridge.

Press, W. H., B. P. Flaner, S. A. Teukolsky, and W. T. Vettering, 1986.   *Numerical Recipes: The Art of Scientific Computing*. Cambridge University Press, Cambridge.

Slamon, D. J., W. Godolphin, L. A. Jones, J. A. Holt, S. G. Wong, D. E. Keith, W. J. Levin, S. G. Stuart, J. Udove, A. Ullrich, and M. F. Press, 1989   Studies of the HER-2/*neu* proto-oncogene in human breast and ovarian cancer. Science 244: 707-712.

Solomon, E., R. Voss, V. Hall, W. F. Bodmer, J. R. Jass, A. J. Jeffreys, F. C. Lucibello, I. Patel, and S. H. Rider, 1987   Chromosome 5 allele loss in human colorectal carcinomas. Nature 328: 616-619.

Sugarbaker, J. P., L. L. Gunderson, and R. E. Wittes, 1985   Colorectal cancer, pp. 800-803, in *Cancer: Principles and Practices of Oncology*, Second Edition, edited by V. T. DeVita, S. Hellman, and S. A. Resenberg. J. B. Lippincott, Philadelphia.

Teich, N. M., 1986   Oncogenes and cancer, pp. 200-228 in *Introduction to the Cellular and Molecular Biology of Cancer*, edited by L. M. Franks and N. M. Teich. Oxford University Press, Oxford.

UNSCEAR (United Nations Scientific Committee on the Effects of Atomic Radiation), 1988   *Radiation Carcinogenesis in Man*. Annex G, pp. 405-544 in *Report to the General Assembly, Sources, Effects, and Risks of Ionizing Radiation*. United Nations Press, New York.

Utsonomiya, J., M. Murata, M. Tanimura, 1980   An analysis of the age distribution of colon cancer in adenomatosis coli. Cancer 45: 198-205.

Vogelstein, B., E. R. Fearon, S. T Hamilton, S. E. Kern, A. C. Preisinger, M. Leppert, Y. Nakamura, R. White, A. M. Smits, J. L. Bos, 1988   Genetic alterations during colorectal-tumor development. N. Engl. J. Med. 319: 525-532.

Vogelstein, B., E. R. Fearon, S. E. Kern, S. R. Hamilton, A. C. Preisinger, Y. Nakamura, and R. White, 1989   Allelotype of colorectal carcinomas. Science 244: 107-211.

Weiss, K. M., 1989   Are the known causes of mortality responsible for the human life span and its evolution? Am. J. Hum. Biol. 1: 307-320.

Weiss, K. M., 1990   The biodemography of variation in human frailty. Demography, in press.

Weiss, K. M., and R. Chakraborty, 1984   Multistage models and the age pattern of familial polyposis coli. Cancer Invest. 2: 443-448.

Weiss, K. M., and R. Chakraborty, 1990   Multistage models and the age-patterns of cancer: does the statistical analogy imply genetic homology? In *Familial Adenomatous Polyposis*, edited by L. Herera. A. R. Liss, New York, in press.

Whitman, M., and D. A. Melton, 1989   Induction of mesoderm by a viral oncogene. In early *Xenopus* embryos. Science 244: 803-806.

Whittemore, A. and J. Keller, 1978   Quantitative models of carcinogenesis. SIAM Review 20: 1-310.

III

GENETICS, DEMOGRAPHY, AND EPIDEMIOLOGY

13

Genetics, Demography, and Epidemiology

PETER E. SMOUSE AND MICHAEL S. TEITELBAUM

While population genetics and demography have different orientations and separate histories, there is an interface between them that is particularly natural, and that is the study of human disease epidemiology. Disease problems and their public health implications are complex, and an interdisciplinary approach is needed; the urge to merge (evident in the chapters of Part III) is motivated by necessity (Bean, Chapter 15). Although no coherent synthesis exists to date, interactions between the two disciplines on this front are becoming progressively more frequent, comfortable, and productive (Schull, Chapter 14).

There are several epidemiological questions that motivate an exchange between genetics and demography, among them: (1) How much of the variation in susceptibility to a particular disease is due to genetic differences among individuals and how can we best use that information? (2) How do we assess the lifetime survival and morbidity impact of that familial variation? (3) How are we to assess the likely impact of public health intervention in the case where most of the relevant risk variables are not amenable to manipulation? (4) How do genetic and environmental risk factors interact, in terms of the usual hazard function methodology? (5) Given that many disorders involve pathogens and their vectors, how does genetic variation in these other organisms relate to that in the human host?

GENETIC EPIDEMIOLOGY

That major improvements in public health have come about as a byproduct of improvements in the human condition (improvements in sanitation, sewage treatment, better nutrition, and inoculation) is now well established (reviewed by Preston, Chapter 22). As environmental (particularly infectious) components of disease are progressively brought under control, we can expect the profile of diseases to shift, with increasing fractions of morbidity and mortality caused by diseases of partial genetic etiology (see discussion in Ott, Chapter 16). The field of genetic epidemiology, while still poorly defined and circumscribed, has been with us since at least the 1940s, when Neel and Schull initiated their studies of the demographic and genetic consequences of radiation in Hiroshima and Nagasaki (Neel and Schull, 1956; Neel, Kato, and Schull, 1974; Schull *et al.* 1981), studies that continue to this day. There followed a whole series of investigations on the mortality and morbidity effects in inbreeding (Schull and Neel, 1965, 1972).

In spite of a long history, however, the discipline is defined more by what people do and the questions asked than by any systematic approach to the subject (Schull, Chapter 14). Current questions of interest are: (1) Can we explain differences in disease pattern among population strata in terms of environmental differences. Are their genetic differences in susceptibility? (2) In the sense of the chapters in Part II of this volume, how does genetic variation alter the frailty distribution (distribution of susceptibility) and the shape of the hazard function? (3) Is the genetic component of a particular disease caused by a single gene, or are there multiple genetic loci involved? How heterogeneous is the genetic etiology? (4) Where are the genes located on the chromosomes, and what is the genetically caused mechanism of disease action? The first two of these questions would seem to be obvious candidates for demographic treatment, but the last two routinely require demographic tools as well. The analysis of simple genetic markers affecting disease traits involves both population and family sampling, and elaborate sampling and estimation methods are required. Even genetically simple disorders such as retinoblastoma (Knudson, 1971) and Huntington's disease (Reed *et al.* 1958; Reed and Neel, 1959) show age-specific onset (incidence) patterns that have to be accounted for in any genetic analysis.

Technical advances in genetics and medicine have contributed to a change in the conventional epidemiologic wisdom, in which genetic etiology once implied that intervention was hopeless. Genetic understanding of erythroblastosis foetalis and phenylketonuria have long since led to effective detection and intervention as routine public health measures (cf. Emery, 1970; Hsia and Holtzman, 1973); an understanding of sickle cell anemia, Tay-Sachs Disease, and myriad other single-gene disorders permits increasingly effective detection with carefully designed population screening and/or prenatal assay (McKusick and Claiborne, 1973). While most of the genetic disorders elucidated so far have been those leading to serious but rare morbidity and mortality, attention is now turning to disorders with more complex etiology (Ott, Chapter 16; Schull, Chapter 14; Weiss, Chapter 12). It is worth remembering that even a small contribution of genetics to certain complex but common diseases may have a substantial impact on the population. Hypertension, diabetes, coronary artery disease, various cancers, and our major psychiatric disorders (depressive syndromes and schizophrenia), currently accounting for a sizable fraction of our public health load, are beginning to yield up genetic/etiological clues that offer hope of effective management in the near future. While the medical benefits of intervention in rare single-gene disorders are encouraging; the public health implications of ameliorating our more common diseases of partial genetic etiology are profound.

Family-linked databases

The first order of business for a useful synthesis of genetics and demography in epidemiological context is a data base with extensive familial and medical record-linking (Weiss *et al.*, 1980; Jorde and Skolnick, 1983; Thompson, Chapter 3; Castilla and Adams, Chapter 4). Such databases range from national twin registries (Christian, 1978; Nance, 1980) to databases on special subpopulations, such as the Hutterites (Hostetler, 1974), to community health studies such as those in Framingham (Dawber, 1980) or Tecumseh (Shreffler *et al.*, 1971; Sing *et al.*, 1971). Perhaps the most useful are collections of interlocking genealogies for a whole population, such as those available for the Laredo Cancer Study (Weiss *et al.* 1980), containing family- and medical-record linked data on over 300,000 people born since 1750, and the Utah Population Database (cf. Bean, Chapter 15), with similar records on over a million individuals born since 1800.

Such databases are put together with both demographic and genetic needs in mind and have some unusual features. Bean (Chapter 15) describes the sorts of features that are needed: (1) Records must be family-linked, with each individual record containing such information as birth date and place, date death, marriage dates, identifiers for parents and children, including birth orders. (2) These demographic records must be linked to detailed medical records. (3) The development of record-linking software is crucial; data entry is an ongoing task. Buchanan *et al.* (1984) and Bean (Chapter 15) have described the implementation strategies for the Laredo and Utah Studies, respectively.

Single gene disorders

The first clue to genetic etiology is a tendency for familial clustering of disease cases. Given family and medical record linkage, it is possible to detect such clustering with contingency analysis or other standard statistical treatments of cases and controls (Chakraborty, Weiss, and Ward 1980; Smouse, Weiss, and Chakraborty 1981; Weiss *et al.*, 1982). This preliminary effort can be followed by studies of familial transmission, the patterns for which have been relatively easy to determine for many of our simpler genetic disorders. Biochemical elucidation of these many disorders has proven more challenging, and represents a major portion of the ongoing agenda for modern human genetics. There are still, however, many genetic disorders that appear to have relatively simple (but not entirely regular) genetic transmission patterns, and for these the analysis is necessarily more complicated, involving elaborate likelihood models of the segregation pattern in multi-generational pedigrees (e.g., Elston and Stewart, 1971; Lange and Elston, 1975; Ott, 1985; Thompson, 1986). In some cases, we are reduced to tracking the familial co-transmission of the disease with a genetic marker thought to be closely linked on the same chromosome (linkage analysis). A critical feature of such analyses is an age-of-onset function, a description of the age pattern of incidence for the susceptible genotype, or—in the parlance of demography—a hazard function (e.g., Elston, 1973; Crowe and Smouse, 1977; Heimbuch, Matthysse, and Kidd, 1980; Risch, 1983; Ott, Chapter 16). In some cases, we need a separate hazard function for each genotype; Weiss (Chapter 12) reviews some of the evidence suggesting that a major difference between the susceptible and nonsusceptible members of the population for developmental disorders is a change in the parameters of their respective hazard functions, resulting in a drastic left-shift and steepening of the curve for the susceptible members of the population. Such alterations in the hazard function are known in a variety of other epidemiologic contexts. This is analogous to different "liability distributions" for the different genotypes, defined in quantitative genetics (cf. Falconer, 1981); threshold models of quasi-continuous genetic disorder theory could be reformulated in terms of such hazard functions (cf. Falconer, 1965, 1967; Carter, 1969).

Complex genetic diseases

Complex diseases—those involving several to many genes and myriad environmental influences—have been more difficult to unravel, but they represent the vast bulk of genetically influenced disease. The genes influencing such traits have been difficult to enumerate, and the sizes of their effects are only now becoming measurable. There are two approaches, the first of which is that of classical quantitative genetics (with unobservable but undeniable genetic heterogeneity). Traditionally, about all we have been able to do is to estimate the fraction of the total variation in the trait ascribable to genetic differences among individuals. The philosophical problem is analogous to that of heterogeneous but unobservable frailty (see the chapters in Part II). Common sense indicates that the genetic variation in susceptibility has to be there on theoretical and empirical grounds, but we are reduced to assumptions about the underlying normal distribution of that variation in the absence of assayable genes (Smith, 1971; Morton and MacLean, 1974; Lalouel *et al.*, 1977; Falconer, 1981).

More recently, we have learned enough about the developmental physiology of some of these disorders that we can begin to identify particular genetic loci that might be viewed, on physiological and developmental grounds, as credible candidates for genetic risk factors. The methods for this candidate gene (or measured genotype) approach are still being developed (e.g., Boerwinkle, Chakraborty, and Sing, 1986; Sing *et al.*, 1988; Boerwinkle and Utermann, 1988), but by being able to specify (and directly assay) particular genotypes, one can stratify the population into potential risk classes, and then determine the fraction of the variation in risk attributable to those loci. This measured genotype approach is analogous to the treatment of heterogeneous frailty; in the parlance of Trussell and Rodríguez (Chapter 8), we convert unobservable z-variables into observable x-covariates. Almost all of the traditional quantitative genetics theory is based on the assumption of a heterogeneous liability (that is, frailty) distribution that is normal, but there is nothing that would prevent one from adopting alternative distributional models more familiar to demographers (e.g., Hoem, Chapter 9; Trussell and Rodríguez, Chapter 8; Wood and

Weinstein, Chapter 11), and there is reason to believe that such an approach would be efficacious (Vaupel, Chapter 10; Weiss, Chapter 12).

THE EPIDEMIOLOGY OF INFECTIOUS DISEASE

To this juncture, we have restricted our genetic-epidemiologic attention to non-infectious diseases. Yet, infectious disease also has an important genetic component, for which demographic treatment is crucial. The usual problems of dealing with disease processes that are physiologically intrinsic to the human organism are greatly exacerbated when the genetics and demography of a disease-causing organism (and sometimes those of a vector as well) must be factored into the picture. Thus, a public health problem is transformed into one of multiple-species genetic ecology. Disease epidemic models have a long history in epidemiology, and the demographic theory has become increasingly sophisticated (reviewed by Bailey, 1975), but we are badly in need of some genetic heterogeneity modeling. Even crude models become complex, but attention to both the genetic and demographic details is crucial if we wish to provide an assessment of strategic alternatives for public health intervention.

To illustrate, Singer (Chapter 17) develops a microsimulation of a three-component system for falciparum malaria (human, mosquito, plasmodium), and then provides a pair of examples concerning the public health aspects of malarial control. The inputs for the simulation are standard demographic schedules for the particular human populations at risk, the life-cycle biology of the mosquito vector and the plasmodium parasite, and standard quantitative genetic models of genetic resistance to the standard pharmacological agents. The first problem he attacks is an evaluation of the efficacy of several alternative field treatments for populations subjected to endemic malaria, the objective being to devise a strategy that delays the genetically inevitable emergence of a resistant strain of *Plasmodium falciparum* as long as possible. The second problem is the spread of malarial infection in complex (compartmentalized) environments, where different sub-populations have different infection rates.

FUTURE DIRECTIONS

A major gap emerges from a perusal of the chapters in this volume. The standard definition of ''demography'' used by many geneticists involves only the gross population effects of total fertility, mortality, and migration. Conversely, the operational definition of ''genetics'' used by many demographers appears to involve only those genetic differences over which it is convenient to average for demographic purposes. It seems obvious that the details of age-specific vital rates deserve more attention from geneticists. On the other hand, a major component of demographic variation within any population is genetically based, a matter that deserves greater attention from demographers. In closing, we mention two areas requiring a synthetic approach to which both geneticists and demographers can contribute: (1) genetic interactions among host, vector, and pathogen, and (2) migrational aspects of infectious disease epidemiology.

Genetic aspects of infectious disease

Although genetic variation in the *Plasmodium* is taken into account in Singer's (Chapter 17) models, both the mosquito vector and the human host are assumed to be genetically uniform for modeling purposes. Human genetic variation in susceptibility to malaria is so widespread and so well known that a mention in passing is sufficient (Livingstone, 1983, 1985), but that in the mosquito is also important, because some control strategies concentrate on eradicating the vector (reviewed in Singer, Chapter 17). Similar considerations apply to strategies of public health control efforts for tuberculosis, trypanosomiasis, onchocerciasis, and AIDS; Singer (Chapter 17) points out that because AIDS compromises the human immune system, the diseases interact in complicated ways. We have a complicated problem in competing risk analysis, with both genetic and demographic components at several different levels. Singer's illustrations are revealing, but modeling efforts of considerably greater demographic and genetic complexity could clearly be deployed to good effect. That work represents a public health opportunity whose modeling surface we have only begun to scratch.

Migrational aspects of genetic disease

That migration has played a prominent role in both demography and genetics should also be evident from the chapters in this volume (Castilla and Adams, Chapter 4; Leslie, Chapter 5; Piazza, Chapter 6). The arrival of low human fertility and high migration rates means that migration will become an ever more important aspect of epidemiology. We need more attention devoted to the migrational component of the spread of infectious diseases (e.g., Sattenspiel, 1987; Sattenspiel and Simon, 1988). Given that genetic differences among individuals affect disease susceptibility, some ethnic groups may be especially vulnerable (e.g., Ward *et al.*, 1980; Weiss, Ferrell, and Hanis 1984; Young *et al.*, 1990). Large-scale movements of people from particular ethnic strata can be expected to have important consequences for the pattern of disease occurrence and for public health intervention strategies, both in demographic and genetic terms.

ACKNOWLEDGMENTS

We thank J. Adams, E. Boerwinkle, A. Hermalin, J. Neel, and K. Weiss for helpful comments on the manuscript. PES was supported by DOE-DE-FG02-87ER60533 and by the New Jersey Agricultural Experiment Station, supported by State funds.

REFERENCES

Bailey, N. T. J., 1975   *The Mathematical Theory of Infectious Diseases and its Applications*. Second Edition. Griffin, London.

Boerwinkle, E., and G. Utermann, 1988   Simultaneous effects of the apolipoprotein E polymorphism on apolipoprotein E, apolipoprotein B, and cholesterol metabolism. Am. J. Hum. Genet. 42: 104-112.

Boerwinkle, E., R. Chakraborty, and C. F. Sing, 1986   The use of measured genotype information in the analysis of quantitative phenotypes in man. I. Models and analytical methods. Ann. Hum. Genet. 50: 181-194.

Buchanan, A. V., K. M. Weiss, R. J. Schwartz, N. L. MacNaughton, M. A. McCartan, AND S. S. Bates, 1984   Reconstruction of genealogies from vital records: the Laredo Epidemiology Project. Comput. Biomed. Res. 17: 326-351.

Carter, C. O., 1969   Genetics of common disorders. Br. Med. Bull. 25: 52-57.

Chakraborty, R., K. M. Weiss, and R. H. Ward, 1980   Evaluation of relative risks from the correlation between relatives: a theoretical approach. Med. Anthropol. 4: 397-414.

Christian, J. C., 1978   The use of twin registries in the study of birth defects. Birth Defects: Original Article Series, Vol. XIV, No. 6a: 167-178.

Crowe, R. R., and P. E. Smouse, 1977   The genetic implications of age-dependent penetrance in manic-depressive illness. J. Psychiat. Res. 13: 273-285.

Dawber, T. R., 1980   *The Framingham Study: The Epidemiology of Atherosclerotic Disease*. Harvard University Press. Cambridge, Mass.

Elston, R. C., 1973   Ascertainment and age of onset in pedigree analysis. Hum. Hered. 23: 105-112.

Elston, R. C., and J. Stewart, 1971   A general model for the genetic analysis of pedigree data. Hum. Hered. 21: 523-542.

Emery, A. E. H., 1970   Antenatal diagnosis of genetic disease, pp. 267-296 in *Modern Trends in Human Genetics, Vol. 1*, edited by A. E. H. Emery. Appleton-Century-Crofts, New York.

Falconer, D. S., 1965   The inheritance of liability to certain diseases, estimated from the incidence among relatives. Ann. Hum. Genet. 29: 51-76.

Falconer, D. S., 1967   The inheritance of liability to diseases with variable age of onset, with particular reference to diabetes mellitus. Ann. Hum. Genet. 31: 1-20.

Falconer, D. S., 1981   *Introduction to Quantitative Genetics*. Second Edition. Wiley, New York.

Heimbuch, R. C., S. Matthysse, and K. K. Kidd, 1980   Estimating age-of-onset distributions for disorders with variable onset. Am. J. Hum. Genet. 32: 564-574.

Hostetler, J. A., 1974   *Hutterite Society*. The Johns Hopkins University Press, Baltimore.

Hsia, D. Y., and N. A. Holtzman, 1973   A critical evaluation of PKU screening, pp. 237-244 in *Medical Genetics*, edited by V. A. McKusick and R. Claiborne. Hospital Practice, New York.

Jorde, L. B., and M. H. Skolnick, 1983   Demographic and genetic applications of computerized record linking: The Utah Mormon genealogy. Sciences Humaines, 56-57: 105-117.

Knudson, A. G., 1971   Mutation and cancer: Statistical study of retinoblastoma. Proc. Natl. Acad. Sci. USA 68: 820-823.

Lalouel, J. M., N. E. Morton, C. J. MacLean, and J. Jackson, 1977   Recurrence risks in complex inheritance with special regard to pyloric stenosis. J. Med. Genet. 14: 408-414.

Lange, K., and R. C. Elston, 1975   Extensions to pedigree analysis: Likelihood computations for simple and complex pedigrees. Hum. Hered. 25: 95-105.

Livingstone, F. B., 1983   The malaria hypothesis, pp. 15-44 in *Distribution and Evolution of Hemoglobin and Globin Loci*, edited by J. E. Bowman. Elsevier, New York.

Livingstone, F. B., 1985   *Frequencies of Hemoglobin Variants, Thalassemia, the Glucose-6-Phosphate Dehydrogenase Deficiency, G6pd Variants, and Ovalocytosis in Human Populations*. Oxford University Press, New York.

McKusick, V. A., and R. Claiborne, 1973   *Medical Genetics*. Hospital Practice, New York.

Morton, N. E., and C. J. MacLean, 1974   Analysis of family resemblance. III. Complex segregation of quantitative traits. Am. J. Hum. Genet. 26: 489-503.

Nance, W. E., 1980   The value of population based twin refistries for genetic and epidemiologic research. Banbury Report 4: 215-234.

Neel, J. V. and W. J. Schull, 1956   *The Effect of Exposure to the Atomic Bombs on Pregnancy Termination in Hiroshima and Nagasaki*. Natl. Acad. Sci.-Natl. Res. Council Publ. 461.

Neel, J. V., H. Kato, and W. J. Schull, 1974   Mortality in the children of atomic bomb survivors and controls. Genetics 76: 311-326.

Ott, J., 1985   *Analysis of Human Genetic Linkage*. The Johns Hopkins University Press, Baltimore.

Reed, T. E., and J. V. Neel, 1959   Huntington's Chorea in Michigan. 2. Selection and mutation. Am. J. Hum. Genet. 11: 107-136.

Reed, T. E., J. H. Chandler, E. M. Hughes, and R. T. Davidson, 1958   Huntington's Chorea in Michigan. 1. Demography and Genetics. Am. J. Hum. Genet. 10: 201-225.

Risch, N., 1983   Estimating morbidity risks with variable age of onset: Review of methods and a maximum likelihood approach. Biometrics 39: 929-939.

Sattenspiel, L., 1987   Population Structure and the spread of disease. Hum. Biol. 59: 411-438.

Sattenspiel, L., and C. P. Simon, 1988   The spread and persistence of infectious diseases in structured populations. Math. Biosci. 90: 341-366.

Schull, W. J., and J. V. Neel, 1965   *The Effects of Inbreeding on Japanese Children*. Harper and Row, New York.

Schull, W. J., and J. V. Neel, 1972   The effects of parental consanguinity and inbreeding in Hirado, Japan. V. Summary and interpretation. Am. J. Hum. Genet. 24: 425-453.

Schull, W. J., M. Otake, and J. V. Neel, 1981   Genetic effects of the atomic bombs: A reappraisal. Science 213: 1220-1227.

Shreffler, D. C., C. F. Sing, J. V. Neel, H. Gershowitz, and J. A. Napier, 1971   Studies on genetic selection in a completely ascertained Caucasian population. I. Frequencies, age, and sex effects, and phenotype associations for 12 blood group systems. Am. J. Hum. Genet. 23: 150-163.

Sing, C. F., D. C. Shreffler, J. V. Neel, and J. A. Napier, 1971   Studies on genetic selection in a completely ascertained Caucasian population. II. Family analyses of 11 blood group systems. Am. J. Hum. Genet. 23: 164-198.

Sing, C. F., E. Boerwinkle, P. P. Moll, and A. R. Templeton, 1988   Characterization of genes affecting quantitative traits in humans, pp. 250-269 in *Proceedings Second International Conference Quantitative Genetics*, edited by B. S. Weir, M. M. Goodman, E. J. Eisen, and G. Namkoong. Sinauer, Sunderland, Mass.

Smith, C. A. B., 1971   Recurrence risks for multifactorial inheritance. Am. J. Hum.

Genet. 23: 578-588.

Smouse, P. E., K. M. Weiss, and R. Chakraborty, 1981   A simple test for the aggregation of disease occurrence in genealogical data. Hum. Hered. 31: 334-338.

Thompson, E. A., 1986   *Pedigree Analysis in Human Genetics*. The Johns Hopkins University Press, Baltimore.

Ward, R. H., P. Raspe, M. Ramirez, R. Kirk, and I. Prior, 1980   Genetic structure and epidemiology: the Tokelau Study, pp. 301-325 in *Population Structure and Genetic Disease*, edited by A. Erikkson, H. R. Forsius, H. R. Nevanlinna, P. L. Workman, and R. K. Norio. Academic Press, New York.

Weiss, K. M., R. Chakraborty, W. J. Schull, D. Rossman, and S. Norton, 1980   The Laredo Epidemiology Project, pp. 267-284 in *Banbury Report No. 4: Cancer Incidence in Defined Populations*, edited by J. Cairns, L. Lyon, and M. Skolnick. Cold Spring Harbor Laboratory, Cold Spring Harbor, New York.

Weiss, K. M., R. Chakraborty, P. P. Majumder, and P. E. Smouse, 1982   Problems in the assessment of relative risk of chronic disease among biological relatives of affected individuals. J. Chron. Dis 35: 539-551.

Weiss, K. M., R. E. Ferrell, and C. L. Hanis, 1984   A new world "syndrome" of metabolic diseases with a genetic and evolutionary basis. Yearbook Phys. Anthropol. 27: 153-178.

Young, T. K., E. J. Szathmary, S. Evers, and B. Wheatley, 1990   Geographic distribution of diabetes among the native population of Canada: A national survey. Soc. Sci. Med. (in press).

14

Some Lessons from Genetic Epidemiology and a Prognosis for Future Work

WILLIAM J. SCHULL

Substantial differences exist within and between developed and developing nations in prenatal and perinatal causes of mortality, the occurrence of fatal disease subsequent to birth, and in the expectation of life. These disparities have a profound effect on public health burdens and the allocation of often limited health care resources. Countless questions, genetic and nongenetic, arise in the search for an understanding of the origin of these differences, and the measurement of the magnitude of their effects. Some of these questions are largely heuristic, but answers to others would have immediate application. Among these latter are: How should we describe the impact of disease or disability? What metric or metrics should we use—the number of cases, years of life lost, quality of life, the economic cost to the individual and his or her society? Do, in fact, any of these measure the totality of the impact? We know, for instance, that among Mexican-Americans one out of every two individuals 35 years of age or older is either a diabetic, will become a diabetic in the subsequent course of life, or no less importantly, is the primary psychological and economic source of support for a diabetic. Case-counting would reflect only a portion of the detriment experienced in these communities from the high prevalence of this disease.

How should the risks be projected to a lifetime, when we know so little of their biological bases and most of the information we have is cross-sectional, describing events in only a portion of the lifetime? How can such projections be used to provide a means for assessing the efficacy of public health interventions? Can they be transported from one population to another, and if so, how? Clearly, not all individuals within a population have the same risk of disease or premature death. Individuals are known to differ in their capacities to repair damage stemming from exposure to ultra-violet or ionizing radiation. How are those individuals with the higher risks to be identified, and these risks ameliorated or prevented? Moreover, as therapeutic regimes grow progressively more specific and the role of genetic variation in response to drugs and their efficacy in

219

individual cases grows more important, iatrogenic disease looms larger. As yet there has been surprisingly little study of these inherited differences in drug response, or for that matter to naturally occurring toxicants in our environment. There is no dearth of questions, just answers.

Demographers have traditionally grappled with some of these problems, albeit in a different guise than the one seen by geneticists. As Livi-Bacci has noted (Chapter 2), demographers ask why do populations grow, why do some survive and others not. With little restatement, these are also genetic concerns. We wonder how genes spread. What determines their persistence or loss? Ultimately, of course, the answer depends upon the intrinsic rates of increase (decrease) of specific genotypes or phenotypes within the population. These rates are a function of age specific probabilities of death and reproduction, but the estimation of these probabilities for specific genotypes or phenotypes is not simple. Demographers are accustomed to large samples, often obtained from a variety of sources, frequently governmentally sponsored censuses or routinely collected vital statistics, but these sources do not normally give the genotype or phenotype of the individuals involved. Thus, the samples the geneticist requires to estimate the probabilities of interest to him must be collected *de novo*, and if the estimates are to have reasonable statistical precision, are commonly too large to be achievable by a single investigator, or even a small group of investigators. It is arguable whether these probabilities of death and reproduction are known with the requisite precision for any inherited disorder. What, for example, does the ''spontaneous'' loss of possibly a third to a half of all conceptuses in a nation such as the United States imply (Wilcox *et al.*, 1988)? Is it fortuitous, devoid of genetic implications? This hardly seems likely, since we know that a disproportionate number of spontaneous abortions involve embryos with chromosomal abnormalities? Is there a limit to the reduction in preparturient death that is possible through the control of maternal illness and the like?

Demographers also strive to understand the implications of migration, within a nation and transnationally, but so do geneticists, for a change in environment can alter the expression of a genotype, and thereby the prevalence or incidence of genetic disease. Migration also increases the opportunities for population admixture, and the diffusion into populations of genes previously infrequent or wholly absent. Even a very small amount of migration between randomly mating subgroups, initially different in their gene frequencies, will readily erase these differences.

Fortunately, my task is not to answer these questions. Rather it is to examine, in general but critical terms, the lessons to be learned from another disciplinary marriage, genetic epidemiology, which might further the dialogue between geneticists and demographers.

WHAT IS GENETIC EPIDEMIOLOGY?

Epidemiology has customarily been defined as the science of epidemics and epidemic diseases; that is, diseases which attack many people in a community or population simultaneously. Commonly, it is supposed that an epidemic involves a disease not continuously present in a particular group of individuals, but rather one brought from the outside through a transportable agent or agents, often bacterial or viral. Charles Creighton's monumental *History of Epidemics in Britain*, for example, deals solely with bubonic plague, cholera, diphtheria, influenza, smallpox, typhoid fever, typhus, and the like. Words such as genetic or inherited do not even occur in this account of twelve centuries of epidemics. If this was an epidemiologist's perception of his discipline, what would it now be when increasingly the focus of public health is upon common, chronic disease? What, indeed, is genetic epidemiology?

Twenty-odd years ago, at the University of Michigan, under circumstances very similar to this meeting, there occurred one of the first attempts to establish a conversation between geneticists and epidemiologists in the study of chronic disease (Neel, Shaw, and Schull, 1965). No specific effort was made to define the discipline aborning, but the proceedings speak to the differences in perspectives and expectations, and the difficulties in establishing a common framework for discussion. Subsequently, others, largely geneticists, have been less timorous about volunteering a definition. Morton (1977), for example, has defined genetic epidemiology as the "science that deals with the etiology, distribution and control of disease in groups of relatives or with the genetic causes of disease in populations." Sing and Moll (1982), however, contend that it is the study of the interaction of environmental and genetic determinants in common diseases. They would agree that its methods are different from those of "traditional" epidemiology. Still others fail to recognize it as either an appropriate or even a necessary conjunction of nouns, and point to the disparate methods and techniques the two disciplines employ. If there is no consensus as to what genetic epidemiology may be, how then are we to learn its lessons or predict its future?

Fortunately a discipline can be defined pragmatically through what its practitioners do, and through the questions they seek to address, among which are the following:

1. What is the role of genetic variation in the etiology of particular disease?

2. Are there significant differences within or among populations in disease prevalence attributable to differences in the frequency of causally related genes?

3. Is the disease in question genetically heterogeneous; that is can the same clinical disorder arise through the segregation of genes at different genetic loci?

4. What is the chromosomal location of the gene(s) involved?

5. What is the mechanism, molecular or cellular, of gene action and how mutable is this process?

Such questions have been addressed in situations as diverse as the investigation of breast cancer (Williams and Anderson, 1984), coeliac disease (Tiwari *et al.*, 1984), the use and abuse of alcohol (Clifford *et al.*, 1984), family resemblance in fasting blood glucose (Laskarzewski, Rao, and Glueck, 1984), immunoglobulin E levels and allergy (Borecki *et al.*, 1985), and even congenital malformations such as cleft lip, with or without cleft palate (Chung, Mi, and Beechert, 1987). Each of the above studies, as well as others that could be cited, has advanced somewhat our understanding of the etiology of the disorder under scrutiny, but this furtherance has come more through the delineation of high risk groups rather than a sharper definition of the genetic processes.

SIMPLY INHERITED GENETIC DISEASE

One of the intellectual appeals of Mendelian inheritance is the preciseness of the predictions which follow from the segregation of genes. Each of the simple modes of inheritance, whether autosomal or sex-linked, leads to a testable statement about the frequency of a specific phenotype in a sibship, the so-called segregation ratio; this ratio varies, of course, as a function of the type of mating and the mode of inheritance postulated. Precise as the predictions of genotypes and phenotypes may be, however, chance can obscure the true mode of inheritance in any given nuclear family, given the small size of human families. Consequently, to test the hypothesis of an autosomal dominant or an autosomal recessive mode of inheritance, for example, we must be able to combine data from a number of different families. These families may vary not only as to the type of mating and the number of children, but also in the manner through which they come to the investigator's attention, i.e., the manner in which they are ascertained. The greater the number of families combined, the less likely are random departures from expectation to obscure the true situation, but it is well to bear in mind that the greater the number of families, the greater the likelihood of introducing genetic heterogeneity which may obscure or confuse.

Analytic methods for coping with the problems posed by ascertainment and the pooling of information from different families, based on the principle of maximum likelihood, have existed for more than a half century, and recent extensions of these, made possible by the advent of the modern computer, have resulted in even more general and powerful statistical techniques. Time does not permit a measured account of these innovations; many will be found in Morton's *Outline of Genetic Epidemiology*, in Thompson's *Pedigree Analysis in Human Genetics*, or cited in Schull and Weiss (1980), and methods appropriate to the analysis of genetic linkage will be found in Ott's *Analysis of Human Genetic Linkage* (but see also Weeks and Lange, 1988).

Correspondence between the simple expectations predicted by the segregation of genes and actual observations are, however, contingent upon a number of assumptions. First, it is usually assumed that one is dealing with a regular phenotype system (see Cotterman, 1953), one where a given genotype is associated with one and only one phenotype, and this holds true for all relevant genotypes. This does not necessarily imply that each genotype has a unique phenotype; two or more genotypes could give rise to the same phenotype as in the case of simple dominance, while still satisfying the condition of regularity stated above. It is tacitly understood that within the range of environmental circumstances compatible with human existence, the phenotype associated with a particular genotype is invariant, or largely so. Alternatively put, there is little or no genotype-environmental interaction. Second, if the phenotype, albeit regular, is not established early in life, cognizance must be taken of the distribution of ages of onset of the condition under scrutiny, and the possible distortion of the expected segregation ratio through competing causes of mortality. This can be, and has been accomplished through a wedding of conventional genetic analysis with life-table methods familiar to the demographer (see, for example, Chidambaram *et al.*, 1988; Risch, 1983). Finally, since it is rarely practicable to identify all of the individuals within a population with a given disease, some sampling strategy is necessary. The latter determines in large measure the probability of inclusion in the study sample, or alternatively the probability of ascertaining a particular affected individual, and in turn affects the observed segregation ratio. For example, if the selection of study cases proceeds through the occurrence of disease, all families so identified will necessarily include at least one affected person. Families which could have produced an affected individual but by chance did not would not be included. Among the families actually ascertained, the expected segregation ratio, if the disease was inherited as a rare, recessive autosomal trait, would not be one-fourth but some greater number. Thus, the model from which one estimates genetic parameters must not only include a model of the manner of gene expression but a model of the sampling process itself.

Specification of the conditional segregation ratio has been termed the "ascertainment problem." Heretofore, this has commonly involved assumptions which often bore little resemblance to the actual sampling process. This statement is not meant to disparage previous work, but rather to call attention to the often complex nature of the sampling itself. Recently, Ewens and his colleagues (see Ewens and Shute, 1986; Shute and Ewens, 1988a,b; see also Hodge, 1988) have described a series of assumption free or nonparametric procedures which appear to hold promise in specific situations. In those instances where the classical ascertainment model is correct, it is still preferable, for it leads to smaller sampling errors of the estimates, but given the uncertainties which often attend genetic sampling, nonparametric methods should prove useful.

The availability of analytic techniques, and the success of geneticists generally in identifying and describing the molecular bases of simply inherited traits has had a generally stultifying effect on the development of models and methods of analysis to cope with more complexly inherited diseases. It has encouraged investigators, first, to compress uncritically into simple molds, disorders whose occurrence rests on an interaction of genetic and environmental forces, and second, to focus largely on those diseases which appear genetically simple, without regard to their public health implications or their contribution to our collective burden of disease and disability. It has also led to, in my view, a profitless debate on the merits of collections of nuclear families as contrasted with solitary, multigenerational ones as the bases for analysis. Obviously both have their strengths and shortcomings. Large multigenerational families can be especially useful in linkage analyses, for example, but provide little or no information on genetic heterogeneity or disease prevalence.

DISEASES OF COMPLEX ETIOLOGY

As has been implied, the common chronic diseases do not generally follow these simple Mendelian principles; their etiology is complex, the culmination of many forces of which genetic constitution is but one. *These are nonetheless the diseases that collectively contribute most to the burden of ill-health.* Studies to identify the role of genetic factors in the origin of such diseases as hypertension or diabetes, where our genetic constitutions and our environments interact, have traditionally been predicated on the general principles and models of quantitative genetics, which are extensions of the simpler, single locus principles. Some of these models are quite elegant mathematically. Two approaches have generally been used.

THE UNMEASURED GENOTYPE

The "unmeasured genotype" approach, as it is sometimes designated, attempts to assign components of variation in the trait to genetic factors, on the one hand, and environmental forces, on the other. In some cases, these studies have sought to determine the presence and possible chromosomal location of a "major" gene, a gene that makes a disproportionate contribution to the total observed variation. Assignment of a component occurs through the examination of the correspondence between a response and the known segregation of genes. Commonly, this approach culminates in statements about the collective importance of genetic variation, often expressed in terms of heritability, and cultural transmission. Studies of this nature are necessary, for they can establish the familial nature of the disease or risk factor, and frequently are all that can be done at a particular point in time. They should, however, be seen as merely first steps, not ends in themselves, because they are essentially phenomenological. Their primary limitations are three. First, since they begin with some statistical preconception of the mode of gene action or, in the case of path analysis, impose on the data a hierarchical cause and effect structure, the results are no more realistic than the model is appropriate, and different models frequently lead to quite different results. Second, they provide no insight into the molecular or cellular bases of the processes that subtend the phenotype itself, although they may be useful in predicting or even assessing the probable impact of strategies of intervention aimed at populations rather than at specific individuals. Third, there is a serious need to reexamine the notion of gene action upon which these models are based in the light of new advances in biochemistry and molecular biology with which the models no longer seem congruent.

THE MEASURED GENOTYPE

The other approach, termed the "measured genotype," employs some of the same statistical tools, but develops somewhat differently. It attempts to decompose variation into components attributable to specific genetic loci rather than the totality of genes involved. For example, it might seek to measure the component of variation in total serum cholesterol assignable to allelic variation at the locus responsible for the apo E lipoprotein (Davignon, Gregg, and Sing, 1988), or the polymorphism in the renin gene to blood pressure (Rapp, Wang, and Dene, 1988). As Sing and his colleagues have shown with regard to serum cholesterol (Davignon, Gregg, and Sing, 1988), this decomposition can provide the basis for projecting the risk of developing coronary artery disease, and thus identifying high risk groups. While this use of genetic studies in concert with more-or-less conventional risk analysis will undoubtedly undergo further refinements, it suggests a paradigm for future analyses of a host of etiologically complex disorders.

At the moment, the principal shortcoming of this approach rests in the fact that its application presupposes the existence of candidate genetic loci at which different genotypes can be identified. Choice of the loci employed must hinge on our knowledge of the presumed disease process. For some diseases, such as atherosclerosis, a substantial body of relevant information exists, but for others, e.g., hypertension, there is little (see, however, Rapp *et al.*, 1988; Sardet, Franchi, and Pouyssegur, 1989). Once candidate genes have been identified, the methods still require sufficient genetic variation to make their application practicable. As a consequence, since the number of genetic loci satisfying these constraints remains small, possibly no more than one or two percent of the entire genome, the situations in which this approach has been applied have been limited. As the number of restriction fragment length polymorphisms grows, however, and the number of genes actually cloned increases, this approach, coupled with other recent developments in molecular biology which make possible the amplification of specific segments of DNA, such as the polymerase chain reaction, will undoubtedly become the method of choice. The merits of this approach rest on the fact that it can provide evidence of the effects of specific genes, and thereby a means of recognition of individuals at high risk. It promises directed intervention through the molecular or cellular processes associated with a particular genetic locus. Perhaps more importantly, the approach may begin to indicate the extent and characteristics of the interaction of genetic and environmental factors contributing to disease. It should be borne in mind, of course, that this method alone cannot address the efficacy of intervention, nor if the risk function is a continuous one will it provide an obvious basis for deciding when intervention should occur, or at what level of risk steps to ameliorate it should be taken. Finally, and perhaps of more demographic interest, directly measuring the frequency and effects of specific genetic variation and investigation of its interaction with local environments may contribute to an understanding of the role of genetic factors in determining differences in their vital rates among separated cultures and populations.

THE FUTURE

Recent advances in biochemistry, cellular, and molecular biology are revolutionizing the ways of measuring genetic diversity, and have already established that the genetic variability that exists in human populations is substantially larger than we have heretofore supposed. Moreover, these new techniques have begun to provide information on the association of specific DNA-characterizable variation with the presence or absence of disease, and have also shown that seemingly similar phenotypes, for instance familial hypercholesterolemia, can have quite different molecular bases, reflecting different mutational events at the same genetic locus. As yet, however, they have seen little application to major public health issues or to the description of population differences of potential importance in understanding the disparate burdens of disease and disability among different segments of society. Presumably this will change with time, but it would be shortsighted to believe these techniques will be the panacea that resolves all of our problems. There will remain the need to understand the dynamic aspects of the interaction between genetic variation and environmental forces, and here population-based studies appear to offer one of the few strategies available to us.

Such studies are tedious, logistically complex, and expensive. They involve a different set of tools, and different strategies. Moreover, it increasingly appears that the process of gene-environment interaction is a dynamic one, and future studies must reflect this fact. If the roots of chronic disease lie in childhood, as seems increasingly likely, then the patterns of life established at that time and their interaction with genetic factors may be more important than subsequent events. If we are to understand the processes of disease and develop intelligent interventions, we must realize that our genetic constitution programs not only the limits of our response to environmental challenges but the rate at which that response occurs, and that the latter may change with the stages of life. Too commonly, complexly distributed phenomena are seen as epigenetic, devoid of any genetic control (see, for example, Dejours, 1987), and that an alteration of the environment automatically foretokens a significant amelioration. No end of conditioning will make a world-class runner of all of us, although we would undoubtedly be somewhat faster after the conditioning than before, nor should we expect that the lowering of cholesterol will necessarily lead to the general elimination of coronary artery disease.

A more profitable avenue of inquiry into mechanisms in this instance is one that focuses on variability in response to a standardized regimen, particularly among those individuals who can be construed as falling outside the customary range of variation. Rates of change, velocities, are often more informative than the end ultimately attained, since the latter can be achieved in a variety of ways. Poehlman and his colleagues (1986; see also Bouchard *et al.*, 1988) have, for example, shown that changes in body fat following short-term (22 days) overfeeding (1,000 calories per day) of monozygotic twins appears to be genetically controlled. The intrapair correlations in adiposity and fat-free mass gains were impressive when viewed in the context of the heterogeneity among sets of twins. Much of the effect was mediated through alterations in the energy expenditure components assessed in the study. More studies of this nature, particularly ones focused on specific sources of genetic variation, are desperately needed, if the massive efforts to change life styles are to prove cost effective and acceptable to a population.

No prescience is needed to see that, as yet, genetic epidemiology has not evolved into the synthesis we seek of the two approaches to the study of chronic disease. Individual studies continue to reflect either one perspective, the genetic, or the other, the epidemiological, but rarely if ever both. Epidemiologists, for example, still favor case-control studies. When they seek evidence of genetic factors, they do so through the study of the occurrence of disease in the relatives of cases as contrasted with controls, positing no model of why or how aggregation occurs. To geneticists, at least, this is not a particularly enlightened approach, for it can and frequently does confound nongenetic bases for familial aggregation with genetic ones. Epidemiologists appear to thrive on model-free hypotheses, geneticists on precisely the opposite. As Murphy (1989) has stated it, epidemiology prospers on badness-of-fit tests, the rejection of the null hypothesis, whereas genetics prospers with goodness-of-fit tests, the capacity to model a set of events suitably. Whether a successful accommodation between these different perspectives will be achieved remains to be seen.

Possibly the most important lesson to be learned from genetic epidemiology is that a dialogue between disciplines does not come easily, however meritorious. It will not grow out of one symposium, or even several, but only through working together in a constructive manner, seeking to learn from one another. This seems more likely to occur if, from the numerous specific areas of mutual interest where we could further one another's aims, we selected a few on which to concentrate. Two possible candidates are cited; obviously there are others.

First, although national censuses and other demographic surveys contain a seemingly limitless store of information on the rates of migration between administrative regions, they seldom include the factors of special interest to geneticists, that is, the relatedness of the migrants, differential migration by phenotype, distribution of migrational distances, and the age-sex distribution of the migrants. It would be useful too if these sources of information distinguished between ''temporary'' and ''permanent'' migration. Another form of migration which is poorly understood and even more poorly recorded is that between social classes within a subdivided population, despite the fact that in many human groups, it may be as potent a force in the dissipation of gene frequency differences as geographic migration. Substantial progress has been made in the development of models to describe the process of homogenization that migration produces, but most models concern themselves with fixed rates of migration and presume migrants to be a randomly drawn sample of the population or community from which they stem. Neither of these suppositions seem realistic. Rates of migration clearly vary enormously as a result of social and economic factors, and most of the evidence suggests that migrants are neither randomly drawn from the communities from which they come, nor do they select their future places of residence haphazardly. Family and social group or ethnic considerations contribute importantly to the decision to leave, as well as where to settle. If the presence or absence of relatives in a particular place exerts a substantial influence on one's decision to emigrate and the selection of one's destination, as the data suggest, migration cannot be Markovian (our usual assumption), for as Kramer (1981) has aptly observed, ''the system remembers its past.'' Thus, our models may be good enough descriptions of the relative impact of migration given the inexact nature of the data available, but not much more. Further progress seems likely to await the accumulation of more information on the migrants themselves so that the idiosyncratic in their behavior can be separated from the systematic.

Second, there is a large class of problems of a theoretical or quasi-theoretical nature, many dealing with the projection of risk in the face of competing causes of mortality, that warrant better solutions than we currently have, or at least alternate ones. Some of these are alluded to in the opening paragraph. Most projection models assume a constancy of risk either in absolute or relative terms, and moreover, that the risk, however expressed, is the same for every member of the population. Models are needed that admit of a heterogeneity in risks arising either from intrinsic, that is, genetic differences, changing treatment modalities, or changes in exposure.

Some may see these remarks as hopelessly pessimistic, or even erroneous. Obviously, I believe they are neither, but if such should be the case one can take solace in the thought that no less distinguished a scientist than Sir Isaac Newton, whose *Principia Mathematica* is one of humanity's greatest intellectual feats, wrote a less well-known book (Newton, 1733) about the fulfillment of the prophecies of Daniel and the Apocalypse of St. John. Daniel (Daniel: 7), you may recall, had a vision of four beasts arising out of the sea, each one more horrendous than its predecessor. The fourth, the most dreadful, had ten horns, but as Daniel gazed upon it, an eleventh appeared. Newton claimed to have discovered that the Church of Rome was this eleventh horn, and computed that it would be erased from the earth between the years 2035 and 2054. You will note that, being the mathematician he was, he provided a confidence interval. While we may chuckle at Newton's folly, some of the scientific shibboleths we now hold may fare no better over time than this fortunately forgotten book. We can console ourselves with the thought that while abilities may differ, we are all created equal in our capacity to err egregiously.

REFERENCES

Borecki, I. B., D. C. Rao, J. M. Lalouel, M. McGue, and J. W. Gerrard, 1985   Demonstration of a common major gene with pleiotropic effects on immunoglobulin E levels and allergy. Genet. Epidemiol. 2: 327-338.

Bouchard, G., L. Perusse, C. Leblanc, A. Tremblay, and G. Theriault, 1988   Inheritance of the amount and distribution of human body fat. Int. J. Obesity 12: 31-41.

Chidambaram, A., A. Chakravarti, R. E. Ferrell, and S. Iyengar, 1988   Estimating the age-at-onset function using life-table methods. Genet. Epidemiol. 5: 255-264.

Chung, C. S., M. P. Mi, and A. M. Beechert, 1987   Genetic epidemiology of cleft lip with or without cleft palate in the population of Hawaii. Genet. Epidemiol. 4: 415-424.

Clifford, C. A., J. L. Hopper, D. W. Fulker, and R. M. Murray, 1984   A genetic and environmental analysis of a twin family study of alcohol, use, anxiety and depression. Genet. Epidemiol. 1: 63-80.

Cotterman, C. W., 1953   Regular two-allele and three-allele phenotype systems. Am. J. Hum. Genet. 5: 193-235.

Creighton, C., 1891   A. *History of Epidemics in Britain.* Volume 1. *From AD 664 to the Extinction of Plague.* Volume 2. 1894. *From the Extinction of the Plague to the Present.* The University Press, Cambridge.

Davignon, J., R. E. Gregg, and C. F. Sing, 1988   Apolipoprotein E polymorphism and atherosclerosis. Arteriosclerosis 8: 1-21.

DeJours, P., 1987   Extreme environments: An overview. pp. 3-11 in *Hypoxia and Cold*, edited by J. R. Sutton, C. S. Houston, and G. Coates. Praeger Publishers, New York.

Ewens, W. J., and N. C. E. Shute, 1986   A resolution of the ascertainment sampling problem. I. Theory. Theor. Pop. Biol. 30: 388-412.

Hodge, S. E., 1988   Conditioning on subsets of the data: Applications to ascertainment and other genetic problems. Am. J. Hum. Genet. 43: 364-373.

Kramer, P. L., 1981   The non-Markovian nature of migration: A case study in the Aland Islands, Finland. Ann. Hum. Biol. 8: 243-253.

Laskarzewski, P. M., D. C. Rao, and C. J. Glueck, 1984   The Cincinnati Lipid Research Clinic Family Study: Analysis of commingling and family resemblance for fasting blood glucose. Genet. Epidemiol. 1: 341-356.

Morton, N. E., 1977   Some aspects of the genetic epidemiology of common diseases, in *Japan Medical Research Foundation: Gene-Environment Interaction in Common Diseases*. University of Tokyo Press, Tokyo.

Murphy, E. A., 1989   Editorial: The basis for interpreting family history. Am. J. Epidemiol. 129: 19-22.

Neel, J. V., M. W. Shaw, and W. J. Schull, (Eds.), 1965   *Genetics and the Epidemiology of Chronic Disease*. Government Printing Office, Washington, D. C.

Newton, I., 1733   *Observations upon the Prophecies of Daniel and the Apocalypse of St. John*. J. Darby and T. Browne, London.

Ott, J., 1985 *Analysis of Human Genetic Linkage*. The Johns Hopkins University Press, Baltimore.

Poehlman, E. T., A. Tremblay, J-P Despres, E. Fontaine, L. Perusse, G. Theriault, and C. Bouchard, 1986   Genotype-controlled changes in body composition and fat morphology following overfeeding in twins. Am. J. Clin. Nutr. 43: 723-731.

Rapp, J. P., S.-M. Wang, and H. Dene, 1988   A genetic polymorphism in the renin gene of Dahl rats cosegregates with blood pressure. Science 243: 542-544.

Risch, N., 1983   Estimating morbidity risks with variable age of onset: Review of methods and a maximum likelihood approach. Biometrics 39: 929-939.

Sardet, C., A. Franchi, and J. Pouyssegur, 1989   Molecular cloning, primary structure and expression of human growth factor-activable Na^+/H^+ antiporter. Cell 56: 271-280.

Schull, W. J., and K. M. Weiss, 1980   Genetic epidemiology: Four strategies. Epidemiol. Rev. 2: 1-18.

Shute, N. C. E., and W. J. Ewens, 1988a   A resolution of the ascertainment sampling problem. II. Generalizations and numerical results. Am. J. Hum. Genet. 43: 374-386.

Shute, N. C. E., and W. J. Ewens, 1988b   A resolution of the ascertainment sampling problem. III. Pedigrees. Am. J. Hum. Genet. 43: 387-395.

Sing, C. F., and P. P. Moll, 1982   Statistical Methods in Human Genetics: A Review of Progress and Problems, pp. 81-97 in *Proceedings X International Congress of Biometrics*, edited by C. G. B. Demetrio and I. D. Geraldi, Embrapa, Brazil.

Thompson, E. A., 1986   *Pedigree Analysis in Human Genetics*. The Johns Hopkins University Press, Baltimore.

Thompson, E. A., 1988   Two-locus and three-locus gene identity by descent in pedigrees. IMA J. Math. Appl. Med. Biol. 5: 261-280.

Tiwari, J. L., H. Betuel, L. Gebuhrer, and N. E. Morton, 1984   Genetic epidemiology of coeliac disease. Genet. Epidemiol. 1: 37-42.

Weeks, D. E., and K. Lange, 1988   The affected-pedigree-member of linkage analysis. Am. J. Hum. Genet. 42: 315-326.

Wilcox, A. J., C. R. Weinberg, J. F. O'Connor, D. D. Baird, J. P. Schlatterer, R. E. Canfield, E. G. Armstrong, and B. C. Nisula, 1988   Incidence of early loss of pregnancy. N. Eng. J. Med. 319: 189-194.
Williams. W. R., and D. E. Anderson, 1984   Genetic epidemiology of breast cancer: Segregation analysis of 200 Danish pedigrees. Genet. Epidemiol. 1: 7-20.

15

The Utah Population Database: Demographic and Genetic Convergence and Divergence

LEE L. BEAN

In 1974 a project was initiated at the University of Utah to construct a population based data set which would provide a resource for genetic epidemiology. Research had clearly demonstrated familial clustering of particular disease entities although much of the previous research had focused on unusual clusters of diseased individuals. The Utah project was based on the assumption that determination of the familial nature of common diseases would be more productive where ascertainment of all cases in a well-defined population was nearly complete and the genetic relationships between all cases were known (Bishop and Skolnick, 1984). Given access to certain unique data sets, it was assumed that the project would provide such opportunities. These records were, first, a large set of family genealogies detailing the historical and contemporary population of a well defined region, Utah, which was populated largely by members of one religious group, the Church of Jesus Christ of Latter-day Saints (LDS, hereafter). In addition to the family records, the Utah Cancer Registry provided exhaustive recording of the incidence of cancers which could be combined with and linked to, the genealogies or family records.

These two sets of records—the genealogies and the Utah Cancer Registry—served as the justification for the creation of the Utah Population Database (UPDB) and, subsequently, the Resource for Genetic Epidemiology (RGE). UPDB represents the computerized sets of records, and in some cases copies of the physical records. This chapter deals primarily with the computerized UPDB files. RGE is a legally constituted entity established by the State of Utah for the management of the database, including development and implementation of procedures to provide record protection, maintenance of confidentiality of individual information, and legal protection for data donors and researchers.

Descriptions of the development of UPDB including the technical details associated with entry and linking of such large data files have been detailed in a variety of publications (See for example, Skolnick *et al.*, 1978; Skolnick *et al.*,

231

1979; Skolnick, 1980; Mineau, Anderton, and Bean, 1989.) Therefore in the second section of this chapter, UPDB is described only briefly.

The development of UPDB has been a long-term project, extending now over a period of 15 years, and the development of the database still continues. UPDB represents not simply a long-term effort, but one that has involved, from the outset, a multidisciplinary effort, merging the interests of demographers and geneticists in the evaluation and exploitation of the data set (Bean, May, and Skolnick, 1978; Jorde and Skolnick, 1983). Multidisciplinary collaboration is evidenced through joint funding, common methodological developments in database management and record linking, and in some cases joint studies. More often, however, the convergence of genetic and demographic interests arise from independent investigations of issues common to researchers representing each field. We demonstrate the independent approaches to the analysis of a common methodological issues, generalizability, in the third section of this chapter.

Many of the research projects which have utilized UPDB are quite independent of one another. At the same time the significance of these independent research projects depends upon the fact that common features of UPDB and its represented population provide specific analytical advantages to both demographers and geneticists. This point is illustrated in the fourth section of the chapter.

In the fifth section of the paper, summaries of selected studies dealing with familial clustering of specific diseases are discussed to demonstrate the common focus on a specific set of variables in both demography and genetics.

THE UTAH POPULATION DATABASE

The assumption that a population database could be developed which included multiple generations in which familial relationships were clearly defined stemmed from the fact that early in the 1970s permission was provided to copy records which are maintained by the Genealogical Society of Utah (GSU hereafter), an organization operated by the LDS church. GSU, located in Salt Lake City, Utah, is the world's largest repository of nominative records. (See Bean *et al.* 1980 for a description of the GSU collection.)

As a religious mandate, dating from the beginning of this century, members of the LDS church have been encouraged to identify their ancestors. To assist members of the church, and others, GSU has operated a continuous program of collecting records which would identify individuals historically: manuscript censuses, vital records, parish records, shipping manifests for migrants, family histories, and others. These records have been used to compile family information and to record it on a form, the family group sheet, similar to, but developed independently of that generated by Louis Henry in France for the purpose of family reconstitution. (The family group sheet has been reproduced in a number of previous publication; for references, see Williams *et al.*, 1979; Hunt *et al.*, 1980; Skolnick *et al.*, 1981.)

The family group sheet includes data on three generations: a husband and wife, their children, and their parents. Each sheet represents a marriage and the following information is recorded: names and place of birth of husband and wife's parents; husband and wife's name, data and place of birth, date and place of death, date and place of marriage, and names of other spouses. If the husband or wife has two or more marriages, separate cross referenced sheets are completed. Children of the couple are listed individually by name, birth order with date and place of birth, date of death, date of marriage, and name(s) of spouse(s). Religious information specific for rites associated with the LDS church is also recorded. In less than half of the records an occupation is listed for the head of the family, but there are no other socioeconomic characteristics listed. Although the number of variables is limited, the data provide information necessary for the analysis of ethnicity, migration, marriage, fertility (including both numbers, timing, and spacing of births), and mortality.

When the UPDB project began, there were approximately nine million family group sheets available at GSU. The set selected for inclusion in UPDB was limited to those records where at least one family member was born or died in Utah or on the pioneer trail followed by the members of the LDS church as they migrated from the midwestern states to Utah. The criterion for selection was specified by the goal of developing as comprehensive a record as possible of the ancestors of the dominant LDS population of the Utah. Based upon the sampling criterion specified, approximately 185,000 family group sheets have now been included in UPDB, representing approximately 1.2 million individuals born after 1800.

The second major set of records which stimulated the development of the UPDB was the Utah Cancer Registry, which was initiated in 1952 in large hospitals, and which became statewide in 1966. All cancers excluding basal and squamous carcinomas of the skin must be reported to the Registry by state law, and the registry has been virtually complete, reporting all cancer cases since 1966 (Lyon *et al.*, 1977). The Utah Cancer Registry is maintained as a separate entity, but linked file structures are created as needed for analysis.

UPDB also includes, at present, the 1880 manuscript census for the state of Utah, which was entered and linked to the genealogy to ascertain the degree to which the genealogy represented the Territory's full population at that date. The file consists of the full information provided in the manuscript census representing approximately 143,000 individuals.

Although the genealogy file is unusually complete for events which occurred in the nineteenth century and the early part of the twentieth century, additional records must be entered to provide fuller coverage of the population during recent decades. Access has been provided to the Utah birth and death records, which became a State registration responsibility in 1904. Death certificates have been entered into UPDB for the period 1934 through 1981, and selected sets have been linked to the genealogies for specific studies discussed below. In 1989 entry of birth certificates will begin.

In summary, the core of UPDB is a set of genealogies representing roughly 1.2 million individuals linked together in various familial structures. Additional record sets include death certificates for approximately 47 years, and the 1880 manuscript census. These records are linked into analysis files with the Utah Cancer Registry. Parallel record sets, while not a constituent part of UPDB, have been linked for specific studies. These include various medical records dealing with cases of myocardial infarction and stroke (Williams *et al.*, 1979; Williams *et al.*, 1984) and birth defects (Fineman and Jorde, 1980.)

ADDRESSING COMMON METHODOLOGICAL ISSUES

Certain advantages of the study population for genetic epidemiological research have been well documented by demographers using the database. In this and the next section of the paper, we demonstrate this linkage through the summary of the arguments and evidence presented in genetic epidemiological studies and through the documentation provided by demographic analysis.

A common issue which has confronted both geneticists and demographers using UPDB is, what does the population represent and can studies based upon this particular population be generalized to other populations? For example, McLellan, Jorde, and Skolnick (1984) note:

> Cultural attributes and past geographic isolation have produced a well-defined population, which is an asset . . . However, these factors also lead to questions regarding the representativeness of this population, particularly in genetic studies in which frequencies of genetic diseases are estimated. Can results obtained from genetic studies of the Mormons be extrapolated to other Caucasian populations?

A priori it can be assumed that the UPDB population is very different from other religious isolates such as the Hutterites, Amish, and Mennonites, which have been the subject of extensive genetic epidemiological studies. The assumption is based upon the evidence of the large founding size of the population, and its diverse origin.

Members of the LDS church who were responsible for the settlement and development of Utah were initially drawn from New England populations when the church was first organized in 1832. As the church community moved increasingly westward, converts were made from a wider range of American communities. LDS missionaries were sent to Canada in 1836 and to England in 1837. Within a few years missionaries had also been sent to other western European nations. Before the migration to Utah in 1847, 4,700 British converts had joined the LDS settlement in Nauvoo, Illinois. The population which settled Utah was drawn therefore from substantial LDS settlements in the Midwest as well as from increasing numbers of converts from European counties. Consequently the population settling Utah drew upon an extensive population pool, both American and European. Data presented below will illustrate that the population settling Utah was dominated by foreign-born migrants, in spite of the common argument that the LDS is the epitome of a native American religion.

Drawing upon the Mormon settlements and widely scattered convert families, the Utah population grew rapidly. In 1847, 1,853 pioneers moved to Utah, followed by 4,633 in 1848 and 7,035 in 1849. From 1850 to 1860 the population increased by 253 % and by 1880 the population was roughly 143,000. The vast majority of the 1880 Utah population—between 85 and 90 % depending upon how one counts "apostates"—were members of the LDS (Bean, Mineau, and Anderton, 1990.)

More important than the significant number of individuals involved in the early settlement was the wide range of different populations from which they

Table 15-1 Distribution (percentages) of MHDP population by place of birth (origin), and birth cohort

Cohort	N	(1)	(2)	(3)	(4)	(5)	(6)	(7)	None	Other
1800-09	3,542	33.1	5.8	15.1	40.0	2.8	0.1	0.1	2.0	1.0
1810-19	6,608	37.7	6.3	18.8	29.9	4.6	0.1	0.1	2.0	0.6
1820-29	10,893	38.1	6.3	23.8	21.9	7.2	0.1	0.1	1.6	0.9
1830-39	15,268	35.1	6.2	24.4	19.0	12.4	0.1	0.1	1.6	1.1
1840-49	19,268	33.8	5.2	24.2	10.3	21.0	0.3	2.2	1.6	1.4
1850-59	30,191	20.9	3.2	20.4	6.4	7.0	0.5	39.1	1.2	1.2
1860-69	41.449	10.1	1.5	13.8	3.9	2.8	2.2	64.0	0.9	0.9
1870-79	53,163	5.1	0.7	8.6	3.3	2.0	4.0	74.4	0.7	1.0
1880-89	69,677	2.6	0.3	4.8	2.8	1.5	10.3	76.0	0.5	1.1
1890-99	80,643	1.1	0.2	2.6	2.6	1.1	14.1	76.1	0.4	1.2

NOTES: (1) England and Wales; (2) Scotland, Ireland; (3) Northern Europe; (4) Canada and other U.S. states not included below; (5) Mormon Trail-Ohio, Mo., Ill., Iowa, Neb., Ind.; (6) Territory-Idaho, Colo., Nev., Ar., Wy., New Mex.; (7) Utah.

were drawn. In Table 15-1 we report the number of individuals recorded in the UPDB by date of birth (birth cohort), and place of birth. Up until the 1850-59 birth cohort, over one-third were individuals born in England and Wales, closer to 40 % if one includes Scotland and Ireland (columns 1 and 2). Among the birth cohorts of 1820-29 through 1860-69 more than 20 % were from Northern European countries. The number of foreign-born migrants falls off sharply beginning with the 1860-69 birth cohort corresponding to the intrusion of the Federal government into the political system. One major consequence of the increasing role of the Federal Government was the declaration as illegal, of the LDS perpetual immigration fund, a funding mechanism that had been used to support migration from Europe. Stripped of the resources to support European migration, the population near the end of the century was increasingly Utah-born. However, most of these individuals were the children of migrants from a number of European nations.

McLellan, Jorde, and Skolnick (1984) have demonstrated the importance of these diverse origins in their study of genetic distances between Utah Mormons and related populations. They summarize their findings as follows:

> Gene frequency data, consisting of six red cell antigen loci, nine electrophoretic systems, and HLA-A and -B are reported for the Utah Mormon population. These are compared statistically to gene frequencies from a U.S. population, 13 European populations, and seven populations from three religious isolates. The Mormon gene frequencies are similar to those of their northern European ancestors. This is explained by the large founding size of the Mormon population and high rates of gene flow. In contrast, the religious isolates (Amish, Hutterities, and Mennonites) show marked divergence from their ancestral populations and each other, due to isolation and random genetic drift.

Further analysis using the "Genealogical Index" (GI) (Skolnick *et al.*, 1981) based upon Malécot's (1948) coefficient of kinship, indicates that because the pioneers were largely unrelated, drawn from several populations as noted above, the coefficient of kinship started low among the earliest cohorts, and then increased, though still remaining very low in contrast to Hutterities, Amish, Mennonites, and other religious isolates. The rise in the coefficient of kinship identified by Bishop and Skolnick (1984) is consistent with Anderton's finding of increasing ethnic intermarriage during the nineteenth century (Anderton, 1986; Jorde, 1989a).

Demographers dealing with the issue of representativeness have approached the problem in two different ways. First an effort was made to compare data from the genealogy with an independent data set. To accomplish this goal, it was decided to enter and link the 1880 manuscript census to the genealogy. The 1880 manuscript census was selected because it was the first U. S. census which identified the relationships of members of a household to the head of the household, providing an additional item of information which might be used in attempting to link the census and the genealogy. The linkage routine developed added two innovations over more common procedures: (1) a phonetic transducer (originally developed by Yves Chiaramella in Grenoble, France and extended with applications to the UPDB data by V. H. Wesley) was used in addition to the Russell-Soundex coding method and (2) family relationships were used in addition to individual indicators (Mineau, Anderton and Bean, 1989). The machine based linkage program generated a match rate of approximately 60 % with an error rate of less than 1%. Hand verification of unlinked cases increased the linkage rate by about 8% among those counties where work has been completed. The unmatched cases represent transitory populations, primarily single males engaged in mining or military service, and other solitary individuals in households: servants and boarders. Female headed households are also underlinked. However, the linkage analysis confirms extensive coverage of those individuals who are likely to have any descendants, and particularly descendants who would be most likely to be present in the contemporary population. In general, however, the linkage rate of between 60% and 78% compares favorably with the percentage of LDS reported in the 1880 census when one considers that many of the individuals counted in the censuses were transitory; migrants who stopped briefly in Utah and then moved to establish Mormon colonies in Arizona, Idaho, Nevada, Wyoming, and elsewhere.

A second demographic approach to the question of representativeness of the population involved a number of analyses which have demonstrated comparability of demographic parameters represented by the UPDB population and a variety of other populations (Anderton and Bean, 1985; Bardet *et al.*, 1981; Lynch, Mineau, and Anderton, 1985; and Mineau, 1988). Similarities are obvious in the case of the relationship between birth spacing and final parity, and patterns of change in levels of fertility over time.

In summary, an analytical advantage for both geneticists and demographers is that the UPDB population is well defined. The population is dominated by a particular religion with religious and ethical standards which distinguish life style, fertility, and mortality. Often regarded as a religious isolate, extensive efforts have been made to demonstrate that the UPDB population does not exhibit unique features restricting any conclusions drawn from the analysis of the data of this particular population. Geneticists and demographers have both addressed the same issue using several different methods, all of which strengthen the initial judgement that the population is ideally suited for studies of genetic epidemiology and historical demography. Moreover specific features of the population provide unusual analytical opportunities.

ANALYTICAL ADVANTAGES

Access to individual records for approximately 1.2 million individuals linked together into kinships, although representing birth cohorts extending over a century and a half, has provided opportunities to investigate fertility change over the course of frontier settlement and development. The population of Utah in general and its LDS population in particular has drawn interest specifically because of its high levels of fertility. The mid and late nineteenth-century levels of fertility are not substantially different from what would be found among other frontier population, although the continuing, relative high levels of fertility are attributed to the profamily and profertility norms of the LDS church. It has also been suggested that high aggregate levels of fertility reflected, in part, the practice of polygyny over a period of approximately half a century. The latter phenomena may have actually reduced overall fertility slightly (Bean and Mineau, 1986) but the practice of polygyny meant that selected males had unusually large numbers of descendants.

The fertility experience of the UPDB women may be summarized in broad terms. Women born in 1800-04 completed child bearing with an average of 7.1 children; women born in 1840-54 had 8.2; and women born in 1895-99 had 4.7 births (Bean, Mineau, and Anderton, 1990.) There were, however, differences in fertility behavior among various subgroups marked generally by levels in fertility rather than trends over time (Bean, Mineau, and Anderton, 1983.) These differences in levels can be demonstrated by comparing age of marriage and mean CEB (Children Ever Born) by birth place (origin) of mothers. Data for selected cohorts are presented in Table 15-2. Note that among the 1810-19 birth cohort, the women born in Great Britain, Canada, and Northern Europe (columns 1 and 2) had fewer children than women born in the United States (columns 3 and 4) reflecting the fact that these women married somewhat later. Among the 1850-59 birth cohort, women born in Great Britain and Ireland, many of whom migrated as children, were married on average at age 20.3 years and produced 8 children. Women born in Utah married earlier, 19.3 years, and ended childbearing with an average of 8.3 children. For women born in the last decade of the nineteenth century, fertility had fallen substantially, to between 4.8 and 5.3 children. In general migrant women continued to have fewer children than native-born Utah women.

Before proceeding with the discussion of the significance of these large family for genetic epidemiology, it is useful to comment on the demographic significance of the above data. At the beginning of this century eugenicists often argued that migrants from Europe were degrading the gene pool of the American native population because they typically had much higher fertility than the native-born. Those often ideologically inspired studies typically compared European migrants with Eastern U.S. urban populations where fertility had already declined substantially. Comparing European migrants with native born Americans among the UPDB population, both groups of which had accepted a religion with strong pronatalist values and settled in the same region, indicates that the Europeans ended childbearing with slightly fewer children. Although the fertility differences are not great, clearly it cannot be argued that the European

Table 15-2 Mean age at marriage (AM) and mean children ever born (CEB) by place of birth (origin) and birth cohort for married women

Birth Cohort		Total	(1)	(2)	(3)	(4)	(5)	(6)	None	Other
1810-19	AM	23.3	23.3	26.1	21.4	20.9	*	*	25.0	*
	CEB	6.7	6.9	5.6	7.1	7.7	*	*	5.7	*
1830-39	AM	22.0	21.9	25.1	19.0	18.7	*	*	23.3	*
	CEB	7.4	7.6	6.3	8.2	8.1	*	*	5.3	*
1850-59	AM	20.2	20.3	22.0	20.2	19.3	19.6	19.3	20.8	22.0
	CEB	7.9	8.0	7.4	7.6	7.9	8.3	8.3	7.3	7.5
1870-79	AM	21.5	21.8	22.7	21.4	21.0	20.9	21.3	21.7	22.8
	CEB	6.7	6.5	6.5	6.6	6.3	7.0	6.7	6.1	6.3
1890-99	AM	21.8	23.0	23.0	21.5	21.1	21.7	21.7	21.6	21.8
	CEB	4.9	4.8	5.0	5.0	4.8	5.3	4.9	4.0	5.5

NOTES:

(For first marriages of women who marry from age 10 through 44 and have complete fertility histories.)

(1) Great Britain and Ireland; (2) Canada and Northern Europe; (3) Other United States; (4) Trail States; (5) Territory outside of Utah; (6) Utah.

*Less than 100 cases.

migrants to the Utah frontier were marked by unusually high fertility relative to native Americans.

The high fertility levels of this population as well as the practice of polygyny produced large sibships for analysis. (See Burt *et al.*, 1985 for a study involving one kindred with over 5,000 members spanning six generations.) The fertility patterns have also produced a significant number of families with two or more same sex siblings for genetic epidemiological research where sex related factors are significant. Support for that argument is provided in two ways; through an examination of family composition, and by reference to a study of early coronary heart disease.

In Table 15-3 we present data from a recent tabulation. This table shows data for selected birth cohorts of 1800-09 through 1860-69 where recording of death dates is most complete. (The addition of death certificates to the UPDB file will extend the coverage of the data.) In the top panel of the table data are presented for women who married only once and who had one or more children. The mean number of children ever born ranges from 6.8 to 8.0. Over two-thirds of all of these families had two or more brothers plus two or more sisters. As a control for family size, the bottom panel provides data for women whose marital experience was disrupted by the death of a spouse followed by a remarriage. Here the mean CEB ranges from 3.5 to 4.4, and the number of families with two or more brothers plus two or more sisters ranges from 29.9% to 41.9%.

In addition to the large numbers of families with two or more same sex siblings, the number of twins available for study, in part associated with large family size and late childbearing, is substantial. Among the 1830-99 birth cohorts of women there were 6,457 pairs of twins of whom 2,344 were dizygotic. The correspondence in mortality patterns among twins is reflected in the fact that for the 1840-59 cohort of twins, both members of the set died under the age of one in 26.5% of the cases, and 25.0% among the 1860-79 cohort. The

Table 15-3 Mean number of children ever born (CEB) and percentage distribution of sets of male (M) and female (F) children in each family (or sibship) by mother's birth cohort

Birth Cohort	Mean CEB	1 M 0 F	0 M 1 F	1 M 1 F	>1 M 0 F	0 M >1 F	>1 M 1 F	1 M >1 F	>1 M >1 F
				Mother Has One Marriage With Children					
1800-09	7.08	2.3	1.9	2.6	3.0	2.5	7.6	7.9	72.2
1820-29	6.86	2.7	2.3	2.9	3.8	2.7	9.3	7.9	68.4
1840-49	8.05	1.3	1.3	1.5	2.4	2.0	7.5	5.9	78.1
1860-69	7.41	1.2	1.4	1.9	2.9	2.7	8.0	7.1	74.7
				Mother Has More Than One Marriage With Children[a]					
1800-09	3.51	9.3	9.3	7.0	9.3	9.3	14.0	11.6	30.2
1820-29	3.76	9.0	10.3	9.9	7.1	5.8	11.5	15.4	31.1
1840-49	4.45	8.1	6.7	8.1	8.7	5.2	11.7	9.6	41.9
1860-69	4.14	7.9	5.2	7.7	9.0	8.5	13.2	13.7	34.8

[a] Each marriage entered separately.

correspondence between non-twin brother-brother and sister-brother pairs is substantially different. Among the 1860-79 cohort 5.6% of the brother-brother pairs died under age 10 and 4.7% of the sister-brother pairs died under age 10. Mortality correspondence among relatives, becomes more significant when related to cause of death.

Williams *et al.* (1979) analyzed premature deaths (after age 35 and before age 55) from coronary heart disease (CHD) among men. For the analysis they employed the UPDB's linked death certificates (1956-75) with cause of death confirmed through an analysis of a sample of death certificates matched to hospital discharge records. When one or more males died from early CHD, familial correlations for age at death are 0.54 for brothers, 0.33 for sisters, and 0.13 for father-son. To confirm the significance of these association, the researchers selected from the UPDB a set of matched controls where deaths between the ages of 35 and 50 were not associated with early CHD. For the controls, the correlations were 0.16 for brothers, 0.11 for sisters, and 0.06 for father-son.

DEMOGRAPHIC VARIABLES IN GENETIC EPIDEMIOLOGY

The justification for the creation of the UPDB was the simultaneous availability of the Utah Cancer Registry and the set of family records, genealogies, which could be linked to investigate familial clustering of cancers. Cancer studies of the Utah population have been extensive, in part stimulated by the low incidence of cancer among the population. Lyon, Gardner, and West (1980) note that the Utah cancer mortality rate was the lowest in the nation. Comparison with the Third National Cancer Survey (TNCS, 1975) finds that the standardized incidence rate for cancer in Utah was 73% of the national incidence rate for all cancers. The low level of cancer in the Utah population reflects the fact that approximately 75% of the population was classified as members of the LDS in the mid-1970s. This low incidence rate of cancer appears to reflect life-style differences, because the LDS, for religious reasons, eschew the use of alcohol, tobacco, coffee, tea and other stimulates. The effect of differences in life-style within Utah is confirmed through analysis of Mormon and non-Mormon cancer rates (Lyon *et al.*, 1976) and cardiovascular mortality (Lyon *et al.*, 1978).

Linking data from the Cancer Registry to the genealogy yields evidence of familial clustering of specific forms of cancer. In these studies kinship relations have been defined by the ''Genealogical index'' which calculates the Malécot coefficients of kinship. Matched controls are selected from the genealogy for each individual with cancer. (For methodological details, see Skolnick *et al.*, 1981.) Skolnick *et al.*, 1981 report:

> The mean kinships of prostate, breast, and colon cancers (ranking first, second, and third, respectively) lie entirely outside their control distributions, suggesting strongly that the observed genetic relationships among these cancer sets did not occur by chance alone.

Prostate cancer, a form of cancer with an incidence rate approximately 23% higher than the national rate, exhibited the highest excess kinship for all major sites studied. The researchers responsible for these studies have, however, been cautious about attributing clustering to genetic factors alone, recognizing that various environmental hypothesis might also account for the observed relationship.

Further studies of familial clustering have proceeded along several lines of investigation (Skolnick *et al.*, 1981). In several of these, the same variables examined in the purely demographic analysis of the genealogies are employed in the further investigation of cancers which appear to cluster within families. In the investigation of prostate cancer described above comparing cancer and control cases, data on marital status, number of children, and age of first marriage are also analyzed. Based upon previous studies, calculations of prostate cancer risk factors reflecting analysis of the demographic characteristics of the population are contradictory. The profamily and pronatal values of the LDS church lead to nearly universal marriage, relatively early age at marriage, and avoidance of ''child-free'' marriages. Analyzing UPDB data, Cannon *et al.* (1982) reported only a slight increase in risk of prostate cancer associated with being ever-married, having offspring, and marrying at a later age. Given the minimal association between these behavioral measures and prostate cancer, research seems appropriately directed towards other, perhaps genetic, causal factors.

Perhaps one of the more significant demographic findings arising from the study of the UPDB population is the evidence of birth spacing as a rational strategy to achieve limited family sizes. The analysis of the UPDB population has demonstrated that even during periods of peak fertility, the population is marked by subcohorts of women who limit family size through the adoption of relatively long birth intervals (Anderton and Bean, 1985). The transition to controlled fertility therefore involves two processes. First, termination of childbearing occurs at increasingly earlier ages (truncation), after a presumably desired number of children are born. Second, during the initial stages of the transition, there is an increasing proportion of women who adopt longer birth intervals.

Evidence on birth spacing suggests that the transition from high to moderate fertility involved multiple strategies, including longer birth intervals across the reproductive years of life in order to achieve a smaller family size. Although this finding is at sharp variance with other historical demographic research (Tolnay and Guest, 1984; Knodel, 1987), attention to this phenomenon is now reflected in the appearance of other studies which document similar phenomena among other populations (David and Sanderson, 1987.)

Regardless of the purely demographic interest in birth spacing, the fact that birth spacing can be well documented among the UPDB population has proved important for other studies. Jorde and Lathrop (1988) have, for example, studied one hypothesis, heterozygote advantage, as an explanation for the high frequency of cystic fibrosis (CF) in Caucasian populations. The study population consisted of 143 grandparents of CF cases taken from the Utah CF registry. From UPDB a control population was selected, consisting of well-match replicate samples. The availability of birth spacing data made it possible to investigate the argument that an elevated gene frequency could be due to shorter birth intervals among the grandparents of carriers. However, no significant differences were found in birth intervals among grandparents and controls, providing no support for the heterozygote advantage hypothesis (Jorde and Lathrop, 1988; Jorde, 1989b).

The ability to document the timing of each birth across the entire life cycle among a sequence of cohorts also means that for this population similar variables are available for epidemiological studies. In the investigation of breast cancer, Hunt *et al.* (1980) studied 236 breast cancer patients drawn from the Cancer Registry and 937 controls drawn from the genealogy matched on year of birth. The study focused on the relationship between reproductive variables (age at first delivery, age at last birth, and number of children) and breast cancer. In this case, consistent with other research, demographic variables reflecting fertility behavior appeared to be strongly associated with breast cancer. The relative odds were elevated to 4.1 where the first birth was delayed and the last birth was before age 35. The findings while significant seem insufficient to account for the degree of familial clustering identified in other studies using UPDB.

CONCLUSIONS

The body of research which has resulted from the creation and continuing development of the UPDB has been substantial, contributing both interesting and significant findings in several fields. The UPDB project was only possible through the extensive collaboration of researchers from a variety of disciplines. The UPDB project is perhaps one of the few where an overt attempt was made to link genetic and demographic studies from the outset. Both groups benefited from the work of the other. There was some collaboration but more often, independent but complementary work. Clearly all researchers using UPDB share common concerns with the coverage and accuracy of the database as well as the software programs which provide easy access to a large file and the subroutines for linkage within and across generation. As we have demonstrated, demographers and geneticists have faced common problems, developing a fuller understanding of the representativeness of the population, but with each group pursuing separate lines of investigation that yield meaningful results to those specializing in one area of research. It should also be recognized, however, that researchers from each field share an interest in similar aspects of the data. The fertility levels and changes are of immediate interest to the demographer, but serve as the bases for large sibships which are analyzed in different forms by the geneticists associated with the project. Variables which clarify the underlying mechanisms involved in the fertility transition among this population—number of children, age at first and last birth, and birth intervals—are subsequently used as independent variables in epidemiological studies.

As the database is developed further, with the input and linkage of the balance of the twentieth-century death certificates and all of the twentieth-century birth certificates, the opportunity will arise for continuing mutually supporting research projects. The linkage between demographers and geneticists will continue because, as 15 years of experience has demonstrated, a large, complex data set may provide great theoretical and empirical advantages to both fields.

ACKNOWLEDGMENTS

Investigators associated with the development of UPDB gratefully acknowledge support provided under the following NIH grants: NCI/CA16573 and 28854 (Skolnick); NICHHD/HD 10267 and 15455 (Bean); NICHHD/HD 16109 (Jorde); NHLBI/HL 21088, 24055, and 21088 (Williams).

REFERENCES

Anderton, D. L., 1986   Intermarriage of frontier immigrant, religious and residential groups: an examination of a macro-structural assimilation hypothesis. Sociological Inquiry 56: 341-366.

Anderton, D. L., and L. L. Bean, 1985   Birth spacing and fertility limitation: a behavioral analysis of a nineteenth century frontier population. Demography 22: 169-183.

Bardet, J. P., K. A. Lynch, G. P. Mineau, M. Hainsworth, and M. Skolnick, 1981   La mortalité maternelle autrefois: Une étude comparée. Annales de Demographie Historique, 31 -49.

Bean, L. L., D. L. May, and M. Skolnick, 1978   The Mormon historical demography project. Historical Methods 11: 45-53.

Bean, L. L., G. P. Mineau, and D. L. Anderton, 1983   Residence and religious effects on declining family size: an historical analysis of the Utah population. Review of Religious Research 25: 91-101.

Bean, L. L., and G. P. Mineau, 1986   The polygyny-fertility hypothesis: a re-evaluation. Pop. Studies 40: 67-81.

Bean, L. L., G. P. Mineau, and D. L. Anderton, 1990   *Fertility Change on the American Frontier: Adaptation and Innovation.* University of California Press, Los Angeles.

Bean, L. L., G. P. Mineau, K. A. Lynch, and J. D. Willigan, 1980   The Genealogical Society of Utah as a data resource for historical demography. Pop. Index 46: 6-19.

Bishop, D. T., and M. Skolnick, 1984   Genetic epidemiology of cancer in Utah genealogies: A prelude to the molecular genetics of common cancers. J. Cell. Physiol. Supplement 3: 63-77.

Burt, R. W., D. T. Bishop, L. A. Cannon, M. A. Dowdle, R. G. Lee, and M. Skolnick, 1985   Dominant inheritance of adenomatous polyps and colorectal cancer. N. Eng. J. Med. 312: 1540-1544.

Cannon, L., D. T. Bishop, M. Skolnick, S. Hunt, J. L. Lyon, and C. R. Smart, 1982   Genetic epidemiology of prostate cancer in the Utah Mormon genealogy. Cancer Surv. 1: 47-69.

David, P. A., and W. C. Sanderson, 1987   The emergence of a two-child norm among American birth-controllers. Pop. Dev. Rev. 13: 1-41.

Fineman, R. M., and L. B. Jorde, 1980   Establishment of a birth defects registry with the aid of a genealogical data base: Utah Registry of Birth Defects and Genetic Disease, pp. 319-331 in *Cancer Incidence in Defined Populations*, edited by J. Cairns, J. L. Lyon, and M. Skolnick. Cold Spring Harbor Laboratory, Cold Spring Harbor, New York.

Hunt, S. C., R. R. Williams, M. H. Skolnick, J. L. Lyon, and C. R. Smart, 1980   Breast cancer and reproductive history from genealogical data. J. Nat. Cancer Inst. 64: 1047-1053.

Jorde, L. B., 1989a   Inbreeding in the Utah Mormons: An evaluation of estimates based on pedigrees, isonymy, and migration matrices. Ann. Hum. Genet. 53: 339-355.

Jorde, L. B., 1989b   Fertility and Cystic Fibrosis. Nature 340: 352.

Jorde, L. B., and G. M. Lathrop, 1988   A test of the heterozygote-advantage hypothesis in Cystic Fibrosis carriers. Am. J. Hum. Genet. 42: 808-815.

Jorde, L. B., and M. H. Skolnick, 1983   Demographic and genetic applications of computerized record linking: The Utah Mormon genealogy. Informatique et Sciences Humaines 56-57: 105-17.

Knodel, J., 1987   Starting, stopping, and spacing during the early stages of fertility transition: the experience of German village populations in the 18th and 19th centuries. Demography 24: 143-162.

Lynch, K. A., G. P. Mineau, and D. L. Anderton, 1985   Estimates of infant mortality on the western frontier: the use of genealogical data. Historical Methods 18: 155-164.

Lyon, J. L., J. W. Gardner, M. R. Klauber, and C. R. Smart, 1977   Low cancer incidence and mortality in Utah. Cancer 29: 2608.

Lyon, J. L., J. W. Gardner, and D. W. West, 1980   *Cancer Among Mormons from 1967-1975*. Banbury Report 4: Cancer Incidence in Defined Populations, pp. 3-30.

Lyon, J. L., M. R. Klauber, J. W. Gardner, and C. R. Smart, 1976   Cancer incidence in Mormons and non-Mormons in Utah, 1966-1970. N. Eng. J. Med. 294: 129-133.

Lyon, J. L., H. P. Wetzler, J. W. Gardner, M. R. Klauber, and R. R. Willimas, 1978   Cardiovascular mortality in Mormons and non-Mormons in Utah, 1969-71. Am. J. Epidemiol. 108: 357-366.

Malécot, G., 1984   *Les Mathématiques de l'Hérédité*. Masson et Cie, Paris.

McLellan, T., L. B. Jorde, and M. H. Skolnick, 1984   Genetic distances between the Utah Mormons and related populations. Am. J. Hum. Genet. 36: 836-857.

Mineau, G. P., D. L. Anderton, and L. L. Bean, 1989   Description and evaluation of linkage of the 1880 manuscript census to family genealogies with implications for Utah fertility research. Historical Methods 22: 144-157.

Mineau, G. P., 1988   Utah widowhood: a demographic profile, pp. 140-163. in *On Their Own: Widows and Widowhood in the American Southwest 1848-1939*, edited by A. Scandron. University of Illinois Press, Chicago.

Skolnick, M., 1980   The Utah genealogical data base: a resource for genetic epidemiology. Cancer Incidence in Defined Populations, Cold Spring Harbor Laboratory, Banbury Report 4: 285-297.

Skolnick, M., L. Bean, D. May, V. Arbon, K. De Never, and P. Cartwright, 1978 Mormon demographic history I. Nuptiality and fertility of once-married couples. Pop. Studies 32: 5-19.

Skolnick, M., L. L. Bean, S. M. Dintelman, and G. P. Mineau, 1979   A computerized family history data base system. Sociol. Soc. Res. 63: 506-523.

Skolnick, M., D. T. Bishop, D. Carmelli, R. Gardner, R. Hadly, S. Hasstel, J. R. Hill, S. Hunt, J. L. Lyon, C. R. Smart, and R. R. Williams, 1981.   A population-based assessment of familial cancer risk in Utah Mormon genealogies, pp. 447-500 in *Genes, Chromosomes, and Neoplasia*, edited by F. E. Arrighi, P. N. Rao, and E. Stubblefield. Raven Press, New York.

Tolnay, S. E., and A. M. Guest, 1984   American family building strategies in 1900: stopping or spacing? Demography 21: 9-18.

Williams, R. R., M. Skolnick, D. Carmelli, A. T. Maness, S. C. Hunt, S. Hasstedt, G. E. Reiber, and R. K. Jones, 1979   Utah pedigree studies: design and preliminary data for premature male CHD deaths, pp. 711-729 in *Genetic Analysis of Common Diseases: Applications to Predictive Factors in Coronary Disease*, edited by C. F. Sing and M. Skolnick. Alan R. Liss, Inc., New York.

Williams, R. R., M. M. Dadone, S. C. Hunt, P. N. Hopkins, J. B. Smith, K. O. Ash, and H. Kuida, 1984   The genetic epidemiology of hypertension: a review of past studies and current results of 948 persons in 48 Utah pedigrees, pp. 419-442 in *Genetic Epidemiology of Coronary Heart Disease: Past, Present, and Future*, edited by D. C. Rao, R. C. Elston, and L. H. Kuller. Alan R. Liss, New York.

16

Genetic Interpretation of Disease Clustering

JURG OTT

Epidemiology focuses on diseases of environmental origin and tries to establish relationships between, for example, disease frequency and occurrence of certain environmental agents. Traditionally, public health policies have assumed that all members of a population are equally susceptible to disease (Harper, 1988). General public health measures such as improved sanitation and diet as well as immunological advances have had a major impact in reducing the incidence of many diseases. Some well-known examples are the elimination of smallpox, control of poliomyelitis, and reduction of tooth caries when fluoride is present (either naturally or given as an additive) in the diet of children.

Knowledge of the classical (Mendelian) genetic principles has led to an understanding of the so-called inborn errors of metabolism (Vogel and Motulsky, 1986), that is, metabolic deficiencies causing disease which are inherited and not (or not primarily) due to environmental agents. The prime example is phenylketonuria (PKU), a disease due to deficiency of the enzyme phenylalanine hydroxylase, which is coded by a gene on chromosome 12. The deficiency causes abnormally high levels of the amino acid phenylalanine in the blood,which, in turn, leads to mental retardation. Once this mechanism was elucidated, a cure for the disease became possible in that a strict diet, free of phenylalanine, completely prevents the disease.

Many abnormal reactions to drugs, for example, suxamethonium sensitivity that can lead to anesthetic accidents (Motulsky, 1970), could also be traced to such inborn errors of metabolism. Research in this area established a branch of genetics called pharmacogenetics which may be seen as constituting the foundation for some of the present successes in human genetics.

A consequence of the decline in frequency of environmentally caused diseases has been an increase in life expectancy, which has resulted in an increase in incidence of diseases with a genetic basis, and of chronic and degenerative diseases that tend to be expressed only later in life. The term ''genetic epidemiology'' (Morton, 1982) has been coined to refer to a branch of science combining prin-

245

ciples of epidemiology and genetics, but there are no sharp distinctions between epidemiology, genetics, and genetic epidemiology. For example, the disease PKU referred to above manifests itself only when phenylalanine is part of the diet, which generally is the case. A recent example falling under the heading of genetic epidemiology is familial Alzheimer's disease, which until recently was seen as a normal "disease of old age." By virtue of genetic linkage to marker loci on chromosome 21, Alzheimer's disease appears to be caused by a single gene, at least in some families (St. George-Hyslop *et al.*, 1987). We may expect that many more diseases will be shown to be caused, or at least be influenced to a varying degree by single genes.

There are several areas in which demography and genetic epidemiology reinforce each other (Cavalli-Sforza and Bodmer, 1977). For one thing, census statistics and other government statistics have frequently provided basic information on disease incidence and prevalence pointing out a need for intervention and measuring success of public health measures. Also, clines (gradients over a large geographical area) in allele frequency can often be correlated with migration and selective pressures in earlier times (Piazza, Menozzi, and Cavalli-Sforza, 1981), or with disease history for genes thought to confer disease resistance. Racial admixture may be determined on the basis of allele frequencies when these are different in the initial populations (Cavalli-Sforza and Bodmer, 1977).

GENETIC MODELS

Vital statistics for diseases are not usually family-based but most often measure disease occurrence on an individual basis only. However, many diseases "run in families," and familial aggregation is one of the first and most direct indicators of a genetic involvement. Once familial aggregation is observed, for example, by way of a significantly increased disease frequency in the siblings of affected individuals as compared with the population prevalence, more refined analysis methods are needed to investigate the nature of the genetic involvement. Some broad classes of genetic models may be characterized as follows.

Polygenic (multifactorial) model

Under the polygenic model, the phenotype (what is observed on an individual) is the result of the concerted action of (infinitely) many genes, each with a small contribution to the phenotype. In addition to these genetic factors, one usually has to allow for other factors as well (see mathematical model, below), at least for a general environmental modifier which generalizes the pure polygenic model to a so-called multifactorial model. For example, body height is generally considered to be a multifactorial trait (Falconer, 1964). A phenotype under polygenic control is, by assumption, normally distributed. In practice, phenotypes often have to be transformed to approximate normality prior to analysis under a multifactorial model.

If the phenotype is qualitative rather than quantitative, an underlying (unobservable) quantity, the so-called *liability*, is postulated and assumed to follow a polygenic mode of inheritance (Curnow and Smith, 1975). If the liability exceeds a certain threshold in a given individual, that individual is affected; that individual is unaffected when the liability stays below the threshold. Multiple thresholds define multiple phenotype classes, and sex-specific disease prevalence may be explained by different thresholds in men and women. An example of such a threshold trait is pyloric stenosis (Vogel and Motulsky, 1986), a disease of the newborn caused by a thickening of the pyloric muscle, preventing release of stomach contents into the duodenum.

The classical mathematical model for a quantitative multifactorial trait may be written as $x_i = \mu + g_i + e_i$, where x_i is the phenotype of the ith individual in a family of size n, μ is an overall mean, g_i stands for the contribution of polygenes, and e_i represents a random environmental deviation. g_i and e_i are assumed to follow the respective normal distributions, $N(0, \sigma_g^2)$ and $N(0, \sigma_e^2)$. The model may be extended by additive terms representing, for example, common family environment (Ott, 1979). The components of x_i are taken to be independent so that the variance, σ_x^2, of x_i is simply the sum of the variances of the components.

The genetic variance, σ_g^2, may be broken into two terms. The so-called additive genetic variance is due to the effects of the single genes if one assumes that they act additively, that is, that a given allele at a gene locus contributes a fixed amount to the phenotype, whether it is present in single or double dose (Crow and Kimura, 1970). The difference between total genetic variance, σ_g^2, and additive genetic variance is called the dominance variance, as it is due to dominance of one allele over another, that is, an allele contributes a different amount to the phenotype depending on which other allele at that locus is present in the genotype. The proportion, σ_g^2/σ_x^2, of the phenotypic variance, σ_x^2, is termed *heritability in the broad sense*, and the proportion of the additive genetic variance is the *heritability in the narrow sense*. In the absence of shared environmental effects, the correlations among pairs of relatives may be shown to be simple functions of these two types of heritabilities (Crow and Kimura, 1970).

As a consequence of the assumptions in the multifactorial model, the phenotypes for the members of a family pedigree jointly follow a multivariate normal distribution, so that the likelihood is simply the multivariate normal density. For qualitative traits, however, the likelihood is the multiple integral over the relevant portions of the multivariate density and is generally obtained by numerical methods on computers.

One of the consequences of the classical multifactorial model is that the variance of a trait among siblings should be the same irrespective of the parental phenotypes or the sibship mean, which may be tested by standard procedures (Williamson *et al.*, 1979). A significant outcome of this test may be interpreted as evidence against the multifactorial model in favor of single-gene models. However, an alternative polygenic model with an infinite number of genes with a graduated effect has been investigated (Matthysse, Lange, and Wagener, 1979), under which the sib variance depends on the parental phenotypes and the sibship mean. Therefore, deviations from the classical polygenic model do not necessarily imply presence of a single gene.

While the multifactorial model has been fruitful in establishing a genetic basis for many traits, the joint action of a large number of loci must be difficult to disentangle. Therefore, more detailed investigations have been possible for traits governed by a small number of genes only.

Single gene models

The blood cholesterol level in normal individuals follows a more or less normal distribution. In some individuals, however, the cholesterol level is generally higher, so that a genetic disease, familial hypercholesterolemia, has been postulated to be determined by a single gene whose presence raises the mean cholesterol level in the blood (Schrott *et al.*, 1972). The resulting mixture of normal distributions may lead to a bimodality of cholesterol levels and can be analyzed at the population level (Ott, 1979) as well in family data (Ott *et al.*, 1974). Pioneering work in this area eventually elucidated the mechanism causing hypercholesterolemia, which won the Nobel prize for Drs. Brown and Goldstein. The gene for familial hypercholesterolemia is now known to reside on chromosome 19. It affects lipoprotein receptor function, resulting in an elevation of cholesterol leading to increased risk of coronary heart disease already at a young age (Vogel and Motulsky, 1986).

Traits following single-gene inheritance may be quantitative, such as the example given above, or qualitative, for example, the ABO or Rh blood groups. In the latter case, the model is easiest represented mathematically by the conditional probability, $P(x_i|g_i)$, that a given genotype, g_i, produces a certain phenotype, x_i, in the ith individual of a family pedigree. $P(x_i|g_i)$ is the so-called penetrance of a phenotype given a certain genotype. In addition to this broad definition of penetrance, some researchers use penetrance in a narrow sense, defining it to be the probability of being affected given a genotype at some disease locus. For a quantitative trait, x, $P(x|g)$ is no longer a probability but represents the density whose mean and variance depend on the given genotype, g. For qualitative and quantitative phenotypes, the likelihood under a single gene model may be written as

$$P(x) = \Sigma P(x|g)P(g) \, ,$$

where x is the vector of phenotypes of the family members, g is a particular vector of genotypes, and the sum is taken over all possible assignments of genotypes to inviduals in a family pedigree. $P(g)$ may be easily evaluated using well-known principles of Mendelian inheritance, but the likelihood, $P(x)$, is a multiple sum containing a large number of terms and generally has to be evaluated numerically for parameter estimation (Elston and Stewart, 1971; Ott, 1974, 1977).

In addition to many well-known blood groups, another example of a qualitative trait following a monogenic mode of inheritance is Huntington's disease. It is not expressed at birth but only later in life (age-dependent penetrance). The gene causing the disease was shown to be genetically linked to a gene locus on chromosome 4 (Gusella *et al.*, 1983). Analysis of genetic linkage (Ott, 1985) has become one of the focal points of single-gene inheritance (see below) in that construction of a complete genetic map of man promises to offer immense research possibilities. Problems of reduced penetrance will be discussed in more detail below, at the end of the section comparing the models.

Table 16-1 Oligogenic models for vitiligo (Majumder, Das and Li, 1988); k = number of disease gene loci, q = population frequency of the disease allele at each locus, s = segregation probability, $ln(L) = log_e$ likelihood plus a constant

k	q	s	$ln(L)$
1	0.068	0.250	-
2	0.260	0.099	32.0
3	0.408	0.055	41.4
4	0.510	0.037	42.4
5	0.584	0.027	41.7

Oligogenic models

Oligogenic models represent a fruitful compromise between polygenic and monogenic modes of inheritance. The assumption is that a rather small number of genes interact to produce a certain phenotype. The interaction among these genes may be additive or multiplicative or a combination of the two, interaction usually being modeled by assuming a particular combination of alleles at the different gene loci conferring susceptibility to disease. For example, two-locus models have been proposed for diabetes (Hodge *et al.*, 1983), and a graphic method for testing two-locus models has been applied to several other diseases (Greenberg, 1981).

A particular class of oligogenic models, multiple homozygosis models, has recently been introduced (Li, 1987). One assumes k unlinked loci, each with two alleles (A,a; B,b; C,c; etc.) with gene frequency q_i at the ith locus. An individual is taken to be affected only when homozygous (a/a, b/b, c/c, etc.) at each locus. Therefore, out of the $3k$ possible genotypes, only one leads to disease. The salient feature of multiple homozygosis models is that even though the gene frequency at each locus may be high, the segregation probability, that is, the probability that an offspring is affected, is generally small. Consequently, despite a strong genetic background, affecteds are mostly singletons, and this is one of the characteristics of, for example, schizophrenia.

So far, multiple homozygosis models have been applied to only very few diseases, for example, to vitiligo families in India (Majumder, Das, and Li, 1988). Vitiligo is a dermatological disorder that tends to become progressive over time. Affected individuals have on their skin pale, milk-white macules. Disease prevalence is 1% in the United States. A likelihood analysis was applied to 298 families in India (Majumder, Das and Li, 1988) where, for simplicity, the allele frequency was taken to be the same at each of the k disease loci. As Table 16-1 shows, the maximum likelihood estimate obtained for the number of loci involved is 4.

COMPARISONS AMONG MODELS

The usual chi-square tests of goodness of fit cannot generally be carried out for family pedigree data as individuals are non-independent. Instead, one relies on one of the following methods that belong in the so-called area of segregation analysis.

One of the simplest ways of comparing whether the data fit one model better than another is to compare the maximum likelihoods achieved under each model. This route has been successfully taken in many investigations. However, it tends to be susceptible to a number of problems. For example, if all candidate models are inadequate, even the best fitting model does not allow for a good fit of the data which is, however, not detected by simply comparing likelihoods achieved under various models. Also, with complex traits, one often has to estimate more than just a few parameters so that the models to be compared are almost saturated and, consequently, the data fit all models almost equally well.

A proper test situation was created by the introduction of the so-called mixed model (Morton, 1982) that allows for a single (so-called major) gene acting on a polygenic and environmental background in producing a quantitative phenotype. This model affords a likelihood ratio test for the presence of a single gene, and variants of this type of approach are often used. The original implementation of the mixed model (Morton and MacLean, 1974) could only be applied to nuclear families (parents and their children) as it involves heavy computation, but newer approximate approaches have alleviated this situation (Hasstedt, 1982; Bonney, 1986; Bonney, Lathrop, and Lalouel, 1988).

The methods discussed so far for finding the appropriate mode of inheritance for the phenotypes in family data all have their merits. However, their results seem to be quite model dependent. For example, a model may postulate a mixture of distributions for the cholesterol levels of unrelated individuals where the component distributions for normal and susceptible genotypes are each assumed to be normal. Skewness is then interpreted as evidence for admixture and presence of at least two genotypes with different mean cholesterol levels. However, this argument is fallacious if the underlying distributions themselves are skewed. Various techniques have been used to transform the data prior to analysis, each presumably leading to somewhat different results. The major drawback of such analyses seems to be that their results are not directly testable, so that they are usually used as preliminary investigations. Also, a significant finding that a single gene is involved in the etiology of a phenotype does not say anything about that gene. Furthermore, there are several pitfalls, such as ascertainment biases (Greenberg, 1986), that may make a single gene appear operating so that results of segregation analyses have to be taken with a grain of salt.

In the last ten years, a method has gained momentum that is extremely powerful at showing involvement of a single gene. The idea is the following. As is well-known, at each (autosomal) locus, each individual has two alleles, one obtained from the mother and the other obtained from the father. When an individual produces a gamete (a sperm or egg cell), at each locus, either the maternal or the paternal allele goes into the gamete with probability $\frac{1}{2}$ each. Looking at two loci jointly, assume that at one of the loci a gamete receives the maternal allele. Generally, again, there is an equal chance for the two alleles at the other locus to occur in the gamete. This situation is termed "free recombination" between the two loci, and the so-called recombination rate or recombination fraction is equal to 50%. However, when two gene loci reside on the same chromosome in close proximity to each other, their inheritance in a family pedigree shows genetic linkage (Ott, 1985). For example, when the maternal allele at a locus is passed on to a gamete, the probability that the maternal allele at a locus nearby will also occur in the same gamete is much higher than 50%, that is, the recombination fraction between neighboring loci is less than 50% and tends to vanish for loci very close to each other.

For disease loci with an unclear mode of inheritance, a reverse argument is used in the following way. One pretends that the disease is caused or at least influenced by the action of a single gene and estimates the recombination fraction between the putative disease locus and a genetic marker locus. If it is possible to unequivocally establish genetic linkage between the two loci, this confirms the presence of a disease gene.

Advances in molecular genetics have made it possible to establish a great number of genetic marker loci on each chromosome (Botstein *et al.*, 1980), and work is in progress leading to a complete map of the human genome at a cost of roughly $660 million over the next five years (U.S. Congress, 1988). These gene markers can now be used as "sign posts" for mapping genes with unknown location. Once a rough map location of a disease gene is known, one hopes to find the gene by molecular genetics techniques and will then be able to elucidate its function in order to understand the mechanism leading to the disease.

Of the many technical and statistical problems involved in human linkage analysis, some are particularly relevant for complex diseases, that is, diseases with uncertain mode of inheritance and unclear distinction between affecteds and unaffecteds. For example, in mental disorders, a gradient of disease definition is sometimes used ranging from strict to intermediate to loose disease definitions. It is common practice to carry out a linkage analysis under each of these disease definitions and to favor the one yielding most evidence for linkage. Other parameters of the genetic model are sometimes also varied in this sense. Such procedures clearly must lead to an increase in the type I error (false positive results), but their effect has not been investigated.

With full penetrance, even in the presence of dominance, the phenotypes of family members often allow an unequivocal determination of their genotypes so that counting recombinants and nonrecombinants yields an efficient estimate of the recombination fraction. When penetrance is reduced, some phenotypes no longer unequivocally exhibit genotypes, which results in a loss of information as measured by the precision of the estimate of the recombination fraction. For simple family types, this loss may be calculated. For example, when full penetrance allows for a direct count of recombination events, a penetrance of 90% under a recombination fraction of 5% increases the variance of the estimate by a factor of 1.66, that is, the estimate has an efficiency of only $1/1.66 = 60\%$ (Ott, 1985, Table 7.2). Therefore, diseases with reduced penetrance require an increased number of observations for valid results of linkage analyses.

If penetrance is reduced for all individuals with susceptible genotypes, its estimation is relatively straightforward. For example, when one of the parents carries a dominant disease gene, one expects that half of the children will be affected. With reduced penetrance, $f < 1$, the proportion of affected will be $f/2$ which may be estimated and tested against the value $\frac{1}{2}$. In many diseases, however, penetrance is strongly age dependent. For example, Huntington's disease usually begins between the ages of 35 and 45. Corresponding "age-of-onset" curves have often beenconstructed as the cumulative distribution of the age at onset of a sample of affected individuals. More sophisticated methods took the pedigree structure and affecteds as well as unaffecteds into account (Elston, 1973). Further refinements also considered the age distribution of the population at large (Crowe and Smouse, 1977; Heimbuch, Matthysse, and Kidd, 1980). Recently, an iterative method of estimating age-of-onset parameters from pedigree data was proposed and applied (Easton *et al.*, 1989).

While good estimates of the age-dependent penetrance seem important in calculating a genetic risk, that is, the probability that an unaffected individual of a certain age is a carrier of the disease gene, practice shows that the estimate of the recombination fraction does not depend too much on the actual form of the age-of-onset curve as long as there is some gradual increase of penetrance with age. Many investigators simply use a straight line for the penetrance as a function of age, starting with zero at a certain young age and rising to its upper limit (perhaps 100%) at some older age, for example, in a linkage study of panic disorder (Crowe *et al.*, 1987).

In the discussion of penetrance, above, the tacit assumption was that individuals not carrying the disease gene have penetrance zero which precludes the occurrence of so-called phenocopies, in human genetics also called false positives or sporadic (nongenetic) cases. For some diseases, such an assumption is unrealistic. For example, in a recent study of osteoarthrosis in Finland (Palotie *et al.*, 1989), a familial form seemed to be segregating in a Mendelian fashion in a large pedigree, but a clear distinction to the relatively common osteoarthrosis of old age could not be made. For linkage analysis purposes, two straight lines were defined as age-of-onset curves, one rising from 0 at age 0 to 1 at age 60 for genotypes involving the disease gene (assumed dominant), and one rising from 0 at age 40 to 0.4 at age 100 (the penetrances at the upper age limits were estimated by maximum likelihood). This parametrization proved to be successful in the linkage analysis with the type II collagen gene on chromosome 12 which was a candidate gene for the disease owing to its leading role for the integrity of the cartilage of the joints.

In most linkage analyses of diseases with age-dependent penetrance, the "age-of-onset curve" used is the distribution function, $F(a)$, of the age at onset, and the age entered in the analysis is present age or age last seen, a'. As pointed out by Elston (1973), for individuals whose age at onset, a, is known, the density function, $f(a)$, rather than $F(a')$ shouldbe used in the likelihood. The practice of using $F(a')$ for all individuals, whether or not their age at onset is known, is often unproblematic since in linkage analysis, one is not interested in the actual value of the likelihood, $L(\Theta)$, with respect to the recombination fraction but only in the ratio, $L(\Theta)/L(0.5)$, where 0.5 represents "free recombination" or no linkage. In simple dominant or recessive situations without phenocopies, the genotype of affecteds is known, irrespective of their age at onset, so that changing their penetrance is equivalent to multiplying the likelihood by a constant which cancels in the likelihood ratio. This argument no longer holds, however, when phenocopies are allowed for. For example, in the study on osteoarthrosis quoted above (Palotie *et al.*, 1989), age of onset was an important discriminator between genetic and nongenetic cases, and using $F(a')$ rather than $f(a)$ in the likelihood for individuals with known age at onset would have left valuable information unused.

IMPLICATIONS OF RECENT GENETIC ADVANCES

Genetic research has demonstrated a genetic basis for many diseases. Even for contagious diseases, susceptibility may vary depending on an individuals' genetic constitution. For example, an autosomal dominant poliovirus sensitivity gene is known to be located on human chromosome 19 (Miller *et al.*, 1974). Consequently, individuals in the population are generally at very different risks for contracting diseases, and this calls for modifications of public health policies (Harper, 1988) (see above). Risk factor control will be more effective when focused on individuals at risk so that perhaps general screening and counseling services will be needed.

Each new technology represents a challenge for society in that it has to adapt to new situations (U.S. Congress, 1988). One of these new situations will be the possibility of determining at birth the carrier status for many diseases. While it may generally be useful to recognize carrier status for a curable genetic disease, this is less clear for nontreatable diseases. Difficult legal questions will also have to be solved. For example, will insurance companies and employers have the right to subject applicants to genetic tests and base their decisions on the outcome of the test? The interests of society will have to be weighed against the interests of single individuals. Even today, with only a small number of known genetic differences between individuals, a gradient of intellectual and/or physical capabilities exists in the population, and society has implemented measures to try to counterbalance some of these inequalities, for example, by levying different percentages of taxes in different income brackets. Advances of molecular genetics will add a new dimension of differences among individuals. Hopefully, ways will be found for society to carry its weaker members and to cope successfully with the problems posed by the new genetics.

REFERENCES

Bonney, G. E., 1986   Regressive logistic models for familial disease and other binary traits. Biometrics 42: 611-625.

Bonney, G. E., G. M. Lathrop, and J.-M. Lalouel, 1988   Combined linkage and segregation analysis using regressive models. Am. J. Hum. Genet. 43: 29-37.

Botstein, D., R. L. White, M. Skolnick, and R. Davis, 1980   Construction of a genetic linkage map in man using restriction fragment length polymorphisms. Am. J. Hum. Genet. 32: 314-331.

Cavalli-Sforza, L. L., and W. F. Bodmer, 1977   *The Genetics of Human Populations*. Freeman, San Francisco, CA.

Crow, J. F., and M. Kimura, 1970   *An Introduction to Population Genetics Theory*. Harper & Row, New York.

Crowe, R. R., R. Noyes, A. F. Wilson, R. C. Elston, and L. J. Ward, 1987   A linkage study of panic disorder. Arch. Gen. Psychiatry 44: 933-937.

Crowe, R. R., and P.E. Smouse, 1977   The genetic implications of age-dependent penetrance in manic-depressive illness. J. Psychiat. Res. 13: 273-285.

Curnow, R. N., and C. Smith, 1975   Multifactorial models for familial diseases in man. J. R. Statist. Soc. 138A: 131-169.

Easton, D. F., M. A. Ponder, T. Cummings, R. F. Gagel, H. H. Hansen, S. Reichlin,

A. H. Tashjian, Jr., M. Telenius-Berg, and B. A. J. Ponder, 1989 The clinical and screening age-at-onset distribution for the MEN-2 syndrome. Am. J. Hum. Genet. 44: 208-215.

Elston, R. C., 1973 Ascertainment and age of onset in pedigree analysis. Hum. Hered. 23: 105-112.

Elston, R. C., and J. Stewart, 1971 A general model for the analysis of pedigree data. Hum. Hered. 21: 523-542.

Falconer, D. S., 1964 *Introduction to Quantitative Genetics.* Oliver and Boyd, Edinburgh and London.

Greenberg, D. A., 1981 A simple method for testing two-locus models of inheritance. Am. J. Hum. Genet. 33: 519-530.

Greenberg, D. A., 1986 The effect of proband designation on segregation analysis. Am. J. Hum. Genet. 39: 329-339.

Gusella, J. F., N. S. Wexler, P. M. Conneally, S. L. Naylor, M. A. Anderson, R. E. Tanzi, P. C. Watkins, K. Ottina, M. R. Wallace, A. Y. Sakaguchi, A. B. Young, I. Shoulson, E. Bonilla, and J. B. Martin, 1983 A polymorphic DNA marker genetically linked to Huntington's disease. Nature 306: 234-238.

Harper, A. E., 1988 Potential contributions of genetic epidemiology. Genet. Epidemiol. 5: 203-206.

Hasstedt, S. J., 1982 A mixed-model likelihood approximation on large pedigrees. Comp. Biomed. Res. 15: 295-307

Heimbuch, R. C., S. Matthysse, and K. K. Kidd, 1980 Estimating age-of-onset distributions for disorders with variable onset. Am. J. Hum. Genet. 32: 564-574.

Hodge, S. E., C. E. Anderson, K. Neiswanger, R. S. Sparkes, and D. L. Rimoin, 1983 The search for heterogeneity in insulin-dependent diabetes mellitus (IDDM): Linkage studies, two-locus models, and genetic heterogeneity. Am. J. Hum. Genet. 35: 1139-1155.

Li, C. C., 1987 A genetic model for emergenesis: In memory of Lawrence H. Snyder, 1901 -86. Am. J. Hum. Genet. 41: 517-523.

Majumder, P. P., S. K. Das, and C. C. Li, 1988 A genetic model for vitiligo. Am. J. Hum. Genet. 43: 119-125.

Matthysse, S., K. Lange, and D. K. Wagener, 1979 Continuous variation caused by genes with graduated effects. Proc. Natl. Acad. Sci. USA 76: 2862-2865.

Miller, D. A., O. J. Miller, V. G. Dev, S. Hashmi, R. Tantravahi, L. Medrano, and H. Green, 1974 Human chromosome 19 carries a poliovirus receptor gene. Cell 1: 167-173.

Morton, N. E., 1982 *Outline of Genetic Epidemiology.* Karger, New York).

Morton, N. E., and C. J. MacLean, 1974 Analysis of family resemblance. III. Complex segregation of quantitative traits. Am. J. Hum. Genet. 26: 489-503.

Motulsky, A. G., 1970 General remarks on genetic factors in anesthesia. Humangenetik 9: 246-248.

Ott, J., 1974 Estimation of the recombination fraction in human pedigrees: Efficient computation of the likelihood for human linkage studies. Am. J. Hum. Genet. 26: 588-597.

Ott, J., H. G. Schrott, J. L. Goldstein, W. R. Hazzard, F. H. Allen, Jr., C. T. Falk, and A. G. Motulsky, 1974 Linkage studies in a large kindred with familial hypercholesterolemia. Am. J. Hum. Genet. 26: 598-603.

Ott, J., 1977 Counting methods (EM algorithm) in human pedigree analysis: linkage and segregation analysis. Ann. Hum. Genet. 40: 443-454.

Ott, J., 1979 Maximum likelihood estimation by counting methods under polygenic and mixed models in human pedigrees. Am. J. Hum. Genet. 31: 161-175.

Ott, J., 1979 Detection of rare major genes in lipid levels. Hum. Genet. 51: 79-91.

Ott, J., 1985   *Analysis of Human Genetic Linkage*. Johns Hopkins University Press, Baltimore, MD.

Palotie, A., P. Väisänen, J. Ott, L. Ryhänen, K. Elima, M. Vikkula, K. Cheah, E. Vuorio, and L. Petlonen, 1989   Predisposition to familial osteoarthrosis is liked to type II collagen gene. Lancet 1(8644): 924-927.

Piazza, A., P. Menozzi, and L. L. Cavalli-Sforza, 1981   Synthetic gene frequency maps of man and selective effects of climate. Proc. Natl. Acad. Sci. USA 78: 2638-2642.

St. George-Hyslop, P. H., R. E. Tanzi, R. J. Polinsky, J. L. Haines, L. Nee, P. C. Watkins, R. H. Myder, R. G. Feldman, D. Pollen, D. Drachman, J. Growdon, A. Bruni, J. F. Foncin, D. Salmon, P. Frommelt, L. Armaducci, S. Sorbi, S. Piacentini, G. D. Stewart, W. J. Hobbs, P. M. Conneally, and J. F. Gusella, 1987   The genetic defect causing familial Alzheimer's disease maps on chromosome 21. Science 235: 885-890.

Schrott, H. G., J. L. Goldstein, W. R. Hazzard, M. M. McGoodwin, and A. G. Motulsky, 1972   Familial hypercholesterolemia in a large kindred: Evidence for a monogenic mechanism. Ann. Intern. Med 76: 711-720.

U.S. Congress Office of Technology Assessment, 1988   *Mapping Our Genes. Genome Projects: How Big, How Fast?* Johns Hopkins University Press, Baltimore, MD.

Vogel, F., and A. G. Motulsky, 1986   *Human Genetics*. Springer-Verlag, New York.

Williamson, J. S., D. T. Bishop, S. Tavaré, and C. J. Tavaré, 1979   On the independence structure that exists within a pedigree with an application to testing for a major gene, pp. 271-276 in *Progress in Clinical and Biological Research*, edited by C. F. Sing and M. Skolnick. Alan R. Liss, New York.

17

The Interface Between Genetics, Demography, and Epidemiology: Examples and Future Directions

BURTON SINGER

An extensive integration of the disciplines of genetics, demography, and epidemiology is long overdue. Although these fields often define separate academic departments where specific narrowly-focused aspects of each area are pursued, a thorough integration of their concepts and methodologies requires a much broader umbrella. One such umbrella is the general area of public health. This is an increasingly rich source of problems requiring the diversity of skills exemplified by these fields for their solution.

The purpose of this chapter is to present some problems from the field of tropical public health that exemplify the integration of ideas and methods from genetics, demography and epidemiology. As part of a detailed discussion of special problems, a diverse array of related public health problems will be indicated that require a close interface between these fields for their solution. The primary focus will be malaria, an arthropod-borne parasitic disease that involves fascinating genetic issues in all three components of an intricate transmission system: the human host, female anopheles mosquitoes, and the malaria parasites. Two public health problems associated with malaria control will be treated in some detail. These are:

1. The specification of anti-malarial drug distribution strategies to maximize the length of time until drug-resistant parasites take over a population;
2. The identification of high-risk subpopulations in migration-initiated malaria epidemics and the development of targeted control strategies.

Malaria is still the most pervasive public health problem in tropical areas of the world; and many attempts at eradication or control utilizing technologically sophisticated interventions (antimalarial drugs and insecticides) have only served to exacerbate the problem. Thus, a wide range of challenging questions are awaiting responses on this topic.

Malaria is by no means unique in its demand for an integration of genetic, demographic, and epidemiological knowledge. This requirement also occurs on such diverse topics as: (1) the diagnosis and treatment of schizophrenia (Risch, 1990; Strauss and Carpenter, 1974) in particular and psychiatric disorders (Merikangas, 1987; Risch and Baron, 1982) in general; (2) strategies for early detection and treatment of breast cancer (Claus, 1988; King and Bailey-Wilson, 1986); (3) assessing the long-term consequences of, and setting regulatory standards for, exposure to low-level ionizing radiation (National Academy of Sciences, 1980); and (4) projecting the deformation of age pyramids for African populations as a consequence of current and projected HIV transmission and a variety of education interventions (Anderson, Mayer, and McLean, 1988).

In the next section the life cycle of malaria parasites and the facets of genetics as they relate to malaria transmission and to infection in the human host are reviewed. The following section focuses on a hybrid genetics-transmission model for malaria as a guide to understanding the development of drug resistance. We highlight a curious ethical dilemma that is associated with the use of antimalarial drugs. If one wants to maximize the length of time that an antimalarial drug is at least capable of reducing—if not eliminating levels of parasitemia in infected humans, then curative doses should be administered very selectively, and less than 30% of the infected population should be treated. Thus, a strategy of maximally effective treatment—curative doses for all infected individuals—in the short term can be deleterious for the community in the longer term.

The penultimate section considers the problem of identifying and characterizing the highest risk individuals who are participants in migration-initiated malaria epidemics. Settlement of the western Amazon region of Brazil, and the complex population mobility at the borders of Thailand, Laos, and Burma provide striking examples of the subtlety of the question: who is at greatest risk of acquiring malaria? Finally, the chapter concludes with a short menu of Public Health problems which have many of the same features as the malaria situation, and where an interface between genetics, demography and epidemiology is essential.

MALARIA: GENETICS, EPIDEMIOLOGY AND DEMOGRAPHY

Parasite life cycle and natural history of infections

Malaria is an acute and chronic infection caused by protozoa of the genus *Plasmodium*. Malarial infections are characterized by fever, enlargement of the spleen, anemia, and, in children, frequently fatal complications. Four species of *Plasmodium* occur naturally in man: *P. falciparum*, *P. vivax*, *P. malariae*, and *P. ovale*. The most virulent species is *P. falciparum*; and in many areas of the world two or three species are present and can simultaneously infect a human host (Young, 1966; Cohen, 1973; Molineaux *et al.*, 1980).

The life cycle of malarial parasites consists of a sexual phase in diverse strains of female anopheles mosquitoes and an asexual phase which multiplies in man. The anopheline phase of the cycle begins when the mosquito ingests human blood containing the sexual forms of the parasite, known as gametocytes. Once inside the mosquito's stomach, the male cell extends and then detaches flagellum-like structures which migrate to, and then fertilize the female cells. The fertilized cells penetrate the wall of the mosquito's stomach and there grow into oocysts containing filament-like structures called sporozoites. Upon reaching maturity, the oocysts rupture, releasing up to several hundred thousand sporozoites to migrate throughout the mosquito's body cavity. Some of these sporozoites reach the salivary glands and there remain dormant until injected into man. For *P. falciparum*, this anopheline phase of the lift cycle lasts between 7 and 14 days.

The sporozoites disappear from the peripheral blood within 30 minutes after injection into man, initiating the exo-erythrocytic stage of the life cycle. The parasites develop in the liver parenchymal cells and reach maturity after about six days. The mature parasites are about 60 μm in their longest diameter and release approximately 40,000 merozoites into the peripheral blood to invade erythrocytes. At this stage of development parasites can first be detected in the blood. A minimum of 10 parasites per mm^3 is normally required for detection by ordinary microscopic examination of a thick blood smear.

The invasion of red blood cells by merozoites initiates the production of the next stage of asexual reproduction, trophozoites. In about 48 hours, each trophozoite releases from 8 to 24 new merozoites for further invasion of red blood cells. After several generations of this process, the parasitized blood cells release male and female gametocytes capable of infecting anopheles mosquitos. A series of trophozoite-gametocyte waves typically follows. As the gametocyte count rises, the trophozoite count falls, and clinical improvement or remission of symptoms frequently occurs. Parasite counts in malaria infections fluctuate markedly. They often show alternating high and low densities on successive days.

Human genetics

Haldane (1949) observed that sickle cell anemia occurred with high frequency almost exclusively in heavily malarious areas of the world. He speculated that malaria was the principal selective agent responsible for balancing the loss, *due to premature death*, of homozygous individuals with type SS hemoglobin (the sickle cell anemics) by increased fitness of heterozygotes with type AS hemoglobin. Subsequently, a variety of epidemiological and *in vitro* investigations (Allison, 1954, 1957; Friedman, 1978) on sickle cell anemia strongly supported this claim. In a major malaria field study conducted in Garki, Kano State, Nigeria (Molineaux and Gramiccia, 1980), sickle cell trait was present in 24% of newborns and 29% of those aged over five years. The hemoglobin S gene-frequency appears to have been maintained by a fitness in heterozygotes (type AS hemoglobin) of 21% over normal homozygotes (type AA hemoglobin). Persons with type SS hemoglobin do not survive to reproductive ages.

The advantage afforded by type AS hemoglobin as protection against *P. falciparum* infections is dependent on the age of the individuals *and* the malaria transmission characteristics. In Garki, Nigeria, the period of advantage started suddenly for infants at approximately 30 weeks of age and continued to 59 weeks. No difference in parasitemia between older children who were type AS, as opposed to type AA individuals, could be demonstrated during wet seasons (high transmission). However, the density of *P. falciparum* asexual stages was lower in type AS individuals up to three years of age during dry seasons (low transmission).

In general, at high levels of transmission, differences between type AS and type AA individuals appear at earlier ages and last a shorter time. This is the result of the rapid acquisition of active immunity under repeated challenge by *P. falciparum* parasites in all individuals. At lower levels of transmission, the advantage for type AS individuals appears at later ages but lasts longer. These interesting genetic, epidemiological, and demographic details have not thus far influenced malaria control strategies primarily because of the high cost (for countries where the disease is endemic) and tedious ongoing data collection that would be required to target for example, drug treatment in type AA persons over prescribed age ranges.

The relationships between malaria and abnormal hemoglobins has recently become still more interesting and perplexing in the light of DNA analysis that makes possible the unequivocal diagnosis of the carrier states of most forms of α-thalassemia (Flint *et al.*, 1986). It is now apparent that this disease occurs at extremely high frequencies in many populations—e.g., in Melanesia—and that it may be the commonest single-gene disorder in the world. Furthermore, epidemiological data (Flint *et al.*, 1986) suggest that there is a selective advantage conferred by α-thalassemia with respect to *P. falciparum* malaria. Thus far, it is not clear what mechanism might be responsible for α-thalassemia to confer protection. Whatever the protective effect is of α-thalassemia against malaria infection, it must be extremely subtle and may not constitute a direct negative effect on parasite development as has been shown for the red cells of individuals with type AS hemoglobin.

Vector genetics

Relatively recent methodological advances such as gas chromatography (Carlson and Service, 1980) and the use of diagnostic allozymes (Mahon, Green, and Hunt, 1976) have revealed that what were once regarded as homogeneous anopheline populations are, in fact, highly heterogeneous. The expression "species complex" has been applied to such populations; and the *A. gambiae* complex of African malaria vectors is among the most intensively investigated heterogeneous anopheles vector groups. Distinct members of a complex represent genetically discrete units; and hybrids are nonfertile. They vary in breeding habits, biting and resting behavior, and competence as host and transmitter of malaria parasites.

A striking example of the importance of understanding the nature of distinct members of a species complex occurred in the previously mentioned Garki, Nigeria malaria project. Several villages received, as control interventions, propoxur spraying on inside walls of huts. With most of the human population sleeping in the huts at night and the local *A. gambiae* vectors biting indoors at night, it would naively appear that the use of insecticides on inside walls of huts could be an effective control measure. However, the morphologically equivalent *A. gambiae* vectors contain two subpopulations (distinct members of the *A. gambiae* complex) one of which is endophilic and the other is exophilic. The endophilic vectors rest on inside walls of huts after taking a blood meal and are, thereby, killed by the insecticide. The exophilic vectors rest outside and survive to produce offpring and continue the transmission process. Thus vector spraying, rather than acting as an effective control measure, merely served as a mechanism for selecting exophilic *A. gambiae* vectors. Of great interest for the present discussion is the fact that the endophilic and exophilic vectors differ by a chromosome inversion. Use of this kind of information to refine malaria control programs lies in the future.

Vector genetics has played havoc with malaria eradication and control efforts on a much larger scale then the village-level selection in the Garki project. Indeed, massive DDT spraying campaigns initiated in the 1950s and repeatedly in the 1960s, which were aimed at malaria eradication, served to demonstrate how rapidly DDT resistant anopheles vectors could dominate local populations. In Sri Lanka, where DDT was once proclaimed to be the principal factor in eradication, the phasing out of routine spraying of DDT revealed a rapid resurgence of anopheles vectors, a new epidemic of malaria, and DDT resistance. Unfortunately the lesson of DDT resistance was not taken seriously enough to prevent the wide introduction of Malathion spraying in Sri Lanka, where once again insecticide resistant strains of anopheles vectors are negating a costly malaria control campaign (Wijesinha, 1985). However, these experiences have finally led to a substantial effort to mobilize people living in the malarious zones to work together in a sustained effort to reduce and control vector breeding sites. Ecologically sound source reduction strategies for vector control, which were frequently effective prior to the development of DDT (Watson, 1953), are once again being introduced and guided, in their design, by increasing knowledge of the interplay between genetics and vector behavior as exemplified within species complexes.

Parasite genetics–drug resistance

Malaria parasites have developed mechanisms of resistance against virtually every drug that has been used against them (Peters, 1982; Molineaux, 1986). The parasites are haploid in the human host and have a brief diploid phase in the mosquito, where recombination events occur. Selection by drugs of resistant strains of parasites takes place in the haploid phase of the parasite life cycle among asexual blood forms and gametocytes, depending on the particular antimalarial drug. Random mating among sexual forms of selected strains takes place in a drug-free environment in the mosquito. Redistribution of genetically-crossed selected variants to the human population follows by injection of sporozoites by female anopheles mosquitoes. The purely selective interpretation of the development of drug resistance is motivated by the fact that no antimalarial drug, to date, is known to be mutagenic. Thus, it is reasonable to presume that resistant strains of parasites are present in the highly heterogeneous parasite populations at very low frequencies prior to the selective action of drugs and that they may have arisen initially by rare random mutations.

Classical genetic experiments (Walliker, 1983) using rodent malarias have demonstrated that the number of gene loci associated with drug resistance is decidedly drug-dependent. Pyrimethamine resistance appears to be a single-locus phenomenon, whereas chloroquine (Beale, 1980; Walliker, 1986) proguanil and primaquine resistance involve multiple loci with the precise number of loci currently unknown. Thus, defensible formal multi-locus models of the development of drug resistance cannot, with the exception of pyrimethamine, currently be formulated. Exploration of drug resistance at the level of DNA is still in a very early phase of development; and it would be premature to speculate on the possible role of information about DNA structure and resistance in the formulation of malaria control strategies.

A SIMULATION MODEL OF TRANSMISSION AND DRUG RESISTANCE

Further quantitative insight about the development of drug resistance can be obtained by specifying models of the malaria transmission process that also incorporate relevant features of parasite genetics. It is possible to calibrate such models with field data on key epidemiological parameters (e.g., mosquito biting rates, incubation periods of parasites in both the mosquito and the human, etc.) and then assess their ability to reproduce prevalence data over time in the presence of prescribed drug distribution schedules. With such models at hand we can ask questions about the rate of development of drug resistance under dosage schedules, and population coverage that are quite different from those in the field settings where the data used for calibration was collected. Thus, the hybrid transmission-genetics model can be used to assess the nature of the development of drug resistance under a wide variety of possible control programs and, hopefully, can be used to guide the design of interventions that maximize the period of effectiveness of given anti-malarial drugs.

In order to focus ideas further, consider the hybrid transmission/genetics model for the development of drug resistance in a closed population outlined schematically in Figure 17-1. Like any model, some simplifying assumptions are made relative to the physical reality; but it is intended that the key features of the transmission process and parasite genetics be retained.

Infected humans and the mosquito population are identified by a log (tolerable dose) distribution for the parasites they are harboring. The tolerable dose of antimalarial drug beyond which given parasites are killed is the *phenotype* introduced for modeling purposes. Discrete gene loci (which vary from one antimalarial drug to another) that are associated with resistance are not formally modeled. Selection of parasites by drugs in the human is represented by truncation and then retention of the upper tail of the log (tolerable dose) distribution. Transfer of gametocytes to mosquitoes is described by the transfer of a log (tolerable dose) distribution from a single human to the "mosquito population." Recombination at the diploid stage in the mosquito is represented by a transformation of this distribution, which increases the mean (corresponding to a tendency toward resistance) and decreases the variance in proportion to the intensity of selection (i.e. depending on the size of the upper tail "selected" by the drug dose). For mathematical details on this kind of quantitative genetics

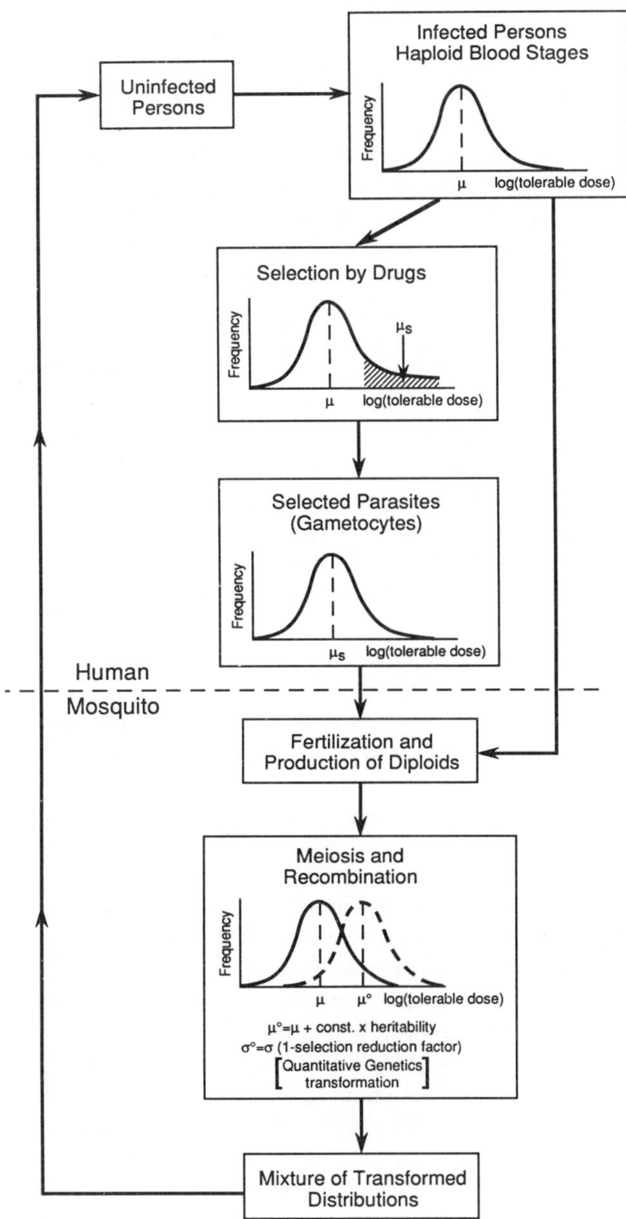

Figure 17-1. Flow chart of micro-simulation model.

model and its relationship to discrete multilocus specifications, the reader should consult Falconer (1981).

The dynamics of transmission and selection of resistant strains by drugs within the model outlined in Figure 17-1 proceeds according to the following algorithm. On each day each human is exposed to the mosquito population depending on the current biting rate. The human acquires a new infection [that is, log (tolerable dose) distribution] with a probability defined to be dependent on the sporozoite rate and, providing he/she is already infected, transmits a distribution to the mosquito population depending on the infectivity of the individual. A newly acquired distribution in the human incubates for seven days after which it either sits alone and is identified with haploid blood stages of the parasite or it is mixed with a previously present distribution, the resulting mixture being viewed as superinfection. The haploid blood stage distribution is also identified with the gametocyte distribution and is thus exposed to the mosquito population.

When drugs are administered they select only from the haploid stage distribution. New gametocytes are defined to be the distribution present before selection for one day and then to be the selected distribution if the individual does not clear parasites in response to the drug dose. If the drug dose clears parasites, that is, selects a sufficiently large fraction of the log (tolerable dose) distribution, then gametocytes defined by a truncated distribution on the day following administration are eliminated. While drugs are active, new distributions that complete the seven-day incubation period are transformed by selection, and the individual either clears parasites or retains some resistant parasites. Precise rules for drug action in this kind of model are given by Cross and Singer (1990).

If an infected human transfers a distribution to the mosquito population following drug administration, then a new distribution corresponding to a new generation of parasite, with increased mean and reduced variance (resulting from truncation selection) is generated corresponding to the processes of fertilization, production of diploids, meiosis, and recombination in the mosquito. If a distribution from an untreated person is transferred to the mosquito population, it is left invariant except for the possibility of being mixed with a small population of new spontaneous mutants, resistant to the drug being administered. The production of drug-resistant mutants is modeled as a spontaneous mutation process independent of the drug administration strategy. Drug action is viewed as a pure selection process on preexisting resistant mutants.

Following genetic transformation induced by selection and/or mixing with a small concentration of resistant mutants, the distribution arising from a single individual is mixed, using weights depending on the immune status and infectivity of the individual, with all other distributions transferred to the mosquito population on the given day. The resulting mixed distribution then incubates for 10 days after which it can be transferred to the human population (sporozoites) only on days 11 and 13 following acquisition by the mosquito population. A mosquito batch is assumed to have a 14-day life span. Its members can pick up distributions from the human population on days 1 and 3 of its life and deposit them (as sporozoites) on days 11 and 13. In any simulation, the mosquito emergence rate and, correlatively, the biting rate are assumed to be governed by rainfall, which of course exhibits substantial seasonal variation in endemic areas.

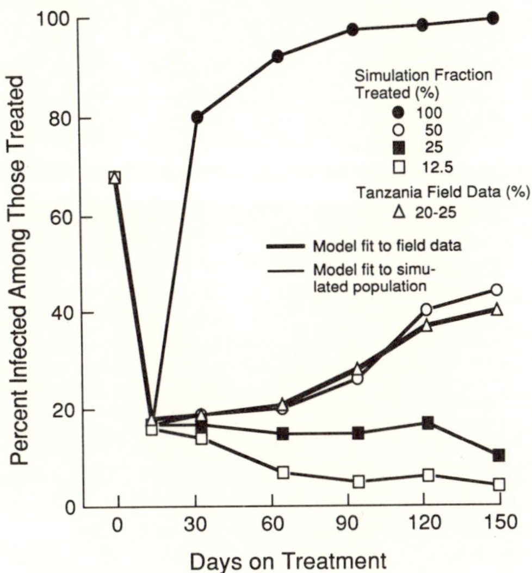

Figure 17-2. Parasite rates nine days post-treatment among those treated with initially-curative dose of pyrimethamine.

The heavy line in Figure 17-2 is based on asexual parasite rates from among the 20-25% sample of people in a village in Tanzania (Clyde and Shute, 1957) treated monthly for six months with what is initially a curative dose of pyrimethamine. The remaining curves show *simulated* parasite rates among the treated population for a variety of dosage and population coverage scenarios. They are *not* based on field data but on simulations corresponding to hypothetical alternative control strategies. The transmission parameters in the model, including the age structure of the population, are chosen to match those in the Tanzanian village where the empirical results were obtained. The qualitative picture is that the drug will remain effective for the longest time on the treated population when coverage is 30% or less and where curative doses are administered. This general pattern persists for simulations of chloroquine distribution in a variety of holoendemic settings. It suggests that close consideration of the potential consequences of the use of different antimalarial drugs in sequence, would be of considerable value for planning new control strategies where there is a desire to treat the maximal number of infected people, and at the same time, retain drug effectiveness. The current widespread dissemination of mefloquine in Thailand in the face of pervasive chloroquine resistance will bear watching in this regard.

MIGRATION-INITIATED MALARIA EPIDEMICS

Dramatic increases in transmission of malaria in Rondônia and Southern Pará, Brazil (Marques, 1987; Ellis, 1988) and in northern Thailand (Singhanetra-Renard, unpublished; Sawyer and Sawyer, 1987, Ellis, 1988) are two instances of an ever increasing problem of rapid rise in disease incidence resulting from migration to frontier-like lands and/or transient interaction by nonindigenous in-

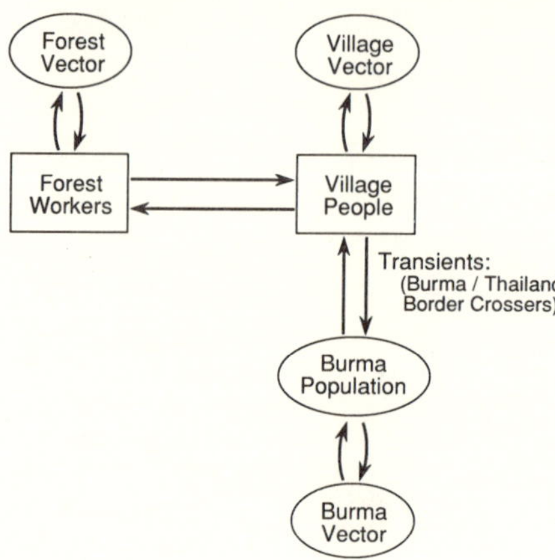

Figure 17-3. Forest malaria: migration and transmission.

fected populations with a less mobile village population through local anopheles vectors. In both cases there is a substantial variation in disease incidence and prevalence among both the highly mobile and the relatively stationary components of the populations. The sources of variation appear to be subtle combinations of socioeconomic characteristics of individuals and ecological features of the settlement areas (Sawyer and Sawyer, 1987). The word *appear* in the previous sentence is of particular importance because the existence of very rich-in-detail household surveys, as in the Machadinho project in Brazil (Sawyer and Sawyer, 1987), present a major data analysis challenge for isolating high-dimensional combinations of conditions that are associated with high prevalance of *P. falciparum* (in Brazil) in a migrant population of nonimmunes.

This challenge is equally if not more severe for northern Thailand in villages that are in close proximity to the borders with Laos and Burma. In some Thai villages, economic conditions are sufficiently desperate that illegal swiddening and the harvesting of forest produce are major activities. This simultaneously improves the economic lot of the villagers and exposes the forest workers to intense contact with anopheles vectors. A caricature of malaria exposure in this setting is shown in Figure 17-3.

Here a village is considered wherein residents are exposed to a vector population within a 2.4–3.2 kilometer radius of breeding sites of the "village vector". Some fraction of the village residents work in the forest and are exposed to a different vector population whose biting rate on man is much higher than that of the village vector. Thus, the forest workers, by virtue of their movement back and forth from the village, provide an increased reservoir of parasites to be transmitted to nonforest workers by the village vector.

Adding to the complexity of this picture is the fact that some Thai villagers, again out of economic necessity, engage in illegal logging, smuggling, and poaching that also involves border crossing into Burma. Laotians and Burmese, already infected with malaria parasites, in turn, cross their borders and make contact with Thai village vectors thereby increasing the local transmission. A

Table 17-1 Typology of circular mobility in northern Thailand regarding activities associated with malaria transmission*

Mobility Pattern	Return After Dark or Late at Night	Overnight - < 1 Week	1 Week - < 1 Month	1 Month - < 1 Year	1 Year or More
			Time		
Rural Village to Foothills	Hunting (frogs) Collecting (bamboo shoots, firewood)		Cultivating[a] (maize, cotton)		
Rural Village to Upland/Forest		Hunting[a] (game)	Logging[a] Hunting (khiat laso) Collecting (honey) Laboring (sawing logs)	Forest[b] Plantation Construction (road, dam) Forest Industry	Mining[a,b] (ores)
Rural Village to Burma, Laos, Cambodia	Smuggling[a,b] (goods)	Visiting[b] (kin)	Trading[a,b] (cattle, goods) Laboring (cowhands)	Cultivating[a,b] (opium) Construction (temples, houses) · Mining (gems)	Mining[a,b] (ores)
Burma, Laos to Forest Settlement in Thailand		Visiting[b] (kin)	Laboring[a,b] (cowhands)	Forest[a,b] Plantation Laboring (farm) Refugees	Mining[b] (ores) Refugees
Burma, Loas to Thai Border Towns	Trading[b] (goods)	Visiting[b] (kin)		Housemaids[b] Shop/ restaurant helpers	

NOTES:

* Adapted from Prothero (1977), Table 1.

a—Exposure to malaria from movement through different ecological zones.

b—Exposure to malaria from movement involving contacts between different groups of people.

typology of the circular mobility in northern Thailand and its relationship to activities which enhance exposure to malaria is presented in Table 17-1.

Malaria control strategies which attempt comprehensive population coverage are virtually impossible to implement in this kind of unstable environment. It might potentially be feasible to target the highest risk subgroups for control efforts, which would involve much more intensive community participation within villages and among those making contact with the diverse array of border crossers. This immediately leads to the problem of succinctly and rapidly characterizing the high risk subpopulations using a variety of forms of evidence, e.g., community participation observations and in-depth interviews, migration surveys, and household surveys (Bilsburrow, Oberai, and Standing, 1984). In order to focus on some analytical issues, the case of malaria in the Amazon, a setting more extensively investigated than northern Thailand, is considered in more detail.

Initial analyses of data from the Machadinho project suggest that those most likely to be found positive for *P. falciparum* are characterized by economic circumstances defined by bundles of commodities owned, nature and location (relative to the forest) of a homesite, nature and location (in the forest or not) of work activity, proximity to the breeding sites of *A. darlingi* (the principal vector of transmission), level of education, background history defined by such variables as "place of origin prior to migration to Rondônia" and work characteristics at the place of origin. Sharp delineation of the combinations of conditions, based on these variables, that characterize the highest risk subpopulations is confounded by the extreme heterogeneity of the migrants. A useful, but as yet underutilized tool, for the characterization of special risk-groups within heterogeneous populations is the family of Grade-of-Membership (GOM) models. We summarize the essential features of GOM constructions and their interpretations, referring the reader to technical references for details about implementation (Manton and Stallard, 1988; Woodbury, Cliver, and Garson, 1978).

Defining subpopulations of a heterogeneous population on the basis of multiple characteristics simultaneously is, on the face of it, a standard problem of cluster analysis. However, when many individuals are identified with multiple risk conditions, no combination of which occurs with high frequency in the full population, then the assignment of individuals to crisply defined categories becomes impossible. A more appropriate strategy is to construct, in a data-intrinsic manner, a small set of "ideal profiles" (or categories), each defined by a multiplicity of conditions, and then associate with each individual a set of "degree-of-similarity scores" or Grades of Membership (GOM scores) relative to these profiles. The GOM scores are interpreted as degrees of similarity of the set of characteristics of an individual to the set of characteristics that define each profile.

For example, suppose that a population of migrants is to be described by their distribution on the following variables:

1.	Economic status (based on commodities owned)	High	Low	
2.	Primary work activity in forest	Yes	No	
3.	Engaged in farming	Yes	No	
4.	Planted coffee trees	Yes	No	
5.	Planted more than 100 coffee trees............	Yes	No	
6.	Planted cacao trees	Yes	No	
7.	Working for someone..........................	Yes	No	
8.	Supervising others............................	Yes	No	
9.	House closed-in by 4 walls....................	Yes	No	
10.	House with protective screening	Yes	No	
11.	House within 1 mile of forest..................	Yes	No	
12.	Number of years in Amazon....................			
13.	Education .... None	0-4 yr	>4 yr	

This is only a partial list of variables that can characterize the degree of exposure of a migrant to the Amazon to malaria. It is presented here to illustrate a methodology. For a more comprehensive discussion of variables associated with exposure to malaria in the Amazon, the reader should consult Sawyer and Sawyer, 1987.

Now consider a profile defined by

I. Low economic status *and* working in the forest *and* house within 1 mile of forest *and* house not closed-in by four walls *and* less than four years of education *and* less than one year in the Amazon

and a second profile defined by

II. High economic status *and* not working in the forest *and* supervising others *and* house closed-in by four walls *and* house with protective screening *and* three years in the Amazon *and* more than four years of education.

Profile I defines a population within which the prevalence rate for *P. falciparum* malaria is very high, while profile II defines a population at minimal risk. Neither of these profiles characterizes many people. Thus there is a need to associate GOM scores with individuals so that their degree of similarity (based on responses to questions 1-13) to each of the profiles may be represented quantitatively. An individual with GOM scores $(g_I, g_{II}) - (7/8, 1/8)$ has most but not all of the characteristics of profile I—that is, he/she loads $7/8$ on profile I and only $1/8$ on profile II. An individual wih GOM scores $(1/2, 1/2)$ is exactly intermediate between the two profiles. If high prevalence is associated with persons for whom g_I exceeds $3/4$, then this defines a high-risk subpopulation to be targeted by a control program.

A Grade-of-Membership representation for our prototype migrant population consists of the logical AND statements defining the profiles—that is, complex combinations of conditions—and a pair of GOM scores for each individual. Of course, depending upon the community, three or four "ideal profiles" might be relevant with individuals having, respectively, 3 or 4 GOM scores indicating their degrees of similarity to the profiles. In highly heterogeneous populations, crisp classification is impossible. For example, in the situation described here, suppose that 35% of the population have $1/4 < g_I < 3/4$; hence, $3/4 > g_{II} > 1/4$. That is, many people cannot be cleanly classified into the categories, I and II, that are best supported by the data. Figure 17-4 shows a prototypical distribution of GOM scores for a 2-profile population of the kind present in Rondônia. The essential feature that clarifies the appropriateness of the Grade-of-Membership concept is that a very small fraction of the population has $g_I = 1$ or $g_I = 0$. In a population where everyone has either $g_I = 1$ or $g_I = 0$, a Grade-of-Membership representation is equivalent to a clustering scheme where every individual is assigned to only one of two populations (profiles).

Grade-of-Membership representations are quite suitable for the rapid ascertainment of high-risk subpopulations that a control program would target in attempting intervention in the presence of high mobility and lack of social cohesion. However, it is important to understand that this kind of representation is much coarser, but nevertheless well suited to rapid assessment, than a more physically detailed transmission model. Indeed, if interest focuses on a deeper scientific understanding of malaria transmission in an environment of unstable

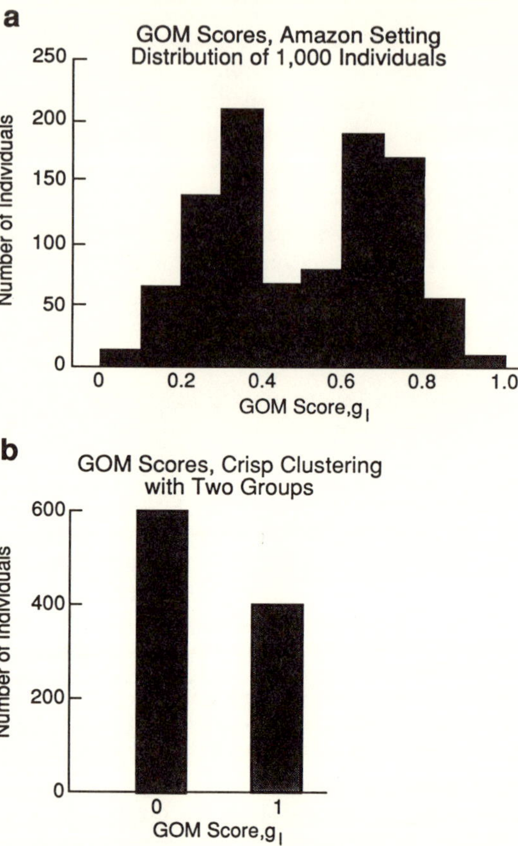

Figure 17-4. Grade-of-Membership score distributions (see text for details.)

population movements such as in the colonization of Rondônia, then a dynamical system corresponding to the schematic diagram in Figure 17-3 represents the minimal level of complexity that must be introduced.

DISCUSSION

The case examples described in the previous sections each require an integration of ideas from genetics, epidemiology and demography to understand a dynamic process and to guide potentially effective disease control strategies. More opportunities for research of this character abound. Three current challenges which promise to remain as such for some time are described below.

1. There is currently a substantial effort being devoted to the development of malaria vaccines (Nussenzweig and Nussenzweig, 1989) and to studying the epidemiological consequences of a diverse array of vaccine characteristics (Halloran, Struchiner, and Spielman, 1989; Struchiner, Halloran, and Spielman, 1989). The development of malaria vaccines is complicated by the great diversity and species-specificity of protective antigens. The antigens are also stage-specific:

they are different for sporozoites, for the stage at which sporozoites undergo rapid multiplication, for the asexual blood stages within erythrocytes, and for gametocytes.

Protective immunity against sporozoites has been achieved by inoculation of irradiated parasites into rodents and primates (including some humans) (Rosenberg *et al.*, 1989). The target antigens consist of single circumsporozoites (CS proteins) that cover the entire surface membrane of the parasite, and are shed when cross-linked by antibodies. Despite the evidence of monoclonal antibodies to CS proteins neutralizing sporozoite infectivity, there is new evidence that, at least for *P. vivax*, a vaccine based on CS proteins will not be universally protective (Cochrane, Nussenzweig, and Nardin, 1980). Indeed, phenotypic heterogeneity in the CS proteins of *P. vivax* is such that over 14% of clinical cases of uncomplicated *P. vivax* malaria at two sites in Thailand produced sporozoites immunologically distinct from any previously characterized. Monoclonal antibodies to the CS protein of other species of human and simian malarias did not bind to these nonreactive sporozoites. Thus, the public health goal of a simple readily administered malaria vaccine is proving to be increasingly elusive with increased knowledge of parasite genetics.

2. Social and economic demography will play a central role in the targeting and delivery of interventions for control of onchocerciasis (River Blindness), African trypanosomiasis, and Chagas' disease. Effective use of Ivermectin (Eckholm, 1989) for onchocerciasis control requires an annual dose of the drug. Education and sustained community compliance in taking repeated doses of Ivermectin at widely spaced times are essential behavioral features of any program which has the hope of long-term reductions in prevalence of onchocerciasis. Community-wide acceptance and correct installation of tse-tse fly traps are central ingredients of African trypanosomiasis control in the Ivory Coast (Laveissiere *et al.*, 1986). Utilization of externally supplied funds for proper house repair and construction, and success in education to prevent infestation by vectors of Chagas' disease is essential for major reductions in incidence and the interruption of transmission (Briceño-Leon, unpublished). In each instance some form of social engineering is a critical feature of a disease control program. Much remains to be done in the design and implementation of these social and economic aspects of tropical public health.

3. The rapid spread of HIV infection with the consequent development of AIDS in Africa has presented new challenges for infectious disease control. In particular, we do not yet understand what the effects are of an HIV compromised immune system in adults who would normally be resistant to severe malaria infection. A resurgence of tuburculosis also seems to accompany the proliferation of HIV infection and AIDS. These disease interactions represent a major public health challenge. Furthermore, it is increasingly important to obtain a quantitative epidemiological and demographic assessment of the impact of HIV contaminated blood supplies on malaria in infants and young children for whom blood transfusions are part of a standard course of treatment.

It is the hope that complex questions such as these, which pervade the domain of tropical public health, will stimulate a more extensive community of scholars to work at the interface of genetics, epidemiology, and demography.

REFERENCES

Allison, A. C., 1954   Protection afforded by sickle cell trait against subtertian malarial infection. Br. Med, J. 1: 190-294.

Allison, A. C., 1957   Malaria in carriers of the sickle cell trait and in newborn children. Exp. Parasitol. 6: 418-447.

Anderson, R. M., R. M. May, and A. McLean 1988   Possible Demographic consequences of AIDS in developing countries. Nature 332: 228-334.

Beale, G. H., 1980   The genetics of drug resistance in malaria parasites. Bull. WHO 58: 799-804.

Bilsburrow, R. E., A. S. Oberai, and G. Standing, 1984,   *Migration Surveys in Low Income Countries*. Croon Helm, London.

Carlson, D. A., and M. W. Service, 1980   Identification of mosquitoes of *Anopheles gambiae* species complex A and B by analysis of cuticular components. Science 207: 1089-1090.

Claus, E., 1988,   *Age of Onset and the Inheritance of Breast Cancer*. Unpublished Ph.D. Dissertation, Department of Epidemiology and Public Health, Yale University, New Haven, CT.

Clyde, D. F., and G. T. Shute, 1957   Resistance of *Plasmodium Falciparum* in Tanganyika to pyrimethamine administered at weekly intervals. Trans. R. Soc. Trop. Med. Hyg. 6: 505-513.

Cochrane, A. H., R. S. Nussenzweig, and E. H. Nardin, 1980,   Immunization against sporozoites, in *Malaria*, Vol. 3, edited by J. P. Kreier. Academic Press, New York.

Cohen, J. E., 1973,   Heterologous immunity in human malaria. Q. Rev. Biol. 48: 467-489.

Cross, A. P., and B. Singer, 1990   Modeling the development of resistance of *Plasmodium Falciparum* to anti-malarial drugs. Trans. R. Soc. Trop. Med. Hyg. (submitted).

Eckholm, E., 1989   Conquering an ancient scourge. New York Times Magazine, January.

Ellis, W. S., 1988   Rondônia's settlers invade Brazil's imperiled rain forest. National Geographic Magazine 174: 772-799.

Falconer, D. S., 1981   *Introduction to Quantitative Genetics*. 2nd Edition. Longman, London and New York.

Flint, J., A. V. S. Hill, D. K. Bowden, S. J. Oppenheimer, P. R. Sill, S. W. Serjeanston, J. Bana-Kori, K. Bhatia, M. P. Alpers, A. J. Boyce, D. J. Weaterall, and J. B. Clegg, 1986   High frequencies of α-thalassaemia are the result of natural selection by malaria. Nature 321: 744-749.

Friedman, M. J., 1978   Erythrocyte mechanisms of sickle cell resistance to malaria. Proc. Natl. Acad. Sci., USA 75 1994-1997.

Haldane, J. B. S., 1949   The rate of mutation of human genes. *Proc. 8th Int. Congr. Genet. (Hereditas*, Suppl. 35): 267-273.

Halloran, M. E., C. J. Struchiner, and A. Spielman, 1989   Modeling malaria vaccines II: population effects of stage-specific malaria vaccines dependent on natural boosting. Math. Biosci. 94: 115-149.

King, M. C., and L. A. C. Bailey-Wilson, 1986   Genetic analysis of human breast cancer: literature review and description of family data. Genet. Epidemiol., Supplement 1: 3-13.

Laveissiere, C., J. P. Hervouet, F. Merouze, and P. Cattand, 1986   La campagne pilote de lutte contre la trypanosomiase humaine dans le foyer de vavoua (Côte d'Ivoire).

Cab Orstom, Ser. Ent. Med. Parasitol. 24: 111-120.

Mahon, R. J., C. A. Green, and R. H. Hunt, 1976   Diagnostic allozymes for routine identification of adults of the *Anopheles gambiae* complex (Diptera Culicidae). Bull. Ent. Res. 66: 25-31.

Manton, K. G., and E. Stallard, 1988   *Chronic Disease Modeling: Measurement and the Evaluation of Risks of Chronic Disease Processes*. Charles Griffin, London.

Marques, A. C., 1987   Human migration and the spread of malaria in Brazil. Parasitol. Today 3: 166-170.

Merikangas, K., 1987 Genetic epidemiology of psychiatric disorders. Ann. Rev. of Psychiatry 6: 625-646, edited by A. J. Francis, and R. E. Hales. Psychiatric Press, Washington, D. C.

Molineaux, L., and G. Gramiccia, 1980   *The Garki Project: Research on the Epidemiology and Control of Malaria in the Sudan Savanna of West Africa*. WHO, Geneva.

Molineaux, L., J. Storey, J. E. Cohen, and A. Thomas, 1980   A longitudinal study of human malaria in the West African savanna in the absence of control measures: relationships between different *Plasmodium* species, in particular *P. falciparum* and *P. malariae*. Am.. J. Trop. Med. Hyg. 29: 725-737.

Molineaux, L., 1986   The epidemiology of drug resistance in malaria: an unfinished conceptual model. Working Paper No. 2.7.1, Informal Consultation on Epidemiology of Drug Resistance of Malaria Parasites, Scientific Working Group on Applied Field Research in Malaria. WHO, Geneva.

National Academy of Sciences, 1980,   *The Effects on Populations of Exposure to Low Levels of Ionizing Radiation*. National Academy Press, Washington.

Nussenzweig, R. S., and V. Nussenzweig, 1989   Antisporozoite vaccine for malaria: experimental basis and current status. Rev. Infect. Dis. 2, Suppl. 3: 5579-5585.

Peters, W., 1982   Anti-malarial drug resistance: an increasing problem. Br. Med. Bull. 38: 187-192.

Prothero, R. M., 1977   Disease and mobility: a neglected factor in epidemiology. Int. J. Epidemiol. 6: 259-267.

Risch, N., 1990   Genetic linkage and psychiatric disorders. Genet. Epidemiol. (in press).

Risch, N., and M. Baron, 1982 X-linkage and genetic heterogeneity in bipolar related major affective illness: reanalysis of linkage data. Ann. Hum. Genet. 46: 153-166.

Rosenbert, R., R. A. Wirtz, D. E. Lanar, J. Sattabongkot, T. Hall, A. P. Waters, and C. Prasittisuh, 1989   Circumsporozoite protein heterogeneity in the human malaria parasite *Plasmodium vivax*. Science 245: 973-976.

Sawyer, D. R., and D. R. T. O. Sawyer, 1987,   *Malaria on the Amazon Frontier: Economic and Social Aspects of Transmission and Control*, CEDEPLAR Technical Report, Federal University of Minas Gerais. Belo Horizonte, Brazil.

Strauss, J., and W. T. Carpenter, 1974   The prediction of outcome in Schizophrenia. Arch. Gen. Psychiatry 31: 37-42.

Struchiner, C. J., M. E. Halloran, and A. Spielman, 1989   Modeling malaria vaccines I: new uses of old ideas. Math. Biosci. 94: 87-113.

Walliker, D., 1983   *The Contribution of Genetics to the Study of Parasitic Protozoa*. Research Studies Press. Ltd., New York.

Walliker, D., 1986 Genetics of the response of malaria parasites to drugs. Working Paper No. 2.2.1, Informal Consultation on Epidemiology of Drug Resistance of Malaria Parasites, Scientific Working Group on Applied Field Research in Malaria, WHO, Geneva, 10-14 November.

Watson, M., 1953   *African Highway: The Battle for Health in Central Africa*. John Murray, London.

Wijesinha, S., 1985   The grass roots approach to malaria control in Sri Lanka. The Ecologist 15: 137-138.

Woodbury, M. A., J. Clive, and A. Garson, 1978   Mathematical typology: A grade of membership technique for obtaining disease definition. Comput. and Biomed. Res. 11: 277-298.

Young, M., 1966   Malaria, pp. 316-359 in *A Manual of Tropical Medicine*, edited by G. W. Hunter, W. W. Frye, and J. C. Swartzweder. W. B. Saunders, Philadelphia.

IV
PERSISTENT ISSUES IN GENETICS AND DEMOGRAPHY

18

Persistent Issues in Genetics and Demography

ALBERT I. HERMALIN AND ANSLEY J. COALE

It is well-recognized that population biology and demography share a common ancestor in Malthus, and in searching for persistent issues and potential points of convergence, it is well to start with him (Malthus, 1798):

> Assuming then, my postulata as granted, I say, that the power of population is indefinitely greater than the power in the earth to produce subsistence for man. Population, when unchecked, increases in a geometrical ratio. Subsistence increases only in an arithmetical ratio. A slight acquaintance with numbers will shew the immensity of the first power in comparison of the second. By that law of our nature which makes food necessary to the life of man, the effects of these two unequal powers must be kept equal.

In the absence of migration, the growth rate of a population or subgroup depends on its fertility and mortality. The Malthusian theory is one of equilibrium or homeostasis, in which population size at first outruns the norms of subsistence, and the resulting impact of lower wages, delayed marriage, and other checks, eventually restores a temporary balance.

During much of the nineteenth century, Malthus' model was debated and developed, mainly within classical economics, in terms of the interrelationships among population growth, wage levels, rates of return, and related factors. The late nineteenth century and early twentieth century witnessed the rise of the neo classical critique of the theory of wages, with greater attention to short run and methodological issues; and the development of demography as a separate discipline through increased empirical research as well as a shift, on the theoretical front, to issues of optimum population size and more importantly, formal demographic modeling, (United Nations, 1973).

This redirection eventually produced demography's most notable methodological orientation, the decomposition of vital rates and other measures into their component elements, particularly, when possible, biological processes. Differences over time or across groups are often "explained" in terms of differences in terms of these components. As Preston (Chapter 22) notes:

> There is no activity more basic to demography or typical of demographers. It's a form of social accounting that, even though it may not provide complete explanations of a phenomenon, helps to clarify what it is that needs explaining.

In highly schematic terms then, demography can be viewed as developing from concerns with overall growth rates, to a focus on the components of these rates—fertility, mortality, and migration—and on modeling how these processes (especially the first two) affect the growth and age structure of a population.

Malthus directly influenced Darwin and Wallace, whose own research was leading them to a similar paradigm (Wallace, 1887):

> ... The most interesting coincidence in the matter, I think, is, that I, *as well as Darwin*, was led to the theory itself through Malthus—in my case it was his elaborate account of the action of 'preventive checks' in keeping down the population of savage races to a tolerably fixed but scanty number. This had strongly impressed me, and it suddenly flashed upon me that all animals are necessarily thus kept down—'the struggle for existence'—while variations, on which I was always thinking, must necessarily often be beneficial, and would then cause those varieties to increase while the injurious variations diminished ...

For Darwin and Wallace, interest centered on the observed variations in flora and fauna, and how new species arise and sustain themselves. The insight they gleaned from Malthus was that different variations would be subject to different growth rates with some dying out because of an unfavorable balance of fertility and mortality rates in a given habitat, and others prospering. From this perspective, population biology resembles demography in focusing on the components of fertility and mortality in order to estimate relative fitness (see Adams, Chapter 1 and Christiansen, Chapter 19). Insofar as selection operates through phenotypic differences, this is a persistent concern of population biology.

POINTS OF DIVERGENCE AND INTERSECTION

Given the common Malthusian ancestry, Lee (1987) notes that:

> It is, then, ironic that the concept of population equilibrium or homeostasis, maintained by density-dependent checks, plays so small a role in human demography and such a dominant one in the study of animal populations. With some exaggeration, we might say that human demography is all about Leslie matrices and the determinants of unconstrained growth in linear models, whereas animal population studies are all about Malthusian equilibrium through density dependence in nonlinear models, with little attention to age distributions and such.

As Lee's quote suggests, a major point of departure between demography and population biology is in the priority given to age structure. While age is the central structural factor of a population to demographers and, along with sex, the basis of classification of almost all data, "the traditional theories of ecology and population genetics have, however, usually ignored the problem of age structure, and have assumed that population can be treated as homogeneous with respect to age." (Charlesworth, 1980). As a result, with some notable exceptions, most population genetic models have not explicitly taken into account age structure (and indeed as Charlesworth, 1980, notes, a full understanding of the *mechanisms* of regulation via population density is rather elusive, despite its importance); whereas models that include age structure, whether dealing with humans or nonhumans, have usually assumed unconstrained growth.

Thus, an area of considerable potential for convergence lies in the development of density dependent and other population-genetic models with explicit attention to age structure. Charlesworth (1980) cites much of the existing work in the area and sets forth many of the challenges, such as identifying the "critical age-group" which most influences the density-dependent components.

The opportunity for more complex modeling is greatly affected by the data base and the rising emphasis on quantitative biology has permitted greater congruence between the data utilized in population biology and human demography. Isomorphism in the structure of the data permits greater commonality in models and methods of analysis, and "vital statistics" and life tables now exist for a number of species. Charlesworth (1980), for example, includes examples for *Drosophila* and the grey squirrel.

Given the great variety of species and their habitats, as well as the different patterns of reproduction, there are considerable difficulties in estimating fertility and mortality data, especially by age-class. Techniques vary by species, ranging from following outcomes in the laboratory, to assuming a stationary population and estimating either an age distribution or ages at death. Christiansen (Chapter 19) makes clear the challenge biologists face when the individual outcomes in terms of offspring and longevity are not easily traced, and he offers one solution in the case of organisms with discrete breeding times.

Beyond common modeling, another incipient area of greater convergence is a common data base. The sources of demographic data have changed substantially in recent years. Surveys have become a key tool for measuring fertility and infant mortality, particularly in developing countries that lack complete or accurate vital registration. Interest in demographic parameters predating modern censuses has led to the development of family reconstitution studies, which rely on ecclesiastical or similar genealogical data to reconstruct birth and death records of families in a delimited territory. This micro approach, as Livi-Bacci (Chapter 2) refers to this strategy, closely parallels the genetic tracing of pedigrees in specified subgroups in order to estimate the degree of inbreeding and other relevant parameters, and the exploration of these connections provides the rationale for Part I of this volume.

As Adams (Chapter 1) cautions, however, many of the questions posed by demographers can be answered by single generation data (parents and numbers of offspring), while those of interest to geneticists require multigenerational data. Nevertheless, the availability of a common data base can be expected to generate opportunities for fruitful interactions.

THE INFLUENCE OF DEMOGRAPHIC PARAMETERS ON EVOLUTION

Demographic data and models also play a role in understanding the sources of genetic variation. The work of Mendel and his successors provided new insights into how traits are inherited and opened up new lines of attack on the questions posed by Darwin and Wallace in terms of the sources of variation in gene frequency and the origins of new genetic material. As Christiansen (Chapter 19) states, the fact that the "substance of inheritance is separate from the expression of the inherited traits," gives rise to the possibility of genetic evolution independent of phenotypic evolution. Of interest here is the way that the demographic properties of a population may influence the path of genetic evolution. One example is presented by Adams and Smouse (1985) in tracing the effect of changes in vital rates and age structure on the genetic structure of a population. They show how the changing pattern of vital rates over the demographic transition (the movement from high birth and death rates, to lower mortality with sustained high fertility, to low birth and death rates) has implications for differential fitness of genotypes (as measured by the net rate of reproduction) by tracing selective coefficients for three diseases with different ages of onset patterns.

Demographic parameters also enter prominently in discussions of the effect of size and levels of contact in determining evolutionary outcomes (Glass, 1971, Wright, 1960). In this regard it is instructive how Christiansen (Chapter 19) examines the competing hypotheses of migration, genetic drift, and natural selection, among others, in accounting for the degree of genetic variation he observed at different locations.

EFFECTIVE, OPTIMAL, AND MINIMUM POPULATION SIZES

Glass (1971) notes that a population completely divided into small isolates runs the risk of the accumulation of deleterious mutations which may result in the extinction of the population. These considerations raise the issue of minimum viable population size from the viewpoint of genetics. Ewens (Chapter 20) pursues this topic in terms of "effective population size," which is the size of a simple population sharing some features with the population under study. He shows that this effective size will depend on the particular feature in question and derives the equations for a number of these, in particular a mutation effective population size which defines a population in terms of its ability to maintain genetic variation of a specified degree. Ewens further notes that effective population size is a genetic concept and any such estimate must be further adjusted to derive an actual population size.

Ewens' differentiation among types of effective population sizes has its socioeconomic echo in concerns with optimum population size in demography. The concept of optimum population size can be traced back to Plato, and to several Renaissance authors as well, but it had its most explicit development in the late nineteenth century, stimulated in part by Malthus and the ensuing debates. While optimum population theory has had its proponents, it has also generated strong criticism mainly in terms of the vagueness of the concept (what should be optimized: economic well being, and if so which measure, or health, conservation of resources, power, etc.), and its static nature—the factors determining the optimum would change while the population was adjusting itself to the original conditions (United Nations, 1973). More recently Sauvy (1969) has used the concept as a tool to investigate the interrelations of population with many other social and economic factors. Concerns with rapid population growth in both developed and developing countries during the 1950s and 1960s and its implications for natural resources, food, energy, and environmental quality led to a further examination of optimum population size (Singer, 1971).

TWO-SEX MODELS

As demography developed into a distinct discipline in the late nineteenth and early twentieth centuries, greater attention was given to modeling the interrelations among demographic factors, in particular the ways in which the processes of population change through fertility and mortality shape the growth rate and age structure of a population. The life table, developed by Graunt in 1662, is a well-known example of such a model, in which a fixed age-specific schedule of mortality and a fixed annual number of births yields a stationary population—one fixed in size and age structure.

A more sophisticated model, the stable population model, which treats the implications of fixed age specific schedules of fertility and mortality, was developed early in this century by Lotka, and has proved widely influential in demographic theory as well as in practical application. One drawback of the model as formulated is that it is a one-sex model; age specific fertility and mortality schedules of either sex may be used to estimate parameters for males *or* females. But if both populations are estimated the implicit growth rates will rarely agree, implying an increasingly unbalanced sex ratio in the total population. Pollak reviews this "two-sex problem" and offers a solution based on a mating rule which is a function of a population's age-sex composition. Insofar as population biologists dealing with sexually reproducing species must take into account mating patterns (see Adams, Chapter 1; Christiansen, Chapter 19), the strategy offered by Pollak may have wider applicability.

DIFFERENCES IN VITAL RATES AND THEIR CAUSES

In empirical research, a major interest of demographers is to trace the differences in vital rates over time and across groups. As noted at the outset, in pursuing these objectives, demographers favor a strategy of decomposition, examining first the levels of fertility, mortality, and migration, and then their components and interrelationships. These investigations have shed considerable light on the ways that social and economic structure have intersected with biological factors in determining the level of population growth. Preston (Chapter 22), for example, shows the importance of a negative auto-correlation in fertility and mortality rates in sustaining a population. For fertility, this operates largely through the mechanism of breastfeeding (and arrangements like postpartum abstinence) whereby high survival among newborns allows longer breastfeeding, and slows down the rate of childbearing in the next period. According to Preston, the negative auto-correlation for mortality operates in part through the differentials in the age-specific probabilities of dying, so that a period of high mortality leads to a period with a higher proportion of survivors at ages with lower mortality risks, thereby reducing the overall rates in the next period.

The nature of the components examined varies for each of the basic processes of change. For mortality, substantial interest in the nineteenth and twentieth century changes centers on the relative role of medical advances, public health activities, individual health practices, and changes in the standard of living. Individual causes of death come into play as means of testing competing hypotheses. Preston describes the lively debate that has developed over these factors, and his analysis leads to new conclusions about the relative importance of each factor.

LEVELS OF FERTILITY IN TRADITIONAL SOCIETIES

It is in the area of fertility that the biological and social factors have been most successfully integrated, and the insights gleaned from recent research are at some variance from earlier thinking. In describing the general pattern of the establishment of modern low birth rates and death rates in the highly industrialized countries of the world, Notestein (1953) explained the high birth rates in traditional societies in the following terms:

> ... They had to be high. We may take it for granted that all populations surviving to the modern period in the face of inevitably high mortality had both the physiological capacity and the social organization necessary to produce high birth rates ... Peasant societies in Europe, and almost universally throughout the world, are organized in ways that bring strong pressures on their members to reproduce ... these arrangements stood the test of experience throughout the centuries of high mortality, are strongly supported by popular beliefs, formalized in religious doctrine, and enforced by community sanctions. They are deeply woven into the social fabric and are slow to change.

This line of argument led many demographers to assume that fertility in premodern societies was close to the biological maximum. In fact, the records and estimates of the rates of childbearing in premodern societies show a surprisingly moderate level of overall fertility, certainly well below the "biological maximum." The biological maximum of human fertility can be approximated by a hypothetical combination of early and universal marriage, (which is found in most of Asia and Africa) with the very high fertility among married women to be found among the Hutterites (Hostetler, 1974) and in the French Canadians in the seventeenth century (Livi-Bacci, Chapter 2). A population subject to this hypothetical combination would experience about 12 births per woman who survived to the highest age of childbearing. In contrast, the rates of childbearing derived from reliable data of premodern populations would produce an average ranging from fewer than 4.5 children per woman in some parts of Northern Europe in the eighteenth century to 5.5 in rural China around 1930, and somewhat over 6 in India around 1900. (See Wood, Johnson, and Campbell, 1985, for an example of moderate fertility in an early agricultural population.) In other words, the actual fertility in traditional societies has ranged from less than 40% to a little over 50% of the biological maximum. There are examples, however, of fertility that would yield from 7 to 8 children per woman: in colonial America, Czarist Russia at the beginning of this century, and in some populations of tropical Africa in the recent past. These populations, however, had recently expanded into newly occupied land, or had recently experienced an increase in fertility through the modification of longstanding customs.

Fertility concepts in genetics and demography

The demographic reason why long-established societies in premodern times had moderate rather than high fertility can be rationalized by considering the combinations of fertility and mortality that yield a rate of increase close to zero. On simple arithmetic grounds, it can be seen that a society that has persisted with more or less fixed customs for a period of centuries must have a moderate rate of increase. For example, if a population were growing at an average of 1% a year, its size would be multiplied by nearly 150 in 500 years, an increase that could not have been sustained within a fixed territory with premodern technology. Thus, the fertility customs that are advantageous in a premodern society are customs that promote moderately high fertility rather than the biological maximum. With a total fertility rate (average number of children born in a lifetime per woman) of 12, a moderate rate of increase could be maintained only if the average duration of life were about 15 years. Such high mortality would be disadvantageous to a society faced with competition with others and vicissitudes such as epidemics and shortages of food. In short, the socially advantageous fertility in a traditional society is moderately high fertility, not extremely high fertility.

Factors maintaining moderate fertility

Preston (Chapter 22) points out that the demographer's penchant for decomposing vital rates into component parts has reached its most elaborate formulation in the study of fertility. Here the interbirth interval is divided into the period of postpartum infecundability, the time to conception, the time added by spontaneous intrauterine mortality, and the period of gestation. These elements plus the proportions marrying and ages at marriage will determine the fertility level of a population. (Bongaarts and Menken, 1983). Differences in fertility across societies can be understood by their differences in terms of these factors.

Fertility at a moderately high level, rather than at the maximal level, has been maintained in traditional societies by a combination of late marriage and a substantial fraction of the population abstaining from marriage (in European populations), and by customs that restrict fertility within marriage, customs characteristic of Asian and African populations. The factors that Europe) restrict fertility within marriage in premodern populations are prolonged nursing, which postpones the resumption of ovulation following birth; customs of postpartum abstinence, in particular during the time when the child is breastfed; health related factors, such as reduced sexual activity in populations subject to chronic fevers, and sterility caused by venereal disease or tuberculosis.

At the same time, the fact that many of the elements of the fertility process are biologically as well as behaviorally determined, identifies the points where genetic elements can play a part. For example, the level of fecundability (the monthly probability that a woman conceives; *effective* fecundability requires that the conception results in a live birth; fertility refers to the actual number of offspring produced) appears to be highly variable; this may be due to behavioral factors (such as coital frequency) or biological factors (for example, level of fetal loss) or both. (See Lam and Smouse, Chapter 7 and Wood and Weinstein, Chapter 11). The implications for further research are strongly dependent on understanding the predominant source of variation, and it has implications for the adequacy of current models in terms of the tenability of assumptions about unobserved heterogeneity. (See Lam and Smouse, Chapter 7.) It may be that further gains in accounting for fertility variations will require demographers to adopt for analytic purposes the geneticists' "narrower" definition of a population as a community of randomly mating individuals in order to better gauge the genetic components of fecundability and its variation across groups (see Adams, Chapter 1).

Moderate fertility and natural selection

The prevalence of moderate rather than maximal fertility in preindustrialized societies, and the idea that moderate fertility might be advantageous to a society and lead it to be more successful in competition with others, appears at first glance inconsistent with the dynamics of natural selection. However, population ecologists have distinguished a series of reproduction strategies that produce relative advantage for different species (May and Rubinstein, 1985). One reproductive strategy is known as *r-selection*, which characterizes insects, and some fish. The species subject to *r-selection* gain an advantage from high reproductivity, which promotes successful survival through the capacity for rapid multiplication when the habitat is sparsely populated. In contrast there is *K-selection* which characterizes species of large body size, a long life span, and slow maturing offspring, who enjoy a stable habitat. The effective strategy here "is to have fewer offspring but to invest more time and energy in raising them" (May and Rubinstein, 1985). Also relevant to reproductive strategies is the way that mating systems and social groupings respond to environmental factors.

The terms *r* and *K* reproductive strategies derive from the key elements of the logistic density dependent model (*r* being the intrinsic rate of natural increase; *K* the carrying capacity of a given environment). This attempt to capture diverse life history patterns as a result of simple selective pressures has been criticized as overlooking the many components constituting a life history and the range of factors, such as trophic position and environmental predictability, that operate on each component (see, for example, Wilbur, Tinkle, and Collins, 1974).

PARITY-RELATED FERTILITY LIMITATION

Whatever the merits of the *r* and *K* distinction, it is nevertheless of interest that traditional societies achieve moderate fertility by practices that are for the most part socially determined, leading to a reproductive strategy little different from the pattern that characterizes whales, elephants, and the giant albatross, species in which moderate fertility has developed genetically through natural selection. The traditional practices that limit marital fertility well below the biological maximum are very different, however, from the practices that limit marital fertility in today's more developed countries. In these latter populations, marital fertility is limited by the widespread practice of contraception, plus resort to induced abortion. A characteristic feature of these practices is that they are undertaken (or at least intensified) by couples who have had as many children as they want. In other words, the practices that limit fertility are used differentially according to the number of children already born. Since the term parity is used to represent the number of children already born to a woman, an important distinction between the modern control of fertility in more developed countries and the traditional limitation of fertility below the biological maximum is that the modern control of fertility is "parity-related."

A puzzling feature of the moderate fertility in traditional societies is the apparent absence of parity-related control of the sort that has become universal in the more developed countries of the world today. The absence of parity-related control in the traditional societies is not, apparently, the result of the mere lack of modern contraceptive technology. Widespread parity-related control developed in some of the departments of France in the late eighteenth and early nineteenth century and was common in most of the departments by the end of the nineteenth century. Strong parity-related control of fertility also existed in certain villages in Hungaryin late eighteenth and early nineteenth century. As recently as just after World War II the most frequent form of effective contraception practiced in France was withdrawal or coitus interruptus, a practice known for centuries in many societies. Therefore the absence of modern technology is not the reason for the virtual absence of fertility regulation related to the number of children already born.

A highly conjectural explanation for the prevalence of customs other than parity-related control to achieve moderate fertility in traditional societies is the relative ease with which social and economic pressures and cultural norms can develop around access to marriage, abstinence, and breastfeeding practices as against practices associated with the cessation of childbearing upon attainment of a certain number. It is also possible, of course, that among certain sub-groups parity-related control did develop and eventually led to a level of fertility too low to sustain the group at then current levels of mortality. This would be an example of the traditional control practices being selected over the parity-related ones. In any event, the importance of socially developed practices to achieve biologically favorable outcomes provides an additional example of how traditional demographic and biological concerns intersect in the understanding of human populations.

ACKNOWLEDGMENTS

The authors wish to thank Julian Adams, Lora Myers, and Peter Smouse for helpful material and comments.

REFERENCES

Adams, J. and P. E. Smouse, 1985   Genetic consequences of demographic changes in human population, pp. 283-299 in *Diseases of Complex Etiology in Small Populations: Ethnic Differences and Research Approaches*, edited by R. Chakraborty and E. J. E. Szathmary. Alan R. Liss, Inc.

Bongaarts, J. and J. Menken, 1983   The supply of children: a critical essay, pp. 27-60 in *Determinants of fertility in Developing Countries* Volume I: *Supply and Demand for Children*. edited by R. A. Bulatao and R. D. Lee. Academic Press, New York.

Charlesworth, B., 1980   *Evolution in age-structured populations*. Cambridge University Press, New York.

Glass, B., 1971   Genetic and evolutionary considerations, pp. 114-126 in *Is There an Optimum Level of Population?* edited by S. F. Singer. McGraw-Hill, New York.

Hostetler, J. A., 1974   *Hutterite Society*. The Johns Hopkins University Press, Baltimore.

Lee, R. D., 1987   Population dynamics of humans and other animals. Demography 24(4): 443-465.

Malthus, T.R., 1798   *The First Essay*.

May, R. M. and D. I. Rubenstein, 1985   Reproductive strategies, pp. 1-23 in *Reproduction in Mammals Book 4: Reproductive Fitness* edited by C. R. Austin and R. V. Short. Second Edition, Cambridge University Press, Cambridge.

Notestein, F. W., 1953   Economic problems of population change. pp. 13-31 in *Proceedings of the Eighth International Conference of Agricultural Economists*. Oxford University Press, Oxford.

Sauvy, A., 1969   *General Theory of Population*. Basic Books, Inc, New York.

Singer, S. F. (ed.), 1971   *Is There an Optimum Level of Population?* McGraw-Hill, New York.

United Nations, 1973   *The Determinants and Consequences of Population Trends: New Summary of Findings on Interaction of Demographic, Economic and Social Factors*. Vol. 1. United Nations, New York.

Wallace, A. R., 1887   Letter from A. R. Wallace to A. Newton. pp. 220-201 in *The Autobiography of Charles Darwin and Selected Letters*, edited by F. DARWIN, Dover Publications Inc., New York.

Wilbur, H. M., D. W. Tinkle, and J. P. Collins, 1974   Environmental certainty, trophic level, and resource availability in life history evolution. Am. Nat. 108: 805-817.

Wood, J. W., P. L. Johnson, and K. L. Campbell, 1985   Demographic and endocrinological aspects of low natural fertility in Highland New Guinea. J. Biosoc. Sci. 17: 57-79.

Wright, S., 1960   Physiological genetics, ecology of populations, and natural selection, in *Evolution after Darwin*, edited by S. Tax, Vol. I. The University of Chicago Press, Chicago.

19

Genetic Comparisons of Life Stages in Natural Populations of *Zoarces viviparus*

FREDDY BUGGE CHRISTIANSEN

Various aspects of the properties of populations viewed as aggregations of individuals are studied in demography, population ecology, and population genetics. The focus on similar individual characteristics as a cause for the dynamic properties of populations ties these subjects together in population biology, and they share the basic demographic description of a population. Population growth is specified in terms of rates of fertility and mortality in the population, and these rates are viewed as originating from the properties of individuals. The simplest model describes the development of a population in terms of birth and death of independently acting individuals. This model produces exponential growth and led Malthus to discuss the properties and implications of unlimited growth. The realization that eventually the Malthusian model population will face resource limitation resulting in competition for resources, decreased fertility, and increased mortality is central to many aspects of population biology.

The work of Malthus was an important inspiration for Darwin in his proposition of natural selection as a mechanism driving biological evolution. Individuals may have different probabilities of survival and show differences in their propensity to breed, and this will result in variation among individuals in the expected proliferation of their offspring. If these differences are associated with variation in a heritable character, then the distribution of the characters in the population will change, in that traits which occur more often in individuals with a high propensity to proliferate will become more common. Differences in survival and fertility are the cause of natural selection on the character in question, and the process may be described in terms of rates of fecundity and mortality as generalized from the individual properties.

Natural selection therefore is described in terms of individual differences in demographic parameters; that is, variation in survival and reproduction throughout the life of the individual. Discussions and evaluations of the effects of fitness are made in terms of the individual demographic characteristics of variants in the population, and likewise fitness measurements should be made in terms of measurements of demographic parameters, as discussed by Bodmer (1968). For human populations these observations are further discussed by Cavalli-Sforza and Bodmer (1971), and the more general applicability of these methods are reviewed by Lewontin (1974), Charlesworth (1980), O'Donald (1983), and Manly (1985).

The individual demographic descriptions are laborious to obtain, or even impractical, in most natural populations of animals. A continuous description of the survival and reproduction of individuals may only be obtained by individual identification. However, for organisms with discrete breeding times, observations of the population at appropriately chosen time points can provide information comparable to information gathered by continuous monitoring of individuals. For instance, the probability of survival of an individual with a given trait from birth to the next breeding time amounts to a sufficient description of natural selection during that period, and this property may be observed as the fraction of surviving individuals carrying this trait.

In Mendelian genetics the substance of inheritance is separate from the expression of the inherited traits; that is, inheritance is indirect. Evolution proceeds by changing the genetic constitution of the population, so the evolutionary implication of natural selection is therefore determined by the degree of genotypic variation induced in the population. Also, indirect inheritance opens the possibility of genetic evolution independent of phenotypic evolution and natural selection. The demographic properties of the population are also important determinants of neutral genetic evolution and of the patterns of neutral genetic variation in recent populations (Kimura, 1983).

Darwinian fitness is described in terms of individual survival and reproduction. The probability of survival or the instantaneous death rate may be viewed as characteristics of the individual in the sense that they are functions of the properties of the individual, its phenotype. In sexually reproducing species, however, the fecundity of a female is a function not only of her own properties, but also of the phenotype of her partner(s). Selection due to variation in fecundity therefore is different from selection due to differential survival, in that fecundity selection acts on pairs of individuals. This will influence the evolutionary interpretation of selection effects, and these effects are more varied than for selection due to variation in survival (Feldman, Christiansen, and Liberman, 1983). Similarly, variation in the propensity to mate or in the ability to find mating partners, usually ascribed to sexual selection, is also a more intricate component in that it results from interactions among individuals (O'Donald, 1980). The last component of selection in sexually reproducing species, gametic selection, works through differential survival of gametes during the haploid phase in the life cycle, or as a deviation from simple Mendelian expectations of inheritance.

This summary of the various aspects of natural selection highlights a general problem for observation of natural populations. The male contribution to breeding will often be difficult to ascertain, but variation in male breeding success is weighed equally with variation in female fecundity in determining the total effect of selection. This problem exists even in populations where individual demographic data are available. Data on individual male breeding may often be impossible to obtain. For humans, for instance, reliable data on male marital

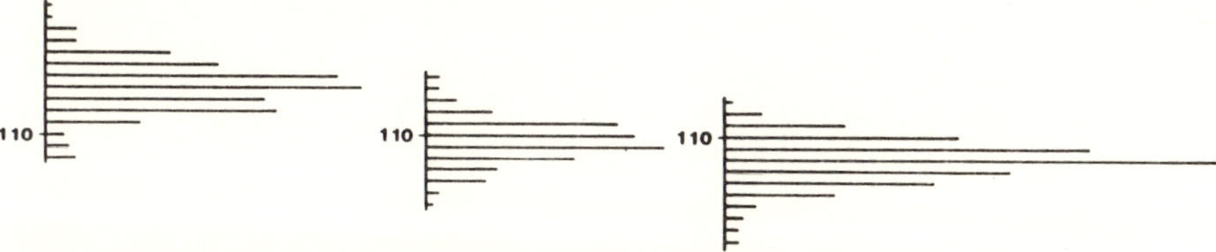

Figure 19-1. Population distribution of the number of vertebrae in *Zoarces viviparus* in three populations in the Ise Fjord Roskilde Fjord complex in Zealand, Denmark. Each bar indicates the number of observations of individuals with a particular number of vertebrae; from left to right, the total number of individuals investigated are 253, 122, and 311. The leftmost histogram shows the population distribution at the mouth of Ise Fjord, and the two histograms to the right describe populations in Roskilde Fjord that opens into Ise Fjord. The rightmost histogram shows the population distribution at the head of Roskilde Fjord. Redrawn after Schmidt (1918).

fertility may be obtained, but the extramarital fertility of the individual male is often hard to assess. One solution is to ascertain genetically the breeding success of the collection of males of a given type in the population; in organisms with discrete breeding times, this approaches optimal information on male breeding selection.

Genetic investigations of natural selection require a description of breeding in terms of the mating and fertility of individuals in the population. The description must be sufficient to give a proper account of the genetic composition among the offspring, the newly produced zygotes in the population. This is difficult in general, but it is possible in organisms with brood protection that breed at discrete times. These requirements are met by *Zoarces viviparus*, a live bearing marine fish abundant in shallow waters along the coasts of northern Europe. Natural populations of this species have been the object of studies on selection influencing protein variation as defined by electrophoresis. Further, the geographical variation among local populations of *Zoarces viviparus* has been studied to provide data on their characteristics and the amount of immigration.

INVESTIGATIONS OF GEOGRAPHICAL VARIATION

Zoarces viviparus was the subject of one of the earliest population genetic investigations of natural populations. Schmidt (1917) comprehensively described the geographical variation of metric characters in populations around Denmark. Figure 19-1 shows an example of the population distribution of the number of vertebrae in three populations about 20 kilometers apart in a Danish estuary. Viviparity allowed Schmidt to study the genetic determination of these characters in natural populations. The fish mate in summer and the females are pregnant until midwinter, when they deliver morphologically fully developed young. From late autumn the characters studied by Schmidt can be evaluated in the fetuses. Investigations of pregnant mothers and their offspring revealed heritabilities of around 80% for the vertebrae counts (Figure 19-2). That is, 80% of the variation in this character is genetically determined; a highly heritable

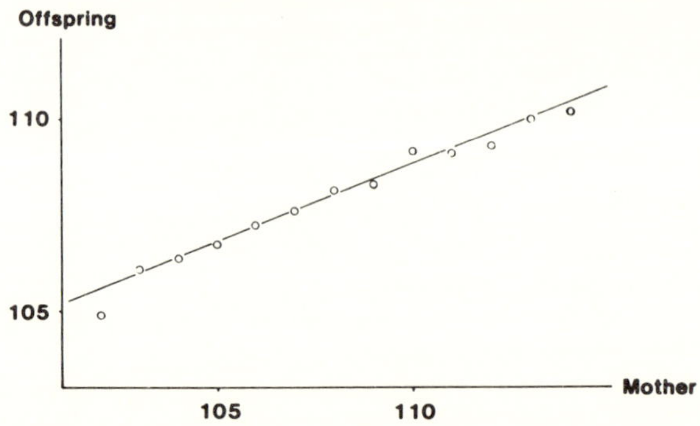

Figure 19-2.   Relation between the number of vertebrae in mothers and in their fetuses in a population sample from the head of Roskilde Fjord. In a sample of 631 mothers, the vertebrae number of each mother is determined, as well as the vertebrae number of five fetuses for each mother. The mothers are sorted according to the number of vertebrae, and the figure shows the mean vertebrae count for the fetuses of the various mother phenotypes. The slope of the indicated mother-offspring regression is 0.40. Redrawn after Smith (1922).

character, although the influence of maternal effects cannot be judged from the data shown in Figure 19-2.

The ability to determine the age of the individual fish allowed investigation of population data structured into cohorts, and these data helped to further evaluate the variation of the character (Figure 19-3). In a given population the mean varied significantly among cohorts (Table 19-1), and the corresponding small variance among cohorts is revealed as heterogeneity among age classes within a sample. Thus, by using the demographic structure of the population, Schmidt was able to argue that the morphological characters are fixed in the individual early in its life. Further, the cohort data clearly show that a small degree of year-to-year variation in the mean count among newborns exists (Table 19-1, and Schmidt, 1921), and this is probably due to environmental influences on the character during development. The year-to-year variation does not suggest a systematic trend, and the populations seem unchanged during the time since Schmidt's investigations (Ege, 1931; Christiansen, Nielsen, and Simonsen, 1988).

The geographical variation in morphological characters showed no obvious general patterns, and although regional differences existed, their importance faded as the many local discrepancies and irregularities in geographical distribution are considered (Schmidt, 1918). A careful comparison between the variation in morphological characters and variation in environmental variables led Schmidt (1917, 1918) to express opposition to naive adaptionist explanations of patterns of variation in natural populations. His conclusions (Schmidt, 1918) from consideration of the data on *Zoarces viviparus* were:

> Our investigations . . . by no means support the hypothesis that the racial characters are determined exclusively by the environment. On the contrary, they seem rather to indicate that differences of environments are not sufficient to explain the structural differences between the races . . .

Table 19-1 Variation in vertebrae counts in *Zoarces viviparus*

	Mean of Squared Deviation	Degrees of Freedom	Variance Component
Within age classes	3.883	4498	
Within cohorts	3.114	18	3.880
Among cohorts	24.683	8	0.037
Within samples	10.423	19	
Among samples	7.927	7	

NOTE:

Analysis of variance of the data depicted in Figure 19-3. A hierarchical analysis of variance of the data structured by age and cohort show that the variation among age classes within cohorts is no more than expected from the within age class variation. The variation among cohorts, however, is highly significant. The last column shows the corresponding components of variance in individual counts. In contrast, a hierarchical analysis of variance of the data structured by age and year of sample show that the variation among age classes within samples is significantly higher than expected from the within age class variation.

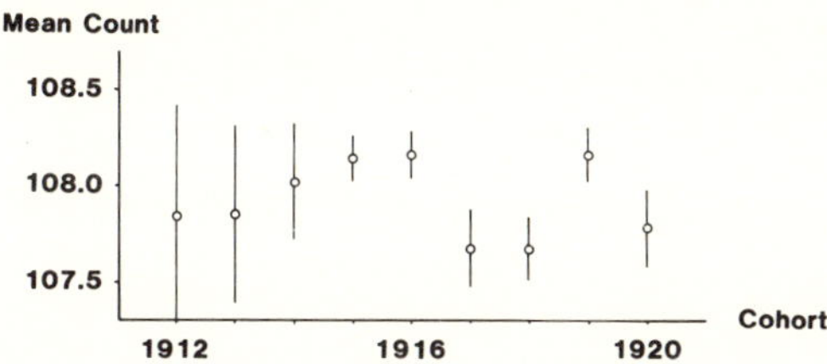

Figure 19-3. Variation in mean vertebrae count among cohorts born in the period 1912-20 in a population at the head of Roskilde Fjord (vertical bars indicate 95% confidence regions). Data from Schmidt, (1921).

However, Schmidt (1917, 1918) found one exception to the picture of erratic variation. In all the major Danish fjords he investigated he saw the same pattern as in Figure 19-1; namely, that the number of vertebrae in fish sampled inside the fjord were lower than the number in samples at the mouth of the fjord.

The geographical patterns in electrophoretically defined genetic variation in *Zoarces viviparus* share some of the aspects of the pattern of morphological variation, but for some loci clear regional patterns override the magnitude of local variation. A polymorphic esterase locus shows large scale clinal variation; that is, a gradual change in a character along geographic lines. From a gene frequency around 0.05 in the Baltic Sea, the frequency of an allele denoted as A_1 increases through the Danish Belts to a maximum around 0.40 in Kattegat, and the frequency again decreases south along the Danish North Sea coast (Frydenberg *et al.*, 1973). A similar pattern is displayed by an allele at a hemoglobin locus where the increase in frequency is from about 0.10 in the Baltic to close to 1.00 in Kattegat, and the cline is parallel to the esterase cline (Hjorth and Simonsen, 1975). In contrast, two polymorphic phosphoglucomutase loci show virtually constant gene frequencies over the same area with small erratic local and regional variations reminiscent of the pattern in the morphological characters (Christiansen *et al.*, 1976). However, the pattern at the esterase locus showed one striking similarity with the pattern of the morphological variation: the larger Danish fjords investigated also showed a cline in the frequency of

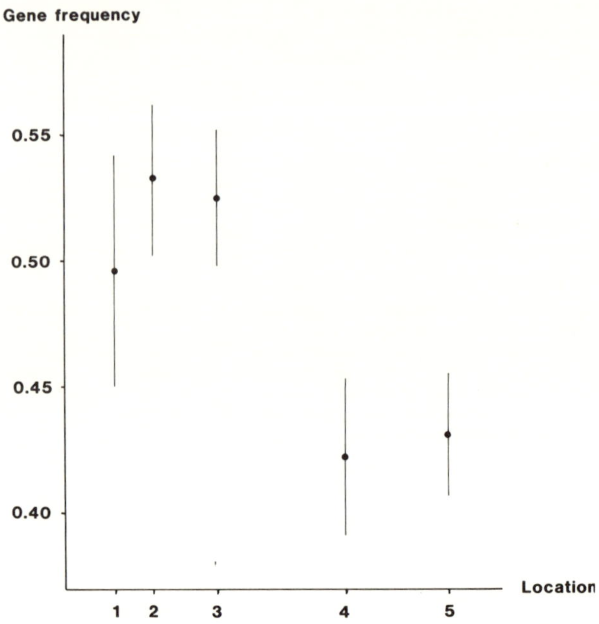

Figure 19-4.   The observed frequency of an esterase allele in populations of *Zoarces viviparus* at five different locations in Mariager Fjord in Jutland, Denmark. Location 1 is at the head of the fjord, location 5 is at the mouth of the fjord, and the locations are depicted at the abscissa showing their approximate relative distances. For each location the mean gene frequency of samples through the years 1969-77 (Frydenberg *et al.*, 1973; Christiansen, Frydenberg and Simonsen, 1984; Christiansen, Nielsen, and Simonsen, 1988) are shown and the 95% confidence regions of the gene frequencies are indicated. The three innermost populations clearly have a higher gene frequency than the two populations at the head of the fjord. The males of one subsample at location 5 show a gene frequency significantly different from that of the females of the same sample. This observation is marked separately (+) and not included in the average of samples at location 5.

one esterase allele, A_1, with a tendency toward a higher frequency inside the fjord than at the mouth. The highest frequency of the esterase allele A_1 was observed inside Mariager Fjord that opens into Kattegat, and the gene frequency difference between the head and the mouth of the fjord (Figure 19-4) is now well documented (Christiansen, Frydenberg, and Simonsen, 1984; Christiansen, Nielsen, and Simonsen, 1988).

The geographical variation (Figure 19-1) in connection with the homogeneity within cohorts (Table 19-1) suggests that very little exchange of individuals occurs per generation between populations about 20 kilometers apart. The constancy over 60 years of the local geographical pattern in morphology adds to the impression of a very sedentary organism (Christiansen, Nielsen, and Simonsen, 1988). The genetic difference between the innermost parts of Mariager Fjord and the populations at the mouth of the fjord further increases this impression of local isolation. The gene frequency difference suggests that exchange of genetic migrants is extremely low between populations separated by a distance that a determined individual *Zoarces viviparus* easily swims in a couple of days.

MOTHER-OFFSPRING INVESTIGATIONS

The access to different life stages allows investigations of variation in demographic parameters and therefore of natural selection in populations of *Zoarces viviparus*. Table 19-2 and Table 19-3 show population data on the esterase polymorphism collected in late autumn in the years 1969 through 1974. Samples of adults are gathered, and for each individual the sex, age, and genotype is determined. At the time of sampling the females breeding in the population that year are recognized as the pregnant mothers, and for each of these the genotype of a randomly chosen offspring fetus is determined and the total number of offspring counted. The data of Table 19-2 and Table 19-3 contain information on the genetic composition of several important stages in the life cycle of *Zoarces viviparus*. The genetic composition of the population of adults structured into cohorts (Table 19-4) is well described, but information on the population of reproducing females and their fecundity is also included. The data on the offspring and on fecundity counts allow estimation of the genetic composition among the population of newly produced offspring any given year, and this population is taken as representing the population of newly produced zygotes.

Comparisons between these discrete demographic stages expose changes in genetic composition supposedly due to natural selection (Christiansen and Frydenberg, 1974, 1976). For example, comparing the population of zygotes and the youngest age class in a cohort measures the effects of variation in survival among the three genotypes. As a result, parameters that provide a full description of selection during that life stage can be estimated. The data contain sufficient information to provide a full description of selection during most stages in the life cycle of *Zoarces viviparus*.

The data also contain information on the breeding males and the matings in the population, in that almost every observation of the genotype of an offspring fetus is an observation of an allele in a transmitted male gamete. The A_1A_1 mothers transmit an A_1 allele to the offspring, so the two offspring classes A_1A_1 and A_1A_2 correspond to observations of eggs fertilized by sperm carrying an A_1 allele and an A_2 allele, respectively (Table 19-3). Similarly, the two offspring classes of the A_2A_2 mothers correspond to observations of the two types of male gametes. The heterozygote A_1A_2 mothers may transmit either an A_1 allele or an A_2 allele to the offspring, so the two classes of homozygote offspring correspond to observations of fertilization by an A_1 allele and an A_2 allele, respectively. The origin of the alleles in the heterozygote offspring from heterozygote mothers cannot be identified. The data of Table 19-3 therefore contain the observation of $305 + 459 + 877 = 1{,}641\ A_1$ alleles and $516 + 877 + 1{,}541 = 2{,}934\ A_2$ alleles observed among $5{,}935 - 1{,}360 = 4{,}575$ sperm, where the male allele is unambiguously determined. The males that participate in the breeding in the population therefore are represented in the mother-offspring sample by

Table 19-2 Genotypic data on *Zoarces viviparus*

Population Section	Genotype			
	A_1A_1	A_1A_2	A_2A_2	Sum
Mothers	821 (13.8)	2696 (45.4)	2418 (40.7)	5935
Other Females	43 (11.8)	161 (44.2)	160 (44.0)	364
Adult Females	1008 (13.7)	3344 (45.3)	3029 (41.0)	7381[a]
Adult Males	693 (13.3)	2332 (44.6)	2201 (42.1)	5226[a]
Adults	1701 (13.5)	5676 (45.0)	5230 (41.5)	12607

NOTE:

The genotype at an esterase locus with alleles A_1 and A_2 is determined in adult individuals sampled during the breeding season in a population in Kaloe Cove just north of Aarhus, Denmark. The adults are sorted according to sex, and the females are sorted into pregnant females and non-pregnant females. The genotypic frequencies are shown as percentages in parentheses. From Christiansen, Frydenberg, and Simonsen (1977).

[a] Includes a sample of adults collected after the breeding season.

Table 19-3 Mother-offspring data on *Zoarces viviparus*

Mother Genotype	Genotype of Offspring			Sum	
	A_1A_1	A_1A_2	A_2A_2		
A_1A_1	305 (37.1)	516 (62.9)		821	(13.8)
A_1A_2	459 (17.0)	1360 (50.4)	877 (32.5)	2696	(45.4)
A_2A_2		877 (36.3)	1541 (63.7)	2418	(40.7)
Sum	764 (12.9)	2753 (46.4)	2418 (40.7)	5935	

NOTE:

The pregnant females of Table 19-2 are classified into seven mother-offspring types according to the genotype of a randomly chosen offspring. The frequency of each mother-offspring class within mother genotypic class is shown as percentages in parentheses. From Christiansen, Frydenberg, and Simonsen (1977).

the sperm they transmit to their offspring (Figure 19-5). This, however, allows only an incomplete description Zof variation in breeding ability of the male genotypes, and the mating in the population is inferred from the mother-offspring combinations. The mother-offspring data also describe the segregation of the two alleles in female heterozygotes, in that Mendelian segregation predicts that half the offspring from heterozygote mothers are heterozygotes (Christiansen

Table 19-4 Cohorts in the data on *Zoarces viviparus*

Age Class	Year of Sample			
	1971	1972	1973	1974
0	71	72	73	74
1	70	-	-	73
2	69	70	71	72
3	68	69	70	71
4	67	68	69	70
5	66	67	68	69
6+	≤ 65	≤ 66	≤ 67	≤ 68

NOTE:
The individuals in Table 19-2 were sampled in the breeding seasons of the years 1969 through 1974. Commencing in the year 1971 the age of the individuals is determined from count of growth zones in otoliths. Ages up to a maximum of 10 years has been observed, but few individuals are more than 5 years old, so older fish are pooled into the class 6+. Maturity usually is reached at age 2, and the adolescent age class 1 is poorly represented in the samples. The table depicts the cohorts in the samples by recording the year of birth of each age class. Entries are the two digits of year of birth.

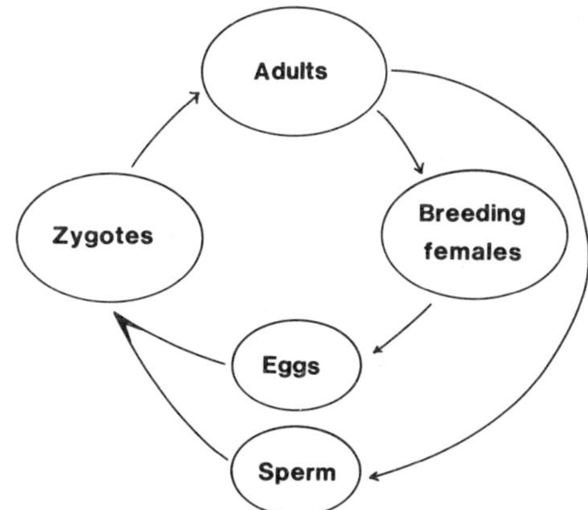

Figure 19-5. Simplified picture of the life stages represented in the population sample including mother-offspring combinations shown in Table 19-2 and Table 19-3.

and Frydenberg, 1973; Ostergaard and Christiansen, 1981). In males the description of the segregation is confounded with the description of male sexual activity.

Comparison between the genotypic frequencies among zygotes and among adults discloses the effect of differential survival, and by similar comparisons between the discrete demographic stages observed (Figure 19-5), the possibility of selection on the observed variation can be investigated through the process of reproduction. Further, comparisons among age classes within cohorts reveal the effects of differential survival of the various types of adults from breeding time to breeding time (Table 19-4). These different components of selection work in quite different ways. Differences in survival remove individuals of the various genotypes from the population in different proportions, so its effect is obvious

Table 19-5 Juvenile zygotic selection in *Zoarces viviparus*

	Genotype		
	A_1A_1	A_1A_2	A_2A_2
Zygotic frequencies	0.129	0.461	0.410
Relative fitness	1.065	1.000	1.037
Frequencies in adults	0.135	0.450	0.415

NOTE:

The table shows the genotypic frequencies among zygotes after random mating among adults, the relative probabilities of survival from zygote to adult, and the genotypic frequencies among adults. The zygotic genotype frequencies are the Hardy-Weinberg proportions corresponding to a gene frequency of allele A_1 of 0.360 ± 0.003. From Christiansen, Frydenberg, and Simonsen (1977).

as a change in genotypic composition of the population. In a parallel way, differential propensity to breed removes adult individuals from the breeding population, but the individuals remain in the population. These processes of sexual selection can only be studied by a characterization of the population of breeders. The procedure of sampling is expected to produce a size bias in the age distribution, in that the traps used are very ineffective in catching small fish. This is seen in the samples as a very low proportion of premature one-year-old fish; actually this age class is represented only in the samples of 1971 and 1974 (Table 19-4). This incomplete information on the age distribution makes it difficult to compare the breeding males (represented as transmitted sperm) with the adult males in the population, because the average gene frequencies among adult males in a sample may be different from the average gene frequency among adult males in the population if differences among cohorts exist (Christiansen and Frydenberg, 1976).

The analysis of the genotypic data of Table 19-2 and Table 19-3 is given by Christiansen *et al.* (1976) and their conclusion is that reproduction occurs by random mating among adults. In particular, no genetic differences are disclosed between adults and breeders, and survival of adults between breeding times is the same for all types. The only significant effect is a difference in genetic composition between the population of zygotes and the adult population, corresponding to differential survival among juveniles (Table 19-5). The selection favors homozygotes, so the prediction will be that the polymorphism is unstable. However, the differential survival seems to vary from year to year. Most cohorts in the investigation show a common pattern, but the first and the last cohort in the samples show clearly deviant patterns (Table 19-6).

The fecundity variation among female genotypes is no more than expected from the variation in fecundity within genotypes, and the influence of the male gamete type on the fecundity of the female is also insignificant. Fecundity is strongly age dependent (Figure 19-6), and the effect of female genotype on fecundity in the two dominating age classes shows the same tendency toward heterozygote inferiority as we saw in the survival component of selection. The variation in fecundity is dominated by effects of female age and size. The youngest age class has a lower fecundity than the rest, but the variation in fecundity among the older fish is due mainly to size (Figure 19-7).

Table 19-6 Cohorts in the data on *Zoarces viviparus*

| | Genotype | | | |
Cohort	A_1A_1	A_1A_2	A_2A_2	Size
66	0.84	1.00	1.85	49
67	1.11	1.00	1.07	762
68	1.03	1.00	1.04	2852
69	1.05	1.00	1.04	4446
70	1.03	1.00	1.00	990
71	0.97	1.00	1.12	365
72	1.13	1.00	0.82	560

NOTE:

The relative probabilities of survival from zygote to adult is calculated within each cohort (Table 19-4). The last column shows the total number of adults observed in the various cohorts. From Christiansen, Frydenberg, and Simonsen (1977).

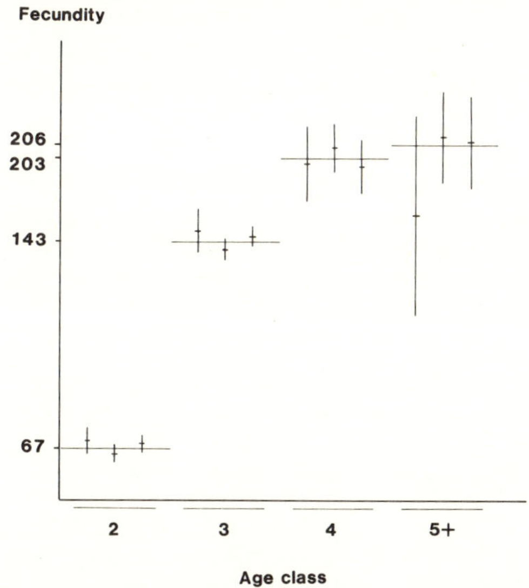

Figure 19-6. The variation in fecundity in *Zoarces viviparus* as a function of age and of genotype at a polymorphic esterase locus. For each age class the average fecundity of the three genotypes are shown with A_1A_1, A_1A_2, and A_2A_2 from left to right. Vertical bars indicate approximate (95%) confidence regions. The fecundity counts have been log-transformed before analysis, and the ordinate axis is accordingly logarithmic.

The insufficient description of the breeding males (Figure 19-5) is a weakness of the investigation of natural selection in *Zoarces viviparus*, and of male sexual selection in particular. Multiple paternities of the broods preclude the collection of more detailed information. The first offspring from a mother usually determines the allele carried by the fertilizing male gamete. Similarly, a second randomly chosen offspring also determines a male gamete, but at the locus considered, this second gamete may carry a gene which is identical to the gene carried by the first gamete and provide no new information. It may alternatively carry a copy of the other gene in the same male and thereby add

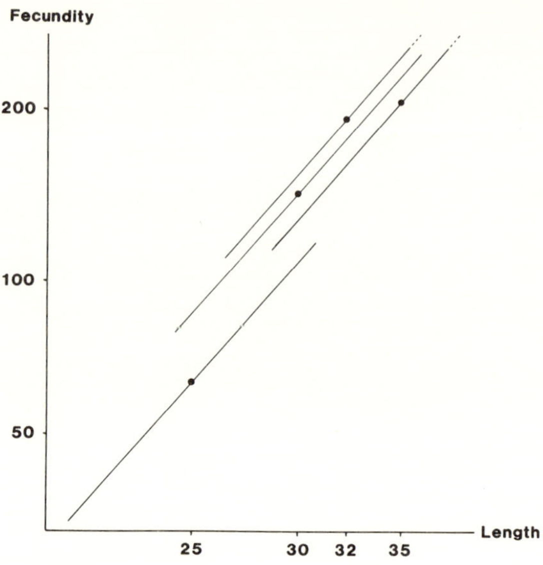

Figure 19-7. Lines of regression of individual fecundity on length for the four age classes of Figure 19-6. For each age class the population mean length and fecundity is shown and the regression line indicates the interval containing 95% of the population Both fecundity counts and lengths have been log-transformed before analysis, both axes in the figure are therefore logarithmic.

information on the genotype of the male. A third possibility exists, namely that the gene in the gamete originates from another male that mated with the female. Thus, if the mothers may carry broods with multiple sires, additional offspring supply very little information on the population of breeding males. However, if the possibility of multiple sired broods is excluded, then mother-offspring combinations with more than one offspring can provide an increase in the qualitative information on the breeding male population (Christiansen, 1980).

Mother-offspring investigations in *Gammarus*

The breeding biology of the marine amphipod *Gammarus oceanicus* allows the collection of population samples of mother-offspring combinations and suggests that each brood is sired by one male only. Mating can occur only at the time when the female molts; before mating the male carries the female around in a precopula. At molting the female releases about 50-100 eggs into her brood pouch, where they are fertilized, and she carries the eggs until they hatch and the juveniles leave the brood pouch. Siegismund (1985) collected mother-offspring combinations of *Gammarus oceanicus*, including five offspring per mother in a study of a mannose phosphate isomerase polymorphic for two alleles denoted B_1 and B_2. The mother-offspring type is characterized in this data by the mother genotype and by the combination of offspring genotypes. As an example, Table 19-7 shows the observed mother-offspring combinations for mothers of a particular homozygote type, B_1, out of a total sample of 361 females with broods:

Table 19-7 Mother-offspring data in *Gammarus oceanicus*

| Genotype of Offspring | | | | |
B_1B_1	B_1B_2	B_2B_2	Observed	Expected
5	0	0	92	92
4	1	0	12	10.2
3	2	0	24	20.3
2	3	0	14	20.3
1	4	0	11	10.2
0	5	0	23	23
Sum			176	176.0

NOTE:

Observed mother-offspring combinations with five offspring per mother where the genotype of the mother is B_1B_1. The expected values in the last column are calculated assuming that the segregating classes are distributed as a doubly truncated binomial distribution with $n = 5$ and $p = 0.5$, corresponding to Mendelian segregation. From Siegismund (1985).

176 are of genotype B_1B_1, 151 of genotype B_1B_2 and 34 of genotype B_2B_2. The offspring of the B_1B_1 mothers always receive the allele B_1 from the mother. The segrègation among the offspring in the genotypes B_1B_1 and B_1B_2 therefore implies that the father transmitted both allele B_1 and allele B_2, so the brood were sired by a heterozygote B_1B_2 male. The two nonsegregating mother-offspring classes, on the other hand, may be sired either by a homozygote male or by a heterozygote male where one of the two offspring types fortuitously did not get represented among the five offspring sampled. The distribution of the various segregating mother-offspring classes is a doubly truncated binomial distribution with $n = 5$ and frequency parameters describing the segregation of the two alleles in the male heterozygote (Nadeau, Dietz, and Tamarin, 1981). Upon the estimation of these segregation frequencies we can estimate the fractions of the nonsegregating mother-offspring classes sired by heterozygote males to obtain estimates of the genotypic frequencies among the males that sired the B_1B_1 females (Siegismund and Christiansen, 1985). A similar analysis may be performed for heterozygote mothers, where the signature of heterozygote males is offspring of both homozygote types within the brood. Thus, the mother-offspring data with several offspring per female contain information to advance a full genotypic description of the breeding process in the males and of the mating in the population (Siegismund and Christiansen, 1985; Siegismund, 1985).

The *Gammarus* data contained five offspring per female even though many more would have been possible. However, as soon as enough offspring are included to guarantee the qualitative information content (three offspring, Christiansen, 1980), the information gained by taking an extra offspring rapidly diminishes with the number of offspring already analyzed. If in addition to the offspring of simple mother-offspring combinations, m further offspring are analyzed, then with Mendelian segregation the probability that both male alleles have been detected is of the order of magnitude $1 - (1/2)^m$. The uncertainty of the evaluation is due to the incomplete information on the mating type, and this uncertainty will lower the probability, so additional offspring is expected to provide more information for broods of heterozygote females (Cooper, 1968). The efficiency of estimation in the heterozygote females is only little more than half the efficiency in the homozygote females, so double as many offspring from heterozygote females would have been desirable.

Discrete breeding times

The key to the selection component analysis discussed here is the population sample of mother-offspring combinations. This allows the description of the breeding components of selection and the genotypic frequencies in the population of offspring formed by the population of parents sampled. In *Zoarces viviparus* this is indeed the population of zygotes for the cohort produced in the year of sample, so the population of fetuses sampled is a natural entity in the population. This identification between the sampled population and the natural initiation of a cohort stems from the discreteness of breeding in the population. Further, the population of mothers sampled in a given year is indeed the population of females that breed in that year, and the number of fetuses in a female reflects her fecundity that year.

In *Gammarus oceanicus*, on the other hand, the identification of the sampled population of offspring with the population of zygotes initiating the next generation is more questionable, because here the breeding is not discrete. The females may breed repeatedly in the final few months of their lives, so the population of breeding females is not well defined in relation to the sample. Females without progeny in their brood pouch may just have released an earlier batch of offspring, they may breed later, or they may not breed at all. In addition, variation in the number of eggs per brood may be a poor representation of the fecundity, as this number may be correlated to the total number of broods produced.

Discrete breeding is therefore important for a comprehensive selection component analysis, based on changes in genetic composition of the population within a generation. Overlap between generations can be handled as long as the age structure of the population can be determined. Continuous breeding through a breeding season is enough to cause difficulties for the analysis, and the solution to these problems is to determine the number of broods of the analyzed individuals. This requires that demographic data based on the individual survival and reproduction be gathered (Bodmer, 1968), and this will often be impractical in natural populations unless individuals can be identified and followed in their breeding (O'Donald, 1983).

DISCUSSION

The mother-offspring investigations disclosed selection during the juvenile development of *Zoarces viviparus*. The mode of selection, however, does not explain the observed polymorphism in the population, because selection favoring both homozygotes, underdominant selection, will eventually lead to monomorphism. The data indicated a resolution of this dilemma in that the underdominant selection seemed to be an episode in time perhaps preceded by an episode favoring genotypes carrying allele A_2, followed by an episode favoring genotypes carrying allele A_1. It is conceivable that variation in relative fitnesses of the genotypes through time may maintain polymorphism (Haldane and Jayakar,

1963; Karlin and Liberman, 1974), but at present we have no support of this contention. As mentioned in the description of the geographical variation in *Zoarces viviparus* the esterase locus shows clinal variation through Danish waters. The Kaloe Cove population, the subject of the mother-offspring investigation, is situated within this cline, so immigration of fish from neighboring populations will influence the composition and the dynamics of the population. Immigration from populations with different gene frequencies will give rise to an excess in the frequency of homozygotes once expected in the population, mimicking the effect of selection against the heterozygote in survival. In population genetics, this is sometimes referred to as the *Wahlund effect* (see, for example, Christiansen, 1988). In the context of the *Zoarces viviparus* population in Kaloe Cove, even unreasonably high rates of immigration from rather distant populations will produce a Wahlund effect of an order of magnitude less than the effect observed (Table 19-5). Immigration from neighboring populations may nevertheless contribute to the stable persistence of the polymorphism in spite of the local selection pressure (Karlin and McGregor, 1972; Slatkin, 1973; Karlin and Richter-Dyn, 1976).

Consideration of the pattern of genetic variation in populations of *Zoarces viviparus* in the fjords led to the conclusion that it is a very sedentary organism with the local population being rather isolated from its neighbors. An immediate question is the cause of the differentiation of the fjord populations (Figure 19-1 and Figure 19-4). Variations in selection pressures among populations may of course produce the observed differentiation, but the independent evolution of virtually isolated populations may also produce the observed differentiation even in the absence of selection (Wright, 1940; Kimura and Maruyama, 1971). The observations of the genetic variation in the populations of Mariager Fjord (Figure 19-4) seem to be consistent with a neutral explanation in terms of differentiation due to random genetic drift. The variation on a regional scale, however, does not seem to be accounted for merely by the evolution of neutral genetic variation in a geographically structured population (Christiansen and Frydenberg, 1974).

Extensive data on the geographical variation exist for two phosphoglucomutase loci (Christiansen *et al.*, 1976), a hemoglobin locus (Hjorth and Simonsen, 1975), and the esterase locus (Frydenberg *et al.*, 1973). The two latter loci show parallel clines through the Danish waters (Figure 19-8) consistent with diffusive migration between the Baltic Sea (low frequencies of both alleles in Figure 19-8) and Kattegat (high frequencies of both alleles in Figure 19-8). Both the Baltic Sea and Kattegat represent regions of fairly constant gene frequencies. Thus, the gene frequency clines at the esterase and hemoglobin loci may be explained entirely by short range migration linking two differentiated regions.

In the same geographical area the two phosphoglucomutase loci are virtually constant, so for these loci we may consider the Baltic Sea and Kattegat as representing distant regions of rather equal gene frequencies with slight clines between the two areas. The two phosphoglucomutase loci show some regional differentiations on a small scale, and the esterase and hemoglobin gene frequencies do not show a perfect linear relationship (Figure 19-8). Nevertheless, the dominant geographical pattern is the marked differentiation between the Baltic Sea and Kattegat in two of the loci and the concurrent homogeneity at the two

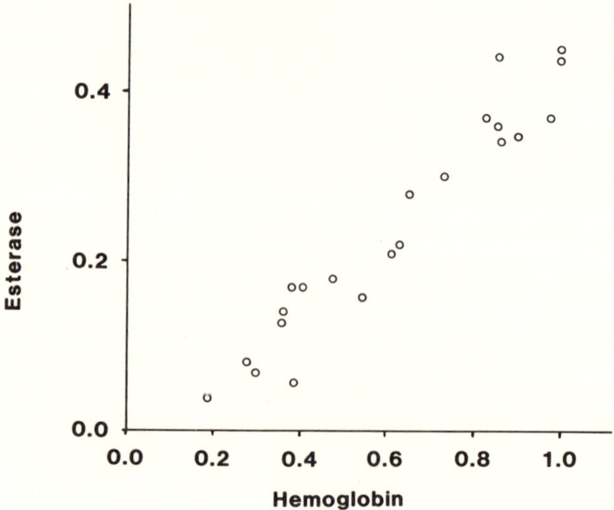#

Figure 19-8. The parallel clines in the polymorphic esterase and hemoglobin loci. For each sample in the region from Western Baltic to Kattegat the points indicate the gene frequencies at the hemoglobin locus and the esterase locus. Redrawn after Christiansen (1977), data from Hjorth and Simonsen (1976).

other loci. Thus, the parallel nature of the four clines, the two steep clines in the esterase and hemoglobin loci and the flat clines in the phosphoglucomutase loci, reduces the observed pattern to the differentiation of two areas, the populations in the Baltic Sea and the populations in Kattegat.

The divergence between two populations may occur as a result of random genetic drift if on the average they exchange considerably less than one gene per locus per generation (Wright, 1940, 1969; Kimura and Maruyama, 1971; Kimura, 1983). In this instance the amount of divergence due to random genetic drift is a function of the time elapsed since the time of separation of the two populations (Wright, 1940, 1969; Kimura, 1955, 1983), and in relation to the pattern of divergence in the four loci in *Zoarces viviparus* any time of separation either make the observed divergence or the observed constancy very unlikely. The parameters of the processes of migration and random genetic drift are independent of the particular loci we consider, so these processes alone are unable to explain both patterns simultaneously.

The seas where these patterns of variation have been observed are less than 10,000 years old, so heterogeneities in differentiation are unlikely to be due to differences in mutation pressure among the loci. Therefore, the conclusion is that selection has to be included as an explanation of the geographical pattern of at least one of the two groups of loci, the divergence or the equality of the two populations in the Baltic Sea and Kattegat (Christiansen and Frydenberg, 1974).

The observation of selection on the esterase locus in Kaloe Cove may suggest that the pattern of the clinal loci is affected by natural selection. However, this conclusion is not founded in the data, because we can only conclude that in Kaloe Cove the esterase polymorphism was influenced by selection during the period when the population was observed. The possibility remains that the esterase alleles showed associated fitness effects (Frydenberg, 1964; Christiansen, 1989) due to linked genetic factors under selection. Similarly, the differentiation between the Baltic Sea and Kattegat might well be independent of the electrophoretic alleles, but due to selection at linked, but unidentified loci (hitch-hiking; Maynard Smith and Haig, 1974; Thomson, 1977) is related to the adaptation of the *Zoarces* populations to the two very different environments (the Baltic Sea is very brackish, whereas Kattegat has close to oceanic conditions). The causal link between natural selection and observed genetic variation influenced by natural selection has to pass through the phenotypic effect of the genetic variation, and the relation of the corresponding phenotype variants to the environment.

ACKNOWLEDGMENTS

The collection of data on fecundity in *Zoarces viviparus* was done in collaboration with O. Frydenberg, A. O. Gyldenholm, and V. Simonsen; F. Abildgaard, A. Frederiksen and J. F. Jorgensen contributed to the analysis of these data. Comments on the manuscript by Al Hermalin and Volker Loeschcke were very helpful.

REFERENCES

Bodmer, W. F., 1968   Demographic approaches to the measurement of differential selection in human populations. Proc. Natl. Acad. Sci. USA 59: 690-699.

Cavalli-Sforza, L. L., and W. F. Bodmer, 1971   *The Genetics of Human Populations.* W. H. Freeman & Co., San Francisco.

Charlesworth, B., 1980   *Evolution in Age-Structured Populations.* Cambridge University Press, Cambridge.

Christiansen, F. B., 1977   Population genetics of *Zoarces viviparus* (L.): A review, pp. 21-47 in *Measuring Selection in Natural Populations*, edited by F. B. Christiansen and T. M. Fenchel. Springer-Verlag, Berlin.

Christiansen, F. B., 1980   Studies on selection components in natural populations using population samples of mother-offspring combinations. Hereditas 92: 199-203.

Christiansen, F. B., 1988   The Wahlund effect with overlapping generations. Am. Nat. 131: 149-156.

Christiansen, F. B., 1989   Population consequences of genetic design in sexually reproducing organisms, in *Genetically-Designed Organisms in the Environment*, edited by H. Money and G. Bernardi. John Wiley & Sons, London. In press.

Christiansen, F. B., and O. Frydenberg, 1973   Selection component analysis of natural polymorphisms using population samples including mother-offspring combinations. Theor. Pop. Biol. 4: 425-445.

Christiansen, F. B., and O. Frydenberg, 1974   Geographical patterns of four polymorphisms in *Zoarces viviparus* as evidence of selection. Genetics 77: 765-770.

Christiansen, F. B., and O. Frydenberg, 1976   Selection component analysis of natural polymorphisms using mother-offspring samples of successive cohorts, pp. 277-

301 in *Population Genetics and Ecology*, edited by S. Karlin and E. Nevo. Academic Press, New York.

Christiansen, F. B., O. Frydenberg, and V. Simonsen, 1977   Genetics of *Zoarces* populations X. Selection component analysis of the *EstIII* polymorphism using samples of successive cohorts. Hereditas 87: 129-150.

Christiansen, F. B., O. Frydenberg, and V. Simonsen, 1984   Genetics of *Zoarces* populations XII. Variations at the polymorphic loci *PgmI*, *PgmII*, *HbI* and *EstIII* in fjords. Hereditas 101: 37-48.

Christiansen, F. B., V. H. Nielsen, and V. Simonsen, 1988   Genetics of *Zoarces* populations XV. Genetic and morphological variation in Mariager Fjord. Hereditas 109: 99-112.

Christiansen, F. B., O. Frydenberg, J. P. Hjorth, and V. Simonsen, 1976   Genetics of *Zoarces* populations IX. Geographic variation at the three phosphoglucomutase loci. Hereditas 83: 245-256.

Cooper, D. W., 1968   The use of incomplete family data in the study of selection and population structure in marsupials and domestic animals. Genetics 60: 147-156.

Felldman M. W., F. B. Christiansen, and U. Liberman, 1983   On some models of fertility selection. Genetics 105: 1003-1010.

Frydenberg, O., 1964   Long-term instability of an *ebony* polymorphism in artificial populations of *Drosophila melanogaster*. Hereditas 51: 198-206.

Frydenberg, O., A. O. Gyldenholm, J. P. Hjorth, and V. Simonsen, 1973   Genetics of *Zoarces* populations III. Geographic variations in the esterase polymorphism *EstIII*. Hereditas 73: 233-238.

Haldane, J. B. S., and S. D. Jayakar, 1963   Polymorphism due to selection of varying direction. J. Genet. 58: 237-242.

Hjorth, J. P., and V. Simonsen, 1975   Genetics of *Zoarces* populations VIII. Geographic variation common to the polymorphic loci *HbI* and *EstIII*. Hereditas 81: 173-184.

Karlin, S., and U. Liberman, 1974   Random temporal variation in selection intensities: case of large population size. Theor. Pop. Biol. 6: 355-382.

Karlin, S., and J. L. McGregor, 1972   Polymorphisms for genetic and ecological systems with weak coupling. Theor. Pop. Biol. 3: 210-238.

Karlin, S., and N. Richter-Dyn, 1976   Some theoretical analyses of migration selection interaction in a cline: a generalized two range environment, pp. 659-706 in *Population Genetics and Ecology*, edited by S. Karlin and E. Nevo. Academic Press, San Francisco.

Kimura, M., 1955   Solution of a process of random genetic drift with a continuous model. Proc. Natl. Acad. Sci. USA 41: 144-150.

Kimura, M., 1983   *The Neutral Theory of Molecular Evolution*. Cambridge University Press, Cambridge.

Kimura, M., and T. Maruyama, 1971   Pattern of neutral polymorphism in a geographically structured population. Genet. Res. 18: 125-131.

Lewontin, R. C., 1974   *The Genetic Basis of Evolutionary Change*. Columbia University Press, New York.

Manly, B. F., 1985   *The Statistics of Natural Selection on Animal Populations*. Chapman and Hall, London.

Maynard Smith, J., and J. Haig, 1974   The hitch-hiking effect of a favourable gene. Genet. Res. (Camb.) 23: 23-35.

Nadeau, J. H., K. Dietz, and R. H. Tamarin, 1981   Gametic selection and the selection component analysis. Genet. Res. (Camb.) 37: 275-284.

O'Donald, P., 1980   *Genetic Models of Sexual Selection*. Cambridge University Press, Cambridge.

O'Donald, P., 1983   *The Arctic Skua*. Cambridge University Press, Cambridge.

Ostergaard, H., and F. B. Christiansen, 1981   Selection component analysis of natural polymorphisms using population samples including mother-offspring combinations, II. Theor. Pop. Biol. 19: 378-419.

Schmidt, J., 1917   Racial investigations I. *Zoarces viviparus* L. and the local races of the same. C. R. Trav. Lab. Carlsberg 13: 277-397.

Schmidt, J., 1918   Racial studies in fishes. I. Statistical investigations with *Zoarces viviparus* L. J. Genet. 7: 105-118.

Schmidt, J., 1921   Racial investigations VII. Annual fluctuations of racial characters in *Zoarces viviparus* L. C. R. Trav. Lab. Carlsberg 14: 1-24.

Siegismund, H. R., 1985   Genetic studies of *Gammarus* IV. Selection component analysis of the *Gpi* and the *Mpi* loci in *Gammarus oceanicus*. Hereditas 102: 241-250.

Siegismund, H. R., and F. Christiansen,   1985, Selection component analysis of natural polymorphisms using population samples including mother-offspring combinations, III. Theor. Pop. Biol. 27: 268-297.

Slatkin, M., 1973   Gene flow and selection in a cline. Genetics 75: 733-756.

Smith, K., 1922   Racial investigations IX. Continued statistical investigations with *Zoarces viviparus* L. C. R. Trav. Lab. Carlsberg 14: 1-42.

Thomson, G., 1977   The effect of a selected locus on linked neutral loci. Genetics 85: 753-788.

Wright, S., 1940   Breeding structure of populations in relation to speciation. Am. Nat. 74: 232-248.

Wright, S., 1969   *Evolution and the Genetics of Populations. Vol. 2. The Theory of Gene Frequencies*. University of Chicago Press, Chicago.

20

The Minimum Viable Population Size as a Genetic and a Demographic Concept

W. J. EWENS

This chapter has three related aims. The first is to investigate the derivation of the frequently-used value 500 for a minimum viable population size (MVPS), and as a result of this investigation to comment on the appropriateness of this value, taking both genetic and demographic factors into account. The second, having noted that the MVPS concept is intimately related both to that of the effective population size and also to the likely amount of genetic variation maintained in a given population, is to introduce a concept of the "mutation effective population size," essentially measuring the size of a population defined by its capacity to maintain genetic variation. Using this quantity, the effects of both demographic and genetic factors on the MVPS can be assessed. The third is to consider two models of catastrophic extinction of populations (or subpopulations) and to calculate, when appropriate, the value of the mutation effective population size, and hence the MVPS, in the case of such extinction.

It is clear that all our considerations center around calculation of an effective population size, so we begin with a review of this concept.

EFFECTIVE POPULATION SIZES

The first substantial analyses of the effects of random changes in gene frequency were made in the 1920s independently by Sewall Wright and R. A. Fisher, and the simplest possible model, introduced independently by them, concerns the evolution in a haploid population of the frequencies of two alleles A_1 and A_2 at some locus [see e.g., Fisher (1958) and Wright (1931, 1969)]. For a fixed population size of N individuals, this model assumes that if there are i A_1 genes in one generation, the probability P_{ij} that there will be j A_1 genes in the following generation is given by

$$P_{ij} = \binom{N}{j} \left(\frac{i}{N}\right)^j \left(1 - \frac{i}{N}\right)^{N-j} \tag{20-1}$$

We are in practice usually more interested in diploid populations, but for most of our investigations the model (20-1) is sufficient to describe these, where the diploid population size is $N/2$. Thus except in certain specific cases where diploidy is an essential component of the analysis and a diploid model is required, we assume a haploid model (as also, for example, do Maruyama and Kimura (1980) in considering questions similar to ours).

The properties of the model (20-1) are known in some detail; here we focus on three of them. The first of these is that if two individuals are taken at random in any given generation, the probability that they have the same parent is $\pi = N^{-1}$, so that $N = \pi^{-1}$. The second is that if we write X_t as the frequency of A_1 in generation t, the variance of X_t is $X_t(1 - X_t)N^{-1}$. The third is that if the transition probabilities P_{ij} are formed into a standard Markov chain transition matrix, the largest nonunit eigenvalue of this matrix, describing in a sense the rate of which genetical variation is lost in the population, is $\lambda = 1 - N^{-1}$, so that $N = (1 - \lambda)^{-1}$.

Now (20-1) is a a very simple model which does not allow for various demographic properties of real populations; for example, variation in population size, geographical structure, the existence of two sexes, and so on. It has nevertheless been used, perhaps largely for historical reasons, as a model against which other more complicated models are gauged. This is most frequently done by the calculation of an "effective population size," which is the size of a simple Wright-Fisher population sharing some feature with the population being studied.

The effective population size depends on what this feature is. If in the population of interest the probability that two individuals have the same parent is π^*, the arguments given above lead to the calculation of an (inbreeding) effective population size N_{ei}, defined by $N_{ei} = (\pi^*)^{-1}$. Similarly if, in the population of interest, the variance of X_{t+1}, given X_t, is $X_t(1 - X_t)N_{ev}^{-1}$, the (variance) effective population size is defined as N_{ev}. Finally, if the largest nonunit eigenvalue of the matrix of transition probabilities of gene frequency in λ^*, the (eigenvalue) effective population size N_{ee} is defined by $N_{ee} = (1-\lambda^*)^{-1}$. Thus the population of interest loses genetic variation, in a certain well-defined sense, at the same rate as a population of size N_{ee} following the simple Wright-Fisher model.

It is interesting to note that whereas the features used as a basis for the calculation of an effective population size are genetic, the reasons for introducing this concept are demographic, and arise because the demographics of real populations are more complex than the demographically-void Wright-Fisher model.

The three effective population sizes defined above are in some cases identical but often are not. Thus, strictly speaking, the expression, "effective population size," is meaningless, and requires a further adjective describing which concept of effective population size is of interest. Of the three concepts described, the one we focus on is the eigenvalue effective size. This is because the original calculations of MVPS used this particular effective population size, as we note later, and also because the variance effective population size need not even exist, since in some populations an equation of the form $\text{Var}(X_{t+1}) = X_t(1-X_t)N_{ev}^{-1}$ does not hold. This is so, for example, for subdivided populations (Rothman, Sing, and Templeton (1974)), where the subpopulation frequencies of A_1 (and not simply the total frequency X_t of A_1) are needed to describe the stochastic behavior of X_{t+1}. Subdivided populations are considered further below.

It is curious, since effective population sizes are often computed in order to assess the likely degree of genetic variation in a population, that no definition of effective population size specifically directed at such an assessment has been put forward (although indirectly Maruyama and Kimura (1980) have done so, using a concept similar to that given below). We therefore now give a formal definition of a mutation effective population size N_{em}.

Suppose genes mutate, at a rate u, in such a way that each new mutant gene is of an entirely novel allelic type, not previously existing in the population. Then genetic variation will persist in the population, and it is a standard result, in the simple Wright-Fisher model based on (20-1), that the stationary probability P that two genes taken at random in any generation are of the same allelic type is given by

$$P = (1 - u)^2[N - (N - 1)(1 - u)^2]^{-1} .$$

Equivalently the population size N is given in terms of P by

$$N = \frac{P^{-1} - 1}{(1 - u)^{-2} - 1} .$$

If in a demographically more complicated model the probability that two genes taken at random are of the same allelic type is P^*, we may define a mutation effective population size $N_{em}(u)$, for a given value of u, by

$$N_{em}(u) = \frac{(P^*)^{-1} - 1}{(1 - u)^{-2} - 1} . \tag{20-2}$$

Usually there will exist a limiting value of $N_{em}(u)$ as $u \to 0$, and if so, we call this the mutation effective population size, denoted N_{em}. Using (20-2), this can be calculated as

$$N_{em} = \frac{1}{2}\frac{dP^{*-1}}{du} , \tag{20-3}$$

where the derivative is calculated at $u = 0$. However, as we note below, it is often more appropriate to use (20-2), in particular when a specified positive value of u is assumed.

The calculations given above assume a haploid population. For monoecious diploid populations the mutation effective population size is found by taking half the values on the right-hand side of (20-2) and (20-3), and for a diploid population allowing two sexes, the mutation effective population size is found to be approximately

$$\frac{4N_m N_f}{N_m + N_f} , \tag{20-4}$$

where N_m is the number of males and N_f the number of females in the population. As is well known, this is essentially equal to the diploid population eigenvalue effective population size, and often differs considerably from the actual population size, a fact which we take up again below.

THE MINIMUM VIABLE POPULATION SIZE

The concept of a minimum viable population size (MVPS) was introduced about ten years ago, the first substantial quantitative attack being that of Franklin (1980). As a result of his calculations, Franklin suggested a value of about 500 for the MVPS. Franklin quite properly qualified his calculation by emphasizing the assumptions on which it was based, but the perhaps natural desire for some specific numerical value has unfortunately led to a reification of this number, and to its use in areas beyond that for which it was originally intended. The two main aims of this chapter are to emphasize, first, that there is no unique numerical value of MVPS, and second that the MVPS, at least as calculated by Franklin, derives from purely genetic arguments and needs further calculation before demographic factors are allowed for. We therefore start our investigation by summarizing the argument which led to the MVPS value of 500.

Franklin's calculations were based on an examination of bristle number variation in *Drosophila*. He calculated that, for bristle number, new mutations add, in each generation, to the genetically caused variation an amount roughly equal to one thousandth of the variation caused by purely environmental effects in a genetically homogeneous population. Making the rough approximation that additive genetic variance σ_a^2 is lost at about the same rate $(1/2N_{ee})$ as heterozygosity, where N_{ee} is the eigenvalue effective population size, a balance between loss and gain of variation is struck when $N_{ee} = 500$. This is, in very broad outline, the derivation of the minimum viable population size of 500.

There are several comments to make about this calculation. Starting from the most narrow of these, the calculation relies on the implicit assumption, pointed out clearly by Lande and Barrowclough (1987), that genetically induced variation should roughly approximate environmentally induced variation, and even though in some cases this may be the case, there is of course no intrinsic reason for equality of the two. If no reference is made to environmentally induced variation, calculation of an explicit value for the MVPS following the above argument largely collapses. If we balance the loss $\sigma_a^2/2N_{ee}$ of additive genetic variance with the gain σ_m^2 through mutation, we obtain, as Lande and Barrowclough point out, the equation

$$\sigma_a^2 = 2N_{ee}\sigma_m^2 \ .$$

Thus, additive genetic variance increases without limit as N_{ee} increases, and conversely, even for small populations, there will be some positive value for σ_a^2. Thus, unless extrinsic and possibly unrealistic assumptions are made, an MVPS cannot be calculated using N_{ee}. An alternative calculation for the MVPS, based on the mutation effective population size, is given below.

The second comment is to note that the calculation of the value 500 relies on a very specific case; namely, bristle number in *Drosophila*. Even for purely genetical questions there is no reason why this number should have any general significance. It appears to be far more appropriate to calculate a value relevant to the case at hand, based on whatever MVPS criteria are decided upon and whatever numerical values apply to that case. The use of a universal fixed numerical value does not appear to be justified, and emphasizing this point is one of the two aims of the paper.

The third comment is that, in calculating the MVPS, one is calculating an effective population size, not an actual population size. The two can differ substantially, as noted above. It is in moving from an effective to an actual population size that demographic factors become relevant. Despite this, an MVPS calculation is sometimes taken to be a calculation of an actual population size, thus losing the entire demographic aspect of the calculation. This point is illustrated in examples given below.

The final point, concerned directly with the second aim of this chapter, is to note that Franklin's calculation relies only on genetical arguments and thus without further modification has no direct connection with demographic questions. Its use in purely demographic contexts is thus inappropriate: the size of a *Drosophila* population needed to maintain a given genetically-induced variation in bristle number has nothing to do with the size of a grizzly bear population necessary to survive some environmental catastrophe.

There is of course a substantial demographic literature on a "minimum size" problem in which numerical calculations sometimes lead to a value of the order of several hundreds, that is to say similar to the Franklin value of 500. An interesting example is the calculation of Hammel, McDaniel, and Wachter (1979), based on incest taboos, where a value of "several hundreds" is reached for the size of a population needed to overcome population size depletion due to these taboos (see also Weiss and Smouse, 1976). Since these calculations start from different premises that those of the genetic MVPS, this similarity must be regarded as a coincidence. We do not pursue demographic calculations here since, having identified Franklin's MVPS as deriving from a genetic requirement, and then being subject to demographic amendment, our aim is to focus on it as a joint genetic-demographic concept. We will, on the other hand, mention briefly examples where the value 500 has been used in purely demographic contexts, and will then reiterate the comment just made, that this is unjustified unless the value 500 is arrived at by an independent argument.

DISCUSSION OF THE MVPS CONCEPT

We now take up numerical examples illustrating the points made concerning MVPS in the previous section. The first of these refers to a calculation of the MVPS based on a requirement of maintenance of genetic variation, using the mutation effective population size. Such a calculation seems to be important in view of the problems associated with the calculation of an MVPS through the eigenvalue effective population size. If in the calculations leading to (20-2) we can take the mutation rate u to be known, we can ask what size diploid population, given this value u, will result in an acceptable value for the probability P of homozygosity. As an example, suppose that a diploid population maintains a sex-ratio of one male to nine females and that the mutation rate u is 10^{-5}. It might be decided that a value of P equal to .8 is sufficiently low for the purposes of maintaining reasonable genetic variation. Using (20-2) and noting the diploid nature of the population, the mutation effective population size is

$$\left(\frac{1}{2}\right)\left(\frac{.8^{-1} - 1}{.99999^{-2} - 1}\right) = 6250 \ .$$

This value is reached by purely genetic arguments. It is now that the demographic factor of sex ratio enters, and (20-4) shows that 6,250 is only 36% of the actual population size, which must therefore be about 17,400 to maintain the genetic variation required.

There are two points to make concerning this calculation, both raised in the previous section. First, it relies on an explicit criterion referring to the extent of genetic variation to be maintained in the population, rather than on the various assumptions leading to the numerical value 500. The second is that the value

Table 20-1 Minimum population sizes for domestic herds recommended by U.S. Congressional Office of Technology Assessment

	Males	Females
Cattle	20	1000
Sheep	20	500
Goats	20	500
Pigs	20	200

6,250, being an effective population size, must be adjusted appropriately to derive an actual population size. This demographic adjustment is crucial since no actual population of size 6,250 is contemplated.

This adjustment is almost always omitted in the literature. Thus in Table 6.2 of the Congressional Office of Technology Assessment publication "Technologies to Maintain Biological Diversity" (1987), numerical values are given on the size of domestic herds below which action might be required to increase genetic variation. These sizes are given in Table 20-1.

The derivation of these numbers is not given in the Congressional report. If they are obtained from the "accepted" MVPS of 500, as the values in the final column suggest, the values in Table 20-1 are self-defeating: a population of 20 males and 500 females has an effective population size of about 83, far less than the possibly intended value of 500. With a 20:500 sex ratio, an actual size of about 3000 is needed for an effective size of 500. This comment emphasizes the demographic adjustment necessary to obtain a true population size, not a conceptual value such as 500.

We turn briefly to purely "demographic" minimum population sizes. The Congressional report, in its first page, refers to ecosystem, species, and genetic diversity and then considers in great detail strategies for the maintenance of all three. Our interest here centers on genetic and species diversity; we have discussed above the size that a genetic population has to be in order to maintain a specified level of variation. So far as species diversity is concerned, a large population size is needed to lower the changes of imminent total extinction of a population: species such as the California condor come immediately to mind. Thus we want large populations for two different reasons in maintaining species and genetic diversity, and this should be remembered in separating clearly genetic and demographic minimum size concepts.

It is clear that the genetic calculations given above and, even if relevant, the MVPS value of 500, have nothing to do with maintaining species diversity. Yet the value 500 has been used for purely demographic or ecological calculations concerning species survival. For example, the Congressional report, in its table 6.1, lists a population of size 500 or less as being in danger of imminent extinction, laying down guidelines for intervention in such cases to conserve the population. This value appears to derive from the same source as the "genetic" MVPS 500, and if so has no direct demographic relevance. Appropriate values should come from the demographic literature such as that referred to above.

Some emphasis has been placed above on separating genetic demographic contributions to the MVPS. Despite this, the problem of species survival is of course in practice not unconnected with genetic calculations, since the loss of genetic variation in small populations can lead to loss of vigor and thus possibly to species extinction. An overall theory, taking demographic and genetic aspects jointly into account, would be most desirable. However, demographers are at a disadvantage so far as their part of a joint theory is concerned: while geneticists can use fairly concrete MVPS criteria, those for demographers are not so clear-cut. Is an acceptable MVPS one which gives a 95% chance of survival for 500 years? A 99% chance of survival for 200 years? Clearly the criteria will vary from case to case, reinforcing the view that even if a joint theory is possible, no fixed specified value for MVPS exists.

CATASTROPHES: A DEMOGRAPHIC MODEL

It is not even certain that, in a purely demographic context, a concept of minimum viable population size is the most appropriate feature to focus on. To illustrate this point, we examine in this section a purely demographic model, involving no genetics, where this might be the case.

Consider a single homogeneous population whose size is determined by birth, death and catastrophic events. We assume linear birth and death rates, so that if the population size is N,

$$Pr[\text{birth} \ \text{in} \ (t, t + \delta t)] = \alpha N(\delta t) \ ,$$
$$Pr[\text{death} \ \text{in} \ (t, t + \delta t)] = \beta N(\delta t) \ ,$$

where we ignore small-order terms throughout. We also assume that a catastrophe might occur in this small time interval. Catastrophes are assumed to arise for reasons extrinsic to the population, and thus the probability that a catastrophe occurs is independent of population size. Thus we assume

$$Pr[\text{catastrophe} \ \text{in} \ (t, t + \delta t)] = \gamma(\delta t) \ .$$

If a catastrophe occurs, each individual independently survives with probability $p, (0 < p < 1)$.

The interpretation of the parameters α, β and γ is immediate: β^{-1} is the mean lifetime of each individual, $\alpha\beta^{-1}$ the mean number of offspring per individual, while γ^{-1} is the mean time between catastrophes.

Clearly the population can survive only if, in some sense, α and p are sufficiently large and β and γ are sufficiently low. The exact criterion for possible survival of the population is given by Ewens et al. (1987), namely

$$\alpha > \beta - \gamma \, log_e \, p \ .$$

If extinction is certain, the mean time for extinction is

$$\frac{log_e(N_0)}{\alpha - \beta + \gamma \, log_e \, p} \ ,$$

where N_0 is the initial population size.

We note two conclusions: first, the criterion for certain eventual extinction of the population is independent of the initial population size, and the second that when eventual extinction is certain, the mean extinction time is only weakly dependent on the initial size. (The variance of the extinction time is similarly weakly dependent on the initial size.) Thus even though this model is a simplified one, it suggests that for large populations, the focus on population size, and the concept of MVPS, could be misplaced: perhaps a more appropriate focus is on α and β. For small populations the probability of extinction of the population is, on the other hand, dependent on the current population size in a significant way.

It would be of great interest to consider models more complex than the above. In particular, no analogous information is yet available on catastrophe models where the birth rate is of the logistic form $\alpha N(K - N)$, where K is a carrying capacity, or on several other methods which can reasonably be constructed to describe population size behavior. It is not clear to what extent the conclusions drawn above would continue to hold in these models.

CATASTROPHES: GENETIC MODELS

In this section we consider joint genetic-demographic calculations of MVPS in subdivided populations subject to catastrophic loss of subpopulations. Since this MVPS is concerned with the amount of genetic variation maintained, we do not consider models (such as that of the previous section) where the entire population can be lost.

Substructured populations have been considered at great length in the literature and a general result, covering the many complex cases considered, is not easily obtained. Here we consider two simple models where an analytic conclusion is possible and which illustrate the points of issue. A more general analysis, covering at least the important cases considered by Maruyama (1972) and Slatkin (1977), would be most desirable.

The first model envisions a haploid population (of size Nn) subdivided into n subpopulations each of size N. Within each subpopulation random mating occurs and in a simple two-allele model equation (20-1) applies within each subpopulation. However, it is assumed that, in each generation, k subpopulations are lost due to catastrophic events and the niches they occupied are replaced by individuals from the remaining subpopulations in such a way that each of the k niches is filled, independently of the other niches, with offspring from individuals from one or other of the $n - k$ surviving subpopulations.

In the spirit of Franklin's calculations we consider first the rate of loss of genetic variation when there is no mutation. In a two-allele model for which (20-1) applies within each surviving subpopulation, genetic variation is lost in the population through two agencies. The first is the random loss of genetic variation within subpopulations (random drift), and the second is the loss of variation through the demographic process of loss of subpopulations. This rate of loss of genetic variation is measured by an eigenvalue λ, which in the case of a random-mating population of size Nn is $1-(Nn)^{-1}$. For the model described above,

$$\lambda = \max[1 - N^{-1}, 1 - q]\,,$$

where the parameter q is determined entirely by the demographic process of loss of subpopulation with replacement by individuals from surviving subpopulations and is given by

$$q = \frac{k(2n - k - 1)}{n(n - 1)(n - k)}\,, \tag{20-5}$$

(Ewens, 1989). Except in extreme cases involving very many subpopulations each of small size, the rate of loss of genetic variation is greater than that for a single random-mating population of size Nn, and the implication for MVPS is that a larger population size is needed in the substructured case if the rate of loss of genetic variation is to be kept at the same value as that for an undivided population.

As stated earlier, it is perhaps easier to consider the mutation effective population size N_{em} (rather than the eigenvalue effective size) and consider the amount of genetic variation maintained in the presence of mutation. In the model considered, N_{em} is given by

$$N_{em} = N + \frac{N(n-1)(1-q)}{q(Nn-1)} , \qquad (20\text{-}6)$$

and except in the extreme cases mentioned in the previous paragraph, this is less than the actual population size Nn.

As an example, suppose that the total population of size 5,000 is divided into 50 subpopulations of equal size, one of which is lost in each generation. Then $N = 100, n = 50, k = 1$, and from (20-5) and (20-6) $N_{em} = 1,300$, about one quarter of the actual population size. In other words, this population maintains about one quarter the amount of genetic variation as an undivided population of the same size. Put in other words, the demographic process of subpopulation loss requires an actual population size about four times as large as is required for a single random-mating population maintaining the same level of genetic diversity.

These calculations are specific to the model considered. A second example is more in the spirit of the demographic model considered earlier in that it allows the total population size to vary, although not in such a way as to allow complete extinction of the population.

We consider again a population divided into n subpopulations, of which k are lost in each generation due to catastrophic events. However, the replacement procedure differs from that of the previous model. Here we assume that the niche previously occupied by a lost subpopulation is filled by a single offspring, drawn at random from the individuals in the surviving subpopulation. At the same time, the number of individuals in each surviving subpopulation increases by unity in each generation. This model is similar to that discussed by Maruyama and Kimura (1980).

The sizes of the various subpopulations in this model are random, but it is easy to see that the mean size of each subpopulation is n/k, so that the mean size of the entire population is n^2/k. Assuming again that all mutant genes are of novel allelic type, reasonable approximations (Ewens (1989)) and simulation (Maruyama and Kimura (1980)) suggest strongly that the mutation effective population size is about $n^2/2k$, that is about half the mean size. In this model, then, the demographic process means that the actual population size must be about twice as large as a single random-mating population maintaining the same level of genetic diversity.

Many other models involving population subdivision and catastrophic extinction can be put forward, some particular cases and important calculations being given by Slatkin (1977). We conclude by emphasizing again that the conclusions on MVPS will differ in these various models (as they do in the two considered above), and as with the value 500, no value or formula should be accepted in advance. An independent calculation must be made for the model deemed most appropriate for the situation at hand.

REFERENCES

Ewens, W. J., P. J. Brockwell, J. M. Gani, and S. I. Resnick, 1987   Minimum viable population size in the presence of catastrophes, pp. 59-68 in *Viable Populations for Conservation*, edited by M. E. Soulé Cambridge University Press, Cambridge.

Ewens, W. J., 1989   The effective population sizes in the presence of catastrophes, pp. 9-25 in *Mathematical Evolutionary Theory*, edited by M. W. Feldman. Princeton University Press, Princeton.

Fisher, R. A., 1958   *The Genetical Theory of Natural Selection*. Dover Publications Inc., New York.

Franklin, I. R., 1980   Evolutionary change in small populations, pp. 135-149 in *Conservation Biology. An Evolutionary-Ecological Perspective*, edited by M. E. Soulé. Sinauer Associates, Sunderland, MA.

Hammel, E. A., C. K. McDaniel, and K. W. Wachter, 1979   Demographic consequences of incest taboos: a microsimulation analysis. Science 205: 972-977.

Lande, R., and G. F. Barrowclough, 1987   Effective population size, genetic variation, and their use in population management," pp. 87-123 in *Viable Populations for Conservation*, edited by M. E. Soulé. Cambridge University Press, Cambridge.

Maruyama, T., 1972   Rate of decrease of genetic variability in a two-dimensional continuous population finite size. Genetics 70: 639-651.

Maruyama, T., and M. Kimura, 1980   Genetic variability and effective population size when local extinction and recolonization of subpopulations are frequent. Proc. Natl. Acad. Sci. USA 77: 6710-6714.

Rothman, E. D., C. F. Sing, and A. R. Templeton, 1974   A model for analysis of population structure. Genetics 78: 943-960.

Slatkin, M., 1977   Gene flow and genetic drift in a species subject to frequent local extinctions. Theoret. Pop. Biol. 12: 253-262.

U. S. Congress Office of Technology Assessment, 1987   *Technologies to Maintain Biological Diversity*. U. S. Government Printing Office, Washington, DC.

Weiss, K. M., and P. E. Smouse, 1976   The demographic stability of small human populations. J. Hum. Evol. 5: 59-73.

Wright, S., 1931   Evolution in Mendelian populations. Genetics 16: 97-159.

Wright, S., 1969   *Evolution and Genetics of Populations, Volume 2: The Theory of Gene Frequencies*. University of Chicago Press, Chicago.

21

Two-Sex Population Models and Classical Stable Population Theory

ROBERT A. POLLAK

Classical stable population theory is inherently a one-sex model of population age structure and growth. The model's essential building blocks are an age-specific fertility or maternity schedule for the female population and an age-specific female mortality schedule. These two schedules define a mapping of the female population vector (by age) in period t into the female population vector in $t + 1$. An equilibrium age structure is one that reproduces itself up to a scale factor, where the scale factor is essentially the equilibrium rate of growth. Because classical stable population theory corresponds to a linear model, the mathematics are straightforward.

Males play virtually no role in classical stable population theory: the mathematics imply either that there are no males and the population reproduces asexually, or that a small number of males can fertilize a large number of females, and females rear their offspring without male assistance. In either case, the number of newborns depends only on the number and age distribution of females and is independent of the number and age distribution of males.

For monogamously mating, age-structured populations, males as well as females can play significant roles in determining age structure and growth. In the case of human populations, a "marriage squeeze" produced by a baby boom provides the standard example. Consider an equilibrium in which the population has been maintaining its age structure and growing at a constant rate, and the marriage pattern is one in which females normally marry older males. Suppose the equilibrium is disturbed by a "baby boom." When the females at the leading edge of the baby boom cohort enter the marriage market, they will find a shortage of "appropriately" older males. Precisely how this short-

317

age affects marriage patterns depends on the willingness of females and males to adjust conventional age-specific marriage patterns in response to changes in the availability of "appropriate" or "suitable" mates. Such adjustments may increase the probability that older males will find mates and the probability that younger males will find mates. These adjustments may also reduce the probability that females in the baby boom cohort will find mates and the probability that older females will find mates. Classical stable population theory admits no such adjustments for females, requiring instead that the fertility of a female of age i remain fixed, independent of the number and age distribution of females and males in the population.

Although the analysis of the equilibrium age distribution and growth rate of age-structured, monogamously mating populations belongs to population biologists and ecologists as well as demographers, in Pollak (1987) I call it "the demographer's two-sex problem." The problem of incorporating males into classical stable population theory is usually regarded as a "puzzle," to borrow Kuhn's terminology. I believe demography's two-sex problem is not merely a puzzle but an "anomaly" symptomatic of a fundamental defect in the classical stable population paradigm.

Classical stable population theory was created by Lotka. Coale (1972), Keyfitz (1968), and Pollard (1973) provide surveys and Arthur (1981, 1982) a recent treatment of the dynamics. Goodman (1953), Fredrickson (1971), Keyfitz (1971), Yellin and Samuelson (1977), Das Gupta (1978), and Schoen (1981) discuss demography's two-sex problem; Das Gupta and Schoen provide extensive references to the literature. My treatment here follows Pollak (1986, 1987, 1990). Feeney (1972), in an unpublished doctoral dissertation, clearly formulates and analyzes a discrete time version of the two-sex problem with the same basic structure as mine, but without a satisfactory proof of the existence of equilibrium.

There are at least two other "two-sex" problems. The most fundamental one is why some species, including our own, reproduce sexually. The second takes sexual reproduction as given and seeks to explain why the sex ratio for a species or a population assumes a particular value. On the first question, see Maynard Smith (1978) and Bernstein *et al.* (1985). R. A. Fisher developed the classical theory of sex-ratio determination; for modern views see Maynard Smith (1980) and Charnov (1982). Demography's two-sex problem takes both sexual reproduction and the sex ratio of newborns as given.

The simplicity of classical stable population theory arises from the *female dominance assumption*—the assumption that the age-specific female fertility rates are constant parameters of the model and, therefore, the same in every period. Because the female rates are parameters of the model, they can be estimated from the vital rates observed in a single period and used to calculate the equilibrium age structure and growth rates of the population. The female dominance assumption implies that any disequilibrium is reflected entirely in male fertility rates. It is instructive to contrast the female dominance model with the mirror image *male dominance* model in which the age-specific male paternity rates are assumed to be constant parameters of the model. The mathematics of the male dominance model is essentially identical to that of the female dominance model; the model yields an equilibrium age structure and an equilibrium growth rate for the male population, and females play no essential role. When actual data are used to calculate the growth rates corresponding to the female dominance and the male dominance models, the calculated rates generally differ, implying that the sex ratio will approach either zero or infinity. The source of the difficulty lies in the assumption that not only the two age-specific mortality schedules but also both the female maternity schedule and the male paternity schedule are fixed parameters of the model. The mathematics allows us to treat either the female or the male fertility schedule as fixed, but not both, at least not so long as we maintain the standard assumption that the age-specific female and male mortality schedules remain fixed. Other than tradition and mathematical convenience, however, there is little justification for assuming that either one of the fertility schedules remains fixed. It is logically possible that age-specific female fertility rates will be unaffected by a marriage squeeze and that the disequilibrium will be reflected entirely in male rates, but it seems improbable. It seems more likely that a marriage squeeze would affect the age-specific fertilities of both females and males. This implies that the fertility rates for both females and males observed during a marriage squeeze are "disequilibrium" rates, and the fertility rates for both females and males depend on the age-sex composition of the population.

In a series of recent papers, Pollak (1986, 1987, 1990), I propose a new model for analyzing the age structure and growth of age-structured, monogamously mating populations. The *birth matrix-mating rule (BMMR)* model generalizes classical stable population theory in a way that permits both female and male fertility rates to depend on the age-sex composition of the population. Because the BMMR model is inherently nonlinear, it is mathematically more complex than classical stable population theory.

The plan of this chapter is as follows. The first two sections describe classical stable population theory and the BMMR model. The next two offer examples and analyze the special cases in which the BMMR model reduces to classical stable population theory. The final section is a brief conclusion.

CLASSICAL STABLE POPULATION THEORY

The mathematics of classical stable population theory is straightforward. Let F denote the female population vector by age, $F = (F_1, \ldots, F_n)$, where n is the greatest age that any individual can attain. Let d denote the female mortality schedule, $d = (d_1, \ldots, d_n)$, where $d_n = 1$, and b the female fertility (maternity) schedule, $b = (b_1, \ldots, b_n)$, where b_i is the number of female offspring born to a female of age i.

These two schedules define a mapping or projection of the female population in period t into the female population in period $t + 1$: the age-specific fertility rates determine the number of newborns in period $t + 1$, and the mortality schedule determines the number in each of the other age categories. Applying the age-specific fertility schedule to the number of females of each age in period t determines the number of newborns in period $t + 1$:

$$F_1^{t+1} = \sum_{i=1}^{n} b_i F_i^t . \qquad (21\text{-}1)$$

Applying the age-specific mortality schedule to the number of females of each age in period t determines the number in the successor category in period $t + 1$:

$$F_i^{t+1} = (1 - d_{i-1})F_{i-1}^t .$$ (21-2)

Matrix notation allows us to express this compactly as

$$F^{t+1} = LF^t$$ (21-3)

where the $n \times n$ projection matrix L is called the "Leslie" matrix:

$$L = \begin{pmatrix} b_1 & b_2 & \ldots & b_{n-1} & b_n \\ (1 - d_1) & 0 & \ldots & 0 & 0 \\ \vdots & \vdots & \ddots & \vdots & \vdots \\ 0 & 0 & \ldots & (1 - d_{n-1}) & 0 \end{pmatrix}$$ (21-4)

Linear algebra provides powerful tools for investigating the existence of equilibrium and dynamic stability in the classical model. An equilibrium is defined as an age distribution, $\hat{F}$, and an equilibrium or intrinsic growth rate, $\hat{r}$, which satisfy the matrix equation

$$(1 + \hat{r})\hat{F} = L\hat{F} .$$ (21-5)

Thus, the equilibrium age distribution is an eigenvector of the Leslie matrix and $(1+\hat{r})$ is the corresponding eigenvalue. Standard demographic terminology calls an age structure that reproduces itself up to a scale factor a "stable age distribution" rather than an "equilibrium age distribution." I have departed from the standard terminology because it blurs the distinction between the existence problem and the dynamic stability problem. Ecologists may find my terminology misleading because the model contains no concept of an equilibrium population size.

A mathematical problem that arises in the one-sex model when the vector $F = 0$ requires special attention because it foreshadows more serious problems in the two-sex model. In mathematical usage it is conventional to say that an equilibrium is a nonzero vector $\hat{F}$ which satisfies (21-5). From a demographic standpoint it is preferable to call $F = 0$, as well as any other vector F that maps into 0, a "trivial" equilibrium. In the one-sex model trivial equilibria are uninteresting both demographically and mathematically. In the two-sex model trivial equilibria are the major obstacles blocking the "natural" fixed-point proof of the existence of a nontrivial equilibrium. In the one-sex model if females beyond a certain age do not reproduce, then population vectors consisting entirely of such females will, after a finite number of periods, map into the 0 vector. In the one-sex model, it is often convenient to drop such nonreproductive females from the model, analyze the reduced model, and reinterpret the results in terms of the original model. In the two-sex model we cannot drop nonreproductive individuals because they may mate with potentially reproductive individuals.

THE BIRTH MATRIX-MATING RULE (BMMR) MODEL

The BMMR model has three essential building blocks: a birth matrix, a mating rule, and the female and male mortality schedules. Thus, the fertility of a female of age i is not specified directly but is derived from the underlying birth matrix and the mating rule and depends on the population's age-sex composition. An element of the birth matrix such as b_{ij} represents the expected number of female offspring born in a period to an "(i, j) union," that is, the union of a female of age i with a male of age j. More precisely, each (i, j) union formed in period t produces b_{ij} female offspring and σb_{ij} male offspring who appear as newborns in period $t + 1$; the parameter σ denotes the secondary sex ratio—the ratio of male to female newborns—and I assume that it is a constant, independent of the population's age-sex structure and independent of the ages of the parents. I denote the birth matrix by $B = \{b_{ij}\}$, the number of (i, j) unions by u_{ij}, and the corresponding unions matrix by U.

The mating rule shows the number of unions of each type—that is, identified by age of female and age of male—as a function of the number of individuals in each age-sex category. A mating rule is thus a mapping, $\mu(F, \ M)$, of the population vector $(F, \ M)$ into the matrix of unions: $U = \mu(F, \ M)$; it is often convenient to write $\mu_{ij} = \mu^{ij}(F, \ M)$, where $\mu^{ij}(F, M)$ denotes the function mapping $(F, \ M)$ into the variable μ_{ij}. Any model of a monogamously mating, age-structured population requires an assumption about the durability of unions. In the BMMR model I assume that unions last for a single period. The advantage of this "Southern California" assumption of serial monogamy is that it exposes the model's logical structure, simplifying substantially both the notation and the analysis. The assumption that matings persist for one period means that the length of the time period plays a double role in the model, as Parlett (1972) points out. Pollak (1987) analyzes the BMMR model with "persistent unions."

The dynamics of the BMMR model are straightforward to describe:

1. The initial population vector, together with the mating rule, determines the number of unions of each type; the number of unions of each type, together with the birth matrix, determines the number of newborns in the next period. The number of newborn females is given by the "newborns" function, $\phi^1(F^t, M^t)$, and newborn males by $\sigma\phi^1(F^t, M^t)$:

$$F_1^{t+1} = \phi^1(F^t, M^t) = \sum_i \sum_j b_{ij} \mu^{ij}(F^t, M^t) ,$$

$$M_1^{t+1} = \sigma\phi^1(F^t, M^t) .$$

(21-6)

Thus, the number of female offspring born to an average female of age i, $B^i(F, M)$—the analogue of the parameter b_i of classical stable population theory—is a function of the population vector (F, M) and is equal to a weighted sum (across male age classes) of the birth matrix parameters $\{b_{i1}, \ldots, b_{in}\}$:

$$B^i(F, M) = \frac{\sum_j b_{ij} \mu^{ij}(F, M)}{F_i} .$$

(21-7)

2. The initial population vector, together with the mortality schedules, determines the number of individuals in each of the other age-sex categories, just as in classical stable population theory:

$$F_i^{t+1} = \phi^{iF}(F^t, M^t) = (1 - d_{i-1}^F)F_{i-1}^t$$

$$M_j^{t+1} = \phi^{jM}(F^t, M^t) = (1 - d_{j-1}^M)M_{j-1}^t .$$

(21-8)

Thus, the BMMR model defines a transformation which maps the population in period t into the population in period $t + 1$:

$$(F^{t+1}, M^{t+1}) = \phi(F^t, M^t) . \tag{21-9}$$

Formally, an equilibrium is defined as an age distribution and a growth rate, $(\hat{F}, \hat{M}, \hat{r})$, that satisfy the equation

$$[(1 + \hat{r})\hat{F}, (1 + \hat{r})\hat{M}] = \phi[\hat{F}, \hat{M}] . \tag{21-10}$$

When the *equilibrium* female and male fertility rates of the BMMR model are used to construct the female dominance model and the mirror-image male dominance model, the two one-sex models are consistent with each other in the sense that they imply identical growth rates for the female and male populations. Furthermore, the equilibrium age structure and growth rates corresponding to these two one-sex models are identical to the equilibrium age structure and growth rate of the BMMR model. Away from equilibrium, however, these two one-sex models and the BMMR model generate different predictions.

Establishing the existence of equilibrium and analyzing dynamic behavior in the BMMR model is more complex than in classical stable population theory because the properties of the mating rule make the BMMR model inherently nonlinear. Any mating rule for monogamous unions must satisfy two accounting requirements: a *nonnegativity* condition ensuring that the number of unions of each type is positive or zero, and an *adding-up* condition ensuring that the number of mated females of each age not exceed the total number of females of that age, with a similar requirement holding for males. This adding-up requirement is the fundamental source of nonlinearity in the BMMR model.

Despite its nonlinearity, we can establish the existence of equilibrium in the BMMR model under suitable assumptions. In addition to the two accounting axioms, I impose three substantive axioms on the mating rule. (1) *Universal scope:* The mating rule must be defined for all nonnegative population vectors. (2) *Continuity:* The function μ must be continuous in (F, M). (3) *Homogeneity:* The function μ must be homogeneous of degree one in (F, M): that is, $\mu(\lambda F, \lambda M) = \lambda \mu(F, M)$ for all $\lambda > 0$. The homogeneity axiom implies that a 1% increase in the number of individuals in every age-sex category results in a 1% increase in the number of unions of every type. Individuals searching for mates in a restricted and increasingly crowded region suggests a "density dependent" mating rule, but the homogeneity axiom precludes dependence of mating on population density. Because the elements of the birth matrix and the mortality schedules are constants, homogeneity of the mating rule guarantees the homogeneity of the mapping $\phi(F, M)$. In classical stable population theory, the corresponding mapping is linear as well as homogeneous. Because we are trying to construct a two-sex model capable of maintaining an unchanging age structure while growing at a constant rate, homogeneity is an attractive assumption. These five axioms on the mating rule, together with a condition that I call *r-productivity*, are sufficient to ensure the existence of equilibrium in the BMMR model.

An equilibrium of the BMMR model is a fixed point of the mapping $\phi(F, M)$ or, more precisely, a fixed point of a related mapping in which the population vector is suitably normalized. A "natural" proof strategy would attempt to apply a fixed-point theorem to this mapping. The difficulty with this strategy is that the mapping carries some points in the domain into $(0,0)$. Although $(0,0)$ and any initial population vector (F^o, M^o) that maps into $(0,0)$ satisfy equation (21-10) and thus are equilibria, this strategy fails to establish the existence of a nontrivial equilibrium. An alternative proof strategy avoids this difficulty by drastically limiting the domain of the mapping, reducing the problem to one dimension. A threshold observation is that the only nonlinear component of the BMMR model is the mating rule. Hence, the only nonlinear component of the mapping $\phi(F, M)$ is the newborns function $\phi^1(F, M)$. To exploit this fact the existence proof decomposes the argument into two parts: the determination of an equilibrium age structure and the determination of an equilibrium growth rate. We begin by disconnecting the birth matrix and the mating rule from the mortality schedules and imagining that the number of newborns grows at a constant rate r. (For definiteness, suppose that in each period storks take away all newborns and bring replacements: in the first period they bring N newborn females and σN males; in the second, $(1+r)N$ females and $(1+r)\sigma N$ males; in the third, $(1+r)^2 N$ females and $(1+r)^2 \sigma N)$ males ...) After n periods the age structure of the population is uniquely determined by the mortality schedules, the secondary sex ratio (σ), the growth rate of newborns (r), and the initial number of newborn females (N). For example, the age structure of the female population at time t is given by

$$F_\tau^t = F_1^1 (1 + r)^{t-\tau} \prod_{k=1}^{\tau-1} (1 - d_k^F) \tag{21-11}$$

where $\tau < n$ and $F_1^1 = N$. For $n = 3$ and no early mortality, this implies $(F_1, F_2, F_3) = [(1+r)^2 N, (1+r)N, N]$. In Pollak (1986) I call a population with this structure an *r-equilibrium*. If the model has a nontrivial equilibrium, $(\hat{F}, \hat{M}, \hat{r})$, then $\hat{F}$ and $\hat{M}$ are r-equilibrium populations corresponding to $\hat{r}$. A *normalized r-equilibrium* female population, $F(r)$, is an r-equilibrium population in which the number of newborn females is 1; the normalized r-equilibrium male population, $M(r)$, is one in which the number of newborn males is σ. Since the newborn females are a cohort of size 1, the one-year-olds are the survivors of a cohort of $1/(1 + r)$, the two-year -olds the survivors of a cohort of $1/(1 + r)^2$, and so on:

$$F_\tau^* = \frac{\prod_{k-1}^{\tau-1}(1 - d_k^F)}{(1 + r)^{\tau-1}} . \tag{21-12}$$

For $n = 3$ and no early mortality, this implies $F(r) = [1, 1/(1+r), 1/(1+r)^2]$.

If the BMMR model has a nontrivial equilibrium, $(\hat{F}, \hat{M}, \hat{r})$, then the equilibrium age structure, $(\hat{F}, \hat{M})$, is an r-equilibrium population for $r = \hat{r}$. Hence, when searching for a nontrivial equilibrium, it suffices to restrict our attention to r-equilibrium populations, thus reducing the problem to a single dimension: an equilibrium of the BMMR model corresponds to a value of r for which the r-equilibrium population satisfies equation (21-10)

$$[(1 + r)F(r) , (1 + r)M(r)] = \phi[F(r), M(r)] . \tag{21-13}$$

By the definition of an r-equilibrium population, the number of individuals in every age-sex category except newborns grows at the rate r. Hence, we need only find a value of r for which the birth matrix and the mating rule imply that

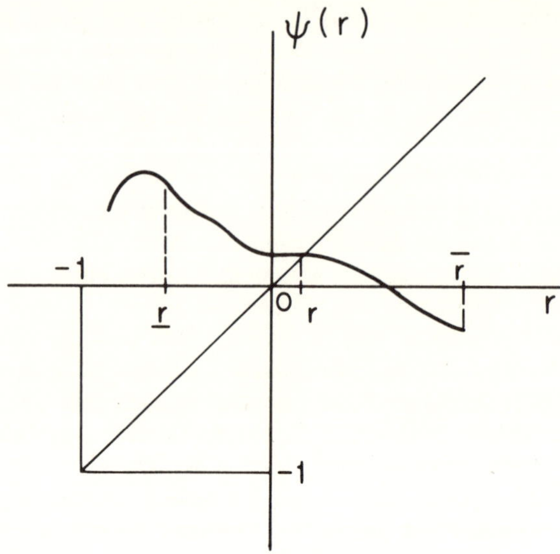

Figure 21-1. Existence of equilibrium in the BMMR Model.

the number of newborn females corresponding to a normalized r-equilibrium population is $(1 + r)$. In the example with no early mortality, this implies that the new female population vector is given by $[(1 + r), 1, 1/(1 + r)]$. Defining the function $\psi(r)$ by

$$\psi(r) = \phi^1[F(r), M(r)] - 1 \tag{21-14}$$

we can express this equilibrium condition as

$$\psi(r) = r . \tag{21-15}$$

In terms of Figure 21-1, establishing the existence of a nontrivial equilibrium requires showing that the function $\psi(r)$ crosses the 45° line. Pollak (1986) proves that the BMMR model has a nontrivial equilibrium provided the model is *r-productive*. The r-productivity condition requires that there exist a value of r for which a normalized r-equilibrium population produces at least $1 + r$ newborn females. The purpose of the r-productivity condition is to ensure that there is some value of r for which the function $\psi(r)$ lies above the 45° line. The fixed birth matrix implies that there is some value of r for which the function ψ lies below the 45° line. For any normalized r-equilibrium population, a larger value of r implies a smaller number of individuals in each age group except newborns; furthermore, as r approaches ∞, the normalized r-equilibrium female population vector approaches $(1, 0, 0, \ldots, 0)$ and the corresponding male vector $(\sigma, 0, 0, \ldots, 0)$. For sufficiently large r, only $(1, 1)$ unions can form, and the adding-up condition implies that the number of such unions cannot exceed min $\{1, \sigma\}$. Thus, for sufficiently large r, the number of newborn females must approach or be less than b_{11} min $\{1, \sigma\}$. For sufficiently large r, this upper bound on the number of newborn females must be less than $1 + r$. Hence, for sufficiently large r, the function $\psi(r)$ lies below the 45° line. [In Pollak (1986) I assume that newborns do not enter unions, so that in the next period the number of newborns approaches 0 and $\psi(r)$ approaches -1 as r approaches ∞.]

Since the function $\psi(r)$ lies above 45° line for some values of r and below it for others, continuity implies that it crosses the 45° line at least once. Hence, the BMMR model has a nontrivial equilibrium.

SPECIAL CASES

Studying nontrivial examples and special cases reveals how the BMMR model works. Because it specifies the mating rule only in a very general way, from the standpoint of applications the BMMR model is not a single model, but a family of models for analyzing growth and age structure in two-sex populations. In this section I discuss examples and special cases that will prove useful in applying the BMMR model. In the next section I distinguish between two senses in which special cases of the BMMR model might coincide with classical stable population theory and characterize the cases in which the two models coincide.

A *female proportional* mating rule is one for which

$$\mu^{ij}(F, M) = \gamma_{ij} F_i \, , \quad \sum_j \gamma_{ij} \leq 1 \qquad (21\text{-}16)$$

in a region of the (F, M) space; I call the region in which (21-16) satisfies the adding-up condition the *relevant region*. A mating rule cannot be female proportional over the entire (F, M) space and satisfy the nonnegativity and adding-up conditions: when the number of males is small, female proportional mating rules imply more unions than there are males. More precisely, since a female proportional mating rule calls for $\sum_i \gamma_{ij} F_i$ unions involving males of age j, it cannot hold unless

$$M_j \geq \sum_i \gamma_{ij} F_i \, . \qquad (21\text{-}17)$$

Furthermore, if not all males of age i are eligible (in a sense I define below), then a female proportional rule can hold only in regions of the (F, M) space for which there are at least $\sum_i \gamma_{ij} F_i$ eligible males of age j.

Female proportional mating rules are consistent with classical stable population theory in the relevant region. For a given female population vector, the set of male population vectors in the relevant region is a subset of those with enough eligible males in each age category to satisfy (21-17). Using equation (21-7), we can verify that, for population vectors in the relevant region, the number of female offspring born to an average female of age i, $\sum_j b_{ij} \gamma_{ij}$, is independent of the age-sex composition of the population.

Schoen's *harmonic mean* mating rule (Schoen, 1981)

$$\mu^{ij}(F, M) = \frac{\alpha_{ij} F_i M_j}{F_i + M_j} \, , \alpha_{ij} > 0, \sum_j \alpha_{ij} \leq 1 \, \forall i \quad \text{and} \quad \sum_i \alpha_{ij} \leq 1 \, \forall j$$

$$(21\text{-}18)$$

provides another example of a mating rule involving only a single parameter for each type of union. Unlike female proportional mating rules, harmonic mean mating rules are "well-behaved" over the entire (F, M) space. To see that the harmonic mean mating rule satisfies the budget constraint, we focus on the requirement that

$$\sum_j \mu^{ij}(F, M) \leq F_i \qquad (21\text{-}19)$$

leaving the analogous demonstration for M_j to the reader. We begin by rewriting the mating rule as

$$\mu^{ij}(F, M) = F_i \mu^{ij}(1, M_j/F_i) = F_i \left[\frac{\alpha_{ij}(M_j/F_i)}{1 + (M_j/F_i)} \right] . \qquad (21\text{-}20)$$

When the number of males approaches infinity, the mating rule requires $\alpha_{ij} F_i$ females of age i:

$$\lim_{M_j \to \infty} \mu^{ij}(F, M) = \alpha_{ij} F_i . \qquad (21\text{-}21)$$

Summing over males of all ages, we see that the required number of females of age i is given by

$$\sum_j \mu^{ij}(F_i, \infty) = F_i \sum_j \alpha_{ij} . \qquad (21\text{-}22)$$

Hence, the adding-up condition for females of age i is satisfied provided $\sum_j \alpha_{ij} \leq 1$. An analogous argument holds for males.

A more general form is the *constant elasticity of substitution* (CES) mating rule

$$\mu^{ij}(F, M) = \left[(\alpha_{ijf})^{-\rho_{ij}} F_i^{-\rho_{ij}} + (\alpha_{ijm})^{-\rho_{ij}} M_j^{-\rho_{ij}} \right]^{\frac{-1}{\rho_{ij}}} \qquad (21\text{-}23)$$

where $\alpha_{ijf} > 0$, $\alpha_{ijm} > 0$, and $\rho_{ij} > 0 \ \forall \ ij$; $\sum_j \alpha_{ijf} \leq 1 \ \forall \ i$, $\sum_i \alpha_{ijm} \leq 1$, $\forall \ j$. Mathematicians call CES forms ''generalized means of order ρ'' (see Hardy, Littlewood, and Polya, 1934). Economists have used CES production functions extensively since their introduction by Arrow, Chenery, Minhas, and Solow (1961). Caswell and Weeks (1986) use the CES mating rule in population biology and Jere Behrman, Samuel Preston, and I are now estimating the CES and other ''marriage functions'' using U.S. and Japanese data. When $\rho_{ij} = 1$ and $\alpha_{ijf} = \alpha_{ijm} = \alpha_{ij}, \forall \ i, j$, the CES mating rule reduces to

$$\mu^{ij}(F, M) = \left[\alpha_{ij}^{-1} F_i^{-1} + \alpha_{ij}^{-1} M_j^{-1} \right]^{-1} \qquad (21\text{-}24)$$

which, after some manipulation, reduces to Schoen's harmonic mean mating rule. The proof that CES mating rules satisfy the adding-up condition parallels that for harmonic mean mating rules; the crucial limit result, (21-21), carries over to the CES case.

Modeling ''spillover effects''—the effects of an increase in the number of females or males of one age on the number of marriages involving females or males of other ages—is a central problem in specifying mating rules. The BMMR model places no restrictions on the pattern of spillover effects, but specifying tractable mating rules that allow for interesting patterns of spillovers has proved difficult. In collaboration with Jere Behrman and Samuel Preston, I am now working on this problem.

CES mating rules, including the harmonic mean, belong to the larger class of *zero spillover mating rules* (Pollak, 1989)—mating rules under which the number of unions involving females of age i and males of age j depends only on F_i and M_j, independent of the number of individuals in the other age-sex categories:

$$\mu^{ij}(F, M) = \mu^{ij}(F_i, M_j) \ . \tag{21-25}$$

Complete random mating

$$\mu^{ij}(F, M) = u^{**} \frac{F_i}{\Sigma F_k} \ \frac{M_j}{\Sigma M_k} \tag{21-26}$$

where

$$u^{**} = \min(\Sigma F_k, \Sigma M_k) \tag{21-27}$$

provides a simple example of a mating rule allowing spillovers, although its symmetric treatment of individuals in all age categories is implausible. When there are more males than females, then $u^{**} = \Sigma F_k$ and complete random mating implies

$$\mu^{ij}(F, M) = F_i \frac{M_j}{\Sigma M_k} \tag{21-28}$$

and

$$\Sigma\Sigma u_{ij} = \Sigma F_k = u^{**} \ . \tag{21-29}$$

That is, when the population contains more males than females, all females mate and each male has an equal probability of mating with each female; hence, the number of (i, j) unions is given by (21-28) and the number of unmated males by $(\Sigma M_k - \Sigma F_k)$. When there are more females than males, the number of (i, j) unions is given by $M_j F_i/\Sigma F_k$, the number of unions by $u^{**} = \Sigma M_k$, and the number of unmated females by $(\Sigma F_k - \Sigma M_k)$.

Some flexibility can be incorporated into a random mating rule by specifying that the number of (i, j) unions is proportional to the number generated by complete random mating:

$$\mu^{ij}(F, M) = \alpha_{ij} u^{**} \frac{F_i}{\Sigma F_k} \frac{M_j}{\Sigma M_k} \ . \tag{21-30}$$

This specification of random mating contains one parameter for each type of union, while complete random mating contains none. An alternative generalization applies the complete random mating formulae, (21-24) and (21-27), not to the entire population of females and males but only to a subset of eligible individuals. Because the distinction between eligible and ineligible individuals is important beyond the context of random mating rules, I introduce it in a general form.

I begin by partitioning the population into "eligible" and "ineligible" individuals. Eligible individuals are potential mates. Ineligible individuals are noncandidates: they are not paired with mates even when there are surplus members of the opposite sex and all members of their own sex in the eligible population have mates. Thus, the ineligible population might include the sick, the very young, and the very old. It may also include various fractions of the individuals in each age-sex category. The classification of an individual as eligible or ineligible, however, must be independent of the age-sex composition of the population. Thus, I assume that the fraction of eligible individuals in each age-sex category is fixed. I denote the vector of females in the eligible population by $f = (f_1, \ldots, f_n)$, of males by $m = (m_1, \ldots, m_n)$, and the proportion of females of age i (males of age j) in the eligible population by $p_i^F (p_j^M)$, so

$$f_i = p_i^F F_i \quad \text{and} \quad m_j = p_j^M M_j \ . \tag{21-31}$$

When $p_i^F = p_j^M = 1$ for all i, j, then $f = F$ and $m = M$, so the eligible population coincides with the full population.

We can generate a "new" mating rule from an old one by replacing the full population vector (F, M) by the eligible population vector (f, m). For example, in the case of Schoen's harmonic mean mating rule (21-18), this replacement yields the mating rule

$$\mu^{ij}(F, M) = \frac{\alpha_{ij} p_i^F p_j^M F_i M_j}{p_i^F F_i + p_j^M M_j} \ , \tag{21-32}$$

which has more parameters than the original rule. In the case of the CES mating rule, new parameters are not introduced or, more precisely, cannot be distinguished empirically from those already appearing in the CES specification. Before returning to random mating rules, it is useful to redefine the maximum feasible number of unions in terms of the eligible population.

I call a mating rule *exhaustive* if it never leaves both unmated males and unmated females in the eligible population. Under an exhaustive mating rule, when there are more males than females in the eligible population, all eligible females are mated; when there are more females than males in the eligible population, all eligible males are mated. Two points require emphasis. First, which sex is in surplus is defined in terms of the eligible population, not the full population. Second, although exhaustive mating rules are analytically convenient, they are unrealistic: in models in which individuals search for mates and search time is limited, some females and some males in the eligible population are likely to be unmated and the number of mated females (males) is likely to depend on the excess of males over females (females over males) in the eligible population. The maximum number of monogamous unions, u^*, in the eligible population is given by

$$u^* = \min\left(\Sigma p_k^F F_k, \Sigma p_k^M M_k\right) = \min\{\Sigma f_k, \Sigma m_k\} \ . \tag{21-33}$$

Under an exhaustive mating rule, this maximum number of unions is always attained.

Random exhaustive mating provides a simple extension of complete random mating. With random exhaustive mating the number of (i, j) unions is given by

$$\mu^{ij}(F, M) = u^* \frac{p_i^F F_i}{\Sigma p_k^F F_k} \ \frac{p_j^M M_j}{\Sigma p_k^M M_k} = u^* \frac{f_i}{\Sigma f_k} \ \frac{m_j}{\Sigma m_k} \ . \tag{21-34}$$

That is, when there are more males than females in the eligible population, all eligible females mate and each eligible male has an equal probability of mating with each eligible female; hence, the number of (i, j) unions is given by $f_i m_j / \Sigma m_k$ and the number of unmated males in the eligible population by $(\Sigma m_k - \Sigma f_k)$. When there are more females than males in the eligible population, the number of (i, j) unions is given by $m_j f_i / \Sigma f_k$ and the number of unmated females in the eligible population by $(\Sigma f_k - \Sigma m_k)$.

The BMMR model simplifies substantially when all males in the eligible population are identical—in the sense that the fertility of an (i, j) union is independent of the male's age—and the mating rule is exhaustive. In Pollak (1986) I call this special case of the BMMR model the *identical males-exhaustive mating rule (IMEX)* model. With identical males, the entries in the birth matrix depend on the female's age but not on the male's age, so the (constant) rows of the birth matrix, $\bar{b}_i$, are the maternity schedule for mated females.

The analysis of the IMEX model treats separately the male surplus and the male deficit cases. Suppose first that males are in surplus. Since the mating rule is exhaustive, all females in the eligible population are mated and, hence, the maternity schedule for mated females becomes the maternity schedule for females in the eligible population. The implied fertility of a female of age i in the full population, b_i, is obtained by multiplying the fertility of a female of age i in the eligible population by the fraction of females of age i who are in the eligible population: $b_i = p_i^F \bar{b}_i$. More formally, any exhaustive mating rule with males in surplus implies

$$\sum_j \mu^{ij}(F, M) = p_i^F F_i \qquad \forall i \ . \tag{21-35}$$

With identical males $(b_{ij} = \bar{b}_i)$, equation (21-7) becomes

$$\sum_j b_{ij} \mu^{ij}(F, M)/F_i = \frac{\bar{b}_i}{F_i} \sum_j \mu^{ij}(F, M) = p_i^F \bar{b}_i = b_i \ . \tag{21-36}$$

The IMEX model with surplus males satisfies the female dominance assumption and thus is consistent with classical stable population theory. The equilibrium age structure and growth rate for the female population coincide with those implied by classical stable population theory, where the b's are given by equation (21-36). Furthermore, provided males remain in surplus in every period, the female population converges to this equilibrium in the manner predicted by the one-sex model. Thus, for the female population, all of the conclusions of classical stable population theory are valid in the IMEX model when males are in surplus. (The mirror-image assumptions—identical females, exhaustive mating, females in surplus—leads to the "male dominance" model.)

THE BMMR MODEL AND STABLE POPULATION THEORY

There are two senses in which special cases of the BMMR model can coincide with classical stable population theory. A special case of the BMMR model coincides weakly with classical stable population theory in a neighborhood (i.e., on an open set in the (F, M) space) if the birth matrix and mating rule are such that

$$\sum_j b_{ij} \mu^{ij}(F, M)/F_i = b_i \tag{21-37}$$

for all (F, M) in the neighborhood. The left hand side of this expression is equation (21-7), the number of female offspring born to an average female of age i; classical stable population theory requires this to be a constant, independent of the age-sex composition of the population. Thus, as the age-sex composition of the population changes, classical stable population theory requires that any increase in births to some types of unions involving females of age i be exactly offset by decreases in birth to other types of unions involving females of age i. The definition refers to a neighborhood in the (F, M) space because a special case of the BMMR model cannot coincide with classical stable population theory for all values of (F, M). The difficulty arises because classical stable population theory implies the number of births to females of age i is independent of the number and age distribution of males, while in the BMMR model the combination of a fixed birth matrix and a monogamous mating rule implies that as the number of males approaches zero, the number of births also approaches 0.

A special case of the BMMR model coincides *strongly* with classical stable population theory if it coincides weakly for all birth matrices in some neighborhood of a birth matrix (b_{ij}). Thus, a special case of the BMMR model coincides weakly with classical stable population theory if it satisfies (21-37) and thus coincides for a particular birth matrix; it coincides strongly if it coincides not only for a particular birth matrix but also when the elements of the birth matrix are free to vary independently in some neighborhood. Because the elements of the birth matrix must be nonnegative, this definition requires modification to accommodate zeros in the birth matrix. One such modification requires that the condition hold only for *admissible* (i.e., nonnegative) variations in elements of the birth matrix in a neighborhood. An alternative modification requires that the condition hold only when the *nonzero* elements of the birth matrix vary independently.

To establish the conditions under which the BMMR model coincides strongly with classical stable population theory, it is convenient to define the function $\bar{\mu}^{ij}(F, M)$ by

$$\bar{\mu}^{ij}(F, M) = \mu^{ij}(F, M)/F_i \; . \tag{21-38}$$

Using these functions, we can rewrite (21-37) as

$$\Sigma b_{ij} \bar{\mu}^{ij}(F, M) = b_i \; . \tag{21-39}$$

The BMMR model coincides strongly with classical stable population theory if and only if the mating rule is female proportional, so that in the relevant region the number of (i, j) unions is proportional to the number of females of age i, and independent of the number of individuals in every other age-sex category:

$$\bar{\mu}^{ij}(F, M) = \mu^{ij}(F, M)/F_i = \gamma_{ij} \; . \tag{21-40}$$

Proof: It is easy to verify that this mating rule implies strong coincidence by substituting into (21-37). To show that it is the only rule that does so, we differentiate (21-39) with respect to b_{ij} and obtain

$$\bar{\mu}^{ij}(F, M) = \frac{\partial b_i}{\partial b_{ij}} \; . \tag{21-41}$$

Strong coincidence implies that the right hand side of (21-41) cannot depend on (F, M), so the left hand side must also be independent of (F, M). Hence, $\bar{\mu}^{ij}(F, M)$ must be a constant, as (21-40) asserts. Q.E.D.

To establish the conditions for weak coincidence, I define the functions $\beta^{ij}(F, M)$ by

$$\beta^{ij}(F, M) = b_{ij}\bar{\mu}^{ij}(F, M) = b_{ij}\mu^{ij}(F, M)/F^i \ . \tag{21-42}$$

Using these functions, we can rewrite (21-37) as

$$\sum_j \beta^{ij}(F, M) = b_i \ . \tag{21-43}$$

Using the β's suppresses the distinction between the birth matrix and the mating rule and represents births per female of age i in the population to (i, j) unions as a function of the population vector.

It is evident from (21-43) that the BMMR model coincides weakly with classical stable population theory if and only if the $\beta^{ij}(F, M)$ functions corresponding to the model sum to a constant (b_i). Examples are easy to construct if we begin by specifying the $\beta^{ij}(F, M)$ functions rather than the birth matrix and the mating rule. The IMEX model with males in surplus coincides weakly with classical stable population theory, as equation (21-36) demonstrates. In general, however, it does not coincide strongly: independent variations in the elements of the birth matrix create differences among males, and unless the $\bar{\mu}^{ij}(F, M)$ are constants, the left hand side of (21-39) depends on the age-sex structure of the population. The IMEX model continues to coincide weakly with classical stable population theory for variations in the birth matrix that satisfy the "identical males" requirement, but not for unrestricted variations unless the mating rule is female proportional.

CONCLUSION

The BMMR model provides a new paradigm for analyzing age structure and growth in monogamously mating populations. The basic building blocks of the model are a birth matrix, a mating rule, and mortality schedules for females and males. The model provides a framework within which the fertility of an average female of age i can depend on the number of competing females of age i, the number of females in other age categories, and the number of males in each age category. Following Pollak (1986, 1990), I have discussed the components of the BMMR model and presented a simple proof of the existence of equilibrium. I have not discussed the uniqueness of equilibrium and its dynamic stability, issues treated in Pollak (1990), nor have I discussed the extension of the BMMR model to "persistent unions," an extension treated in Pollak (1987).

The traditional model for analyzing age structure and growth—classical stable population theory—is a one-sex model in the sense that males play no essential role. Classical stable population theory is clearly appropriate for one-sex populations but it is appropriate for monogamously mating two -sex populations only under very restrictive conditions. In this paper I characterize the special cases of the BMMR model that coincide with classical stable population theory.

The male surplus case of the IMEX (identical males, exhaustive mating) model is an example of a special case of the BMMR model that coincides "weakly" with classical stable population theory. The coincidence is weak in the sense that it holds only when values of the elements of the birth matrix are precisely aligned. The class of special cases of the BMMR model that coincide weakly with classical stable population theory, although broader than the IMEX model, is highly restrictive.

The female proportional mating rule—a mating rule in which the number of (i, j) unions is proportional to the number of females of age i and independent of the number of males and of the number of females in other age categories—implies a BMMR model that coincides "strongly" with classical stable population theory. The coincidence is strong in the sense that it continues to hold when the elements of the birth matrix are subjected to independent perturbations. The female proportional mating rule is the only rule for which the BMMR model coincides strongly with classical stable population theory. This result implies that the conclusions of classical stable population theory hold only under highly restrictive conditions and, hence, confirms the importance of the added generality of the BMMR model.

ACKNOWLEDGMENTS

This research was supported in part by the National Institutes of Health, the National Science Foundation. and the Population Council's Research Awards Program on Fertility Determinants funded by USAID. The first two sections of the chapter are drawn from Pollak (1990). Parts of the third section are drawn from Pollak (1986). I am grateful to Beth Allen, W. Brian Arthur, Jere R. Behrman, Joel Cohen, George Farnbach, Griffith Feeney, Joseph Felsenstein, Robert Fogel, Marc Nerlove, Samuel Preston, Peter Smouse, Susan Watkins, Joel Yellin, and anonymous referees for helpful conversations and to Judith Farnbach for editorial assistance.

REFERENCES

Arrow, K. J., H. B. Chenery, B. S. Minhas, and R. M. Solow, 1961   Capital-labor substitution and economic efficiency. Rev. Econ. Stat. 43: 225-250.

Arthur, W. B., 1981   Why a population converges to stability. Am. Math. Monthly 88: 557-563.

Arthur, W. B., 1982   The ergodic theorems of demography: A simple proof. Demography 19: 439-445.

Bernstein, H., H. C. Byerly, F. A. Hopf, and R. E. Michod, 1985   Genetic damage, mutation, and the evolution of sex. Science 229: 1277-1281.

Caswell, H., and D. Weeks, 1986 Two-sex population models: chaos, extinction, and other dynamic consequences of sex. Am. Nat. 128: 707-735.

Charnov, E. L., 1982   *The Theory of Sex Allocation.* Princeton University Press, Princeton.

Coale, A. J., 1972   *The Growth and Structure of Human Populations.* Princeton University Press, Princeton.

Das Gupta, P., 1978   An alternative formulation of the birth function in a two-sex model. Pop. Studies 32: 367-379.

Feeney, G. M., 1972   *Marriage Rates and Population Growth: The Two-Sex Problem in Demography*, Ph.D. dissertation, University of California, Berkeley.

Fredrickson, A.G., 1971   A mathematical theory of age structure in sexual populations:

random mating and monogamous marriage models. Math. Biosci. 10: 117-143.

Goodman, L. A., 1953   Population growth of the sexes. Biometrics 9: 212-225.

Hardy, G. H., J. E. Littlewood, and G. Polya, 1934   *Inequalities*. Cambridge University Press, Cambridge.

Keyfitz, N., 1968   *Introduction to the Mathematics of Population*.   Addison-Wesley Publishing Co., Reading, MA.

Keyfitz, N., 1971   The mathematics of sex and marriage, pp. 89-108 in *Proceedings of the Sixth Berkeley Symposium on Mathematical Statistics and Probability*, edited by L. M. Le Cain, J. Neyman, and E. L. Scott. University of California Press, Berkeley.

Kuhn, T. S., 1962; 1970   *The Structure of Scientific Revolutions*. Second Edition, enlarged. University of Chicago Press.

Lotka, A. J.,   The stability of the normal age distribution. Proc. Natl. Acad. Sci. USA 8: 339-345.

Maynard Smith, J., 1978   *The Evolution of Sex*. Cambridge University Press, Cambridge.

Maynard Smith, J., 1980   A new theory of sexual investment. Beh. Ecol. Sociobiol. 7: 247-251.

Parlett, B., 1972   Can there be a marriage function? pp. 107-135 in *Population Dynamics*, edited by T. N. E. Greville. Academic Press, New York.

Pollak, R. A., 1986   A reformulation of the two-sex problem. Demography 23: 247-259.

Pollak, R. A., 1987   The two-sex problem with persistent unions: A generalization of the birth matrix-mating rule model. Theoret. Pop. Biol. 32: 176-187.

Pollak, R. A., 1990   Two-sex demographic models. J. Political Econ. (in press).

Pollard, J. H., 1973   *Mathematical Models for the Growth of Human Populations*. Cambridge University Press, Cambridge.

Schoen, R., 1981   The harmonic mean as the basis of a realistic two-sex marriage model. Demography 18: 201-216.

Yellin, J., and P. A. Samuelson, 1977   Comparison of linear and nonlinear models for human population dynamics. Theoret. Pop. Biol. 11: 105-126.

22

Sources of Variation in Vital Rates:
An Overview

SAMUEL H. PRESTON

Demography is often called an observational science, somewhat analogous to astronomy. Rather than working by deductive logic, its principal approach is to observe and then account for what it observes. What catches its eye is variation: changes in birth rates and death rates, differences in fertility or marriage behavior from one group to the next.

Genetic change or variation is not capable of accounting for much of the observed variation in aggregate indices over the time spans that demographers typically examine, and for this reason demography is not at present making much use of genetics. Most changes in rates have been too fast, and differences among groups at a moment in time too large, to be explicable in genetic terms. It was not always so. Enthusiasm for the new discipline of genetics in the late nineteenth and early twentieth centuries led to a much more prominent role for genetics in demographic explanation. Different European ethnic groups were treated as separate races and were believed to have widely differing susceptibilities to disease. Black Americans were thought to have exceptionally high mortality because of such afflictions as a smaller, "tropical lung" (Hoffman, 1896). Such ascriptions were ultimately found inadequate as explanations of mortality variation, the dominant sources of which were shown to be environmental and behavioral.

What demographers have set their sights on is undoubtedly somewhat arbitrary. Instead of asking why life expectancy at birth for the world as a whole has doubled in this century, they might have asked why no one has ever been recorded as living past age 120. Instead of asking why childbearing is increasingly delayed in the United States, they might have asked why the reproductive age span for women is essentially confined to ages 15 to 50. The choice of questions reflects many factors, but perhaps above all the fact that demography is a social science and recruits its practitioners predominantly from other social sciences. The social sciences have two basic operating principles that direct at-

335

tention to variation. The first one is that societies matter, that social institutions and arrangements affect individual behavior and well-being. The second one is that societies are improvable. Neither of these principles is advanced by the study of constants, whether genetic or not.

Though genetics forms a largely unexamined scaffolding for demography, biology in general is playing a much more active role. The processes of birth and death are obviously biological processes, and all social influences must work through these to affect rates. In the past several decades there has been increased recognition that misunderstanding and misinterpretation can result from failure to account for the biological components of these processes. By integrating biological and social factors, demographers have recently added elegance, precision, and concreteness to the models and measurement devices used to understand demographic variation.

This work has achieved its most elegant form and scored its greatest successes in the analysis of birth intervals, work associated especially with the names of Louis Henry, Robert Potter, Mindel Sheps, Jane Menken, and John Bongaarts (e.g., Bongaarts and Potter, 1983; Sheps and Menken, 1973). In this body of work, the completed fertility of a woman is viewed as the quotient of her length of exposure to intercourse during her reproductive life and the length of time it takes her to go from one birth to the next. This interbirth interval is then broken up into components that represent time of exposure to the risk of conception, the period of pregnancy itself, and an insusceptible period following birth. The length of these periods can in turn be related to biosocial variables representing fecundability, use and effectiveness of contraception, spontaneous and induced abortion, the length of breastfeeding, and postpartum sexual behavior. Marriage patterns appear in the numerator as influences on the lifetime duration of exposure to intercourse.

Bongaarts (1982) has shown that the bulk of variation in completed fertility from one population to the next can be explained in terms of the basic components of this model. One especially useful result is the demonstration of the importance of length of breastfeeding in accounting for fertility variation, especially among populations not practicing contraception. Other important results from this body of research demonstrate the synergistic relation between contraception and abortion; important nonlinearities in the relation between fertility and contraceptive effectiveness; and the importance of variances as well as means in explaining group-processes in terms of individual-level processes.

In developing these models, it was not enough to know that there is a biological connection between two variables. Nor could laboratory research shed much light on their relationship. It was necessary to examine empirically the relationship between them in human populations, for example, the effect of length of breastfeeding on the return of ovulation. These relationships are conventionally summarized in "model" age or duration schedules, empirical representations of the typical patterns found in human populations. Here investigators have followed the lead of Coale and Demeny (1966) in constructing model life tables. Genetic factors sometimes make cameo appearances in applications of these models, for example, in explaining the earlier age at menarche for blacks than whites in the United States, despite blacks' poorer economic and nutritional conditions (Menken, 1987).

Birth interval models represent the demographers' classic decompositional approach to studying vital rates. First, variation is observed, then a model or explanatory scheme is developed that breaks down the variation into constituent parts, and then differences in some overall index are assigned to the various components. There is no activity more basic to demography or typical of demographers. It is a form of social accounting that, even though it may not provide complete explanations of a phenomenon, helps to clarify what it is that needs explaining. Mortality analysis has several similar decompositional devices (e.g., Mosley and Chen, 1984), though none are as elegant or compelling as birth interval models because the biomedical influences on mortality are far more complex. Because it is not a biological process, migration lacks these procedures. Partly as a result, migration studies have lost status within demography in the past several decades, although not in the social sciences more generally.

SHORT-TERM FLUCTUATIONS IN VITAL RATES

The devices of demography have been applied to the study of both long-term and short-term variation in vital rates. One noteworthy product of the latter studies is the demonstration of the power of genetic and social forces to cushion the effects of the stochastic elements to which human populations have been subject. Until a few hundred years ago, of course, humans lived in highly uncertain environments. Food supplies were subject to the vagaries of weather and fire and crop disease; human disease had a partially independent cycle of its own.

If genetic and social factors had encouraged a positive autocorrelation in demographic time series (that is, if negative deviations from trend in one period had been associated statistically with negative deviations from trend in the next period), then there would be an enhanced danger that one of these threatening episodes would send a population into a tailspin that led to extinction. Instead, a variety of mechanisms introduced negative autocorrelation into these time series, whereby negative deviations in one period are associated with positive deviations in the next. Most obvious on the genetic side is the lactation-ovulation mechanism that links fecundity to child survival. In good times, more children survive, hence more women are breastfeeding at a moment in time, and as a result fewer children are born nine months later. In bad times, more children die and more women are restored to the fecund state, to replace them earlier. The biosocial models of reproduction have clarified the great importance of breastfeeding in regulating the *volume* of fertility, but its role in distributing births over time was perhaps equally important for group survival.

A second genetic mechanism that provided such a cushion is the age pattern of mortality. Comparing high and low mortality populations, the biggest differences in one-year survival rates occur under age 5 and above age 50 (Coale and Demeny, 1966). The smallest differences tend to occur in the late childhood ages where women have their highest reproductive value. As a result, an episode of excess mortality would tend to increase the fraction of the population who were in the reproductive ages and permit a quicker restoration of population numbers (cf. Watkins and Menken, 1985). The evolution of systems of acquired immunity to infectious disease also produced negative autocorrelation in time series of death rates.

How important these mechanisms were is suggested by the effort that most societies made to reinforce them. Most Asian and African societies developed devices that linked the resumption of intercourse to the survival of the previous child. Postpartum taboos on intercourse tended to be long and tied to the survival of the previous child (Nag, 1968). In the extreme, as in many parts of India, they involved physical separation of husband and wife. In Europe these taboos have been much less common, but a similar cushioning outcome was obtained by a unique marriage system that tied intercourse to the availability of land. Nubile women were held in reserve until such time as death or social incapacity opened up reasonable prospects for supporting a new family (Wrigley, 1987).

How the combination of these mechanisms operated to mute the effect of crisis is illustrated in Figure 22-1, drawn from work carried out in collaboration with Ann Janetta, an historian at the University of Pittsburgh. Using temple records of deaths by age over two centuries, we have reconstructed vital rates in a Japanese village of about 1,200 people. This village seems to have had remarkably little trend in vital rates over this period and its population was nearly stationary in size. That stationarity was achieved in part through negative autocorrelation in the demographic time series. For example, after a subsistence crisis in which a quarter of the population died between 1831 and 1836, the death rate fell well below average over the next 20 years and the birth rate rose to above-average values over the next decade. It is likely that a European-style marriage system was in place in rural Japan during this period, which would account for the prolonged fertility response. Obviously, other forms of heterogeneity besides that associated with age (e.g., variation in constitutional frailty) may also be implicated in the mortality response.

LONG-TERM TRENDS IN VITAL RATES

Human populations eventually gained much greater control of these sources of instability. By 1800, techniques in agriculture and transportation in Europe had improved to the point where crises were less severe and there was no longer a strong positive correlation between the price of grain and mortality (Flinn, 1981). An even greater breakthrough has occurred within the past century as populations have gained control of the infectious diseases from which the majority of people were dying.

I will focus on studies of the sources of this decline in mortality in order to illustrate the range of things that demographers do to study long-term variation in vital rates. The quantitative dimensions of the changes are reasonably clear. Life expectancy at birth for the world as a whole at the beginning of the twentieth century was in the high twenties. It is now, according to UN estimates, 61 years (United Nations, 1986:114). Life expectancy in the United States and other industrial countries was about 49 years at the turn of the century, compared to about 75 years at present. Since prehistoric life expectancies were evidently in the low twenties, more than half of the progress since prehistoric times in both industrial and non-industrial countries has occurred during this century.

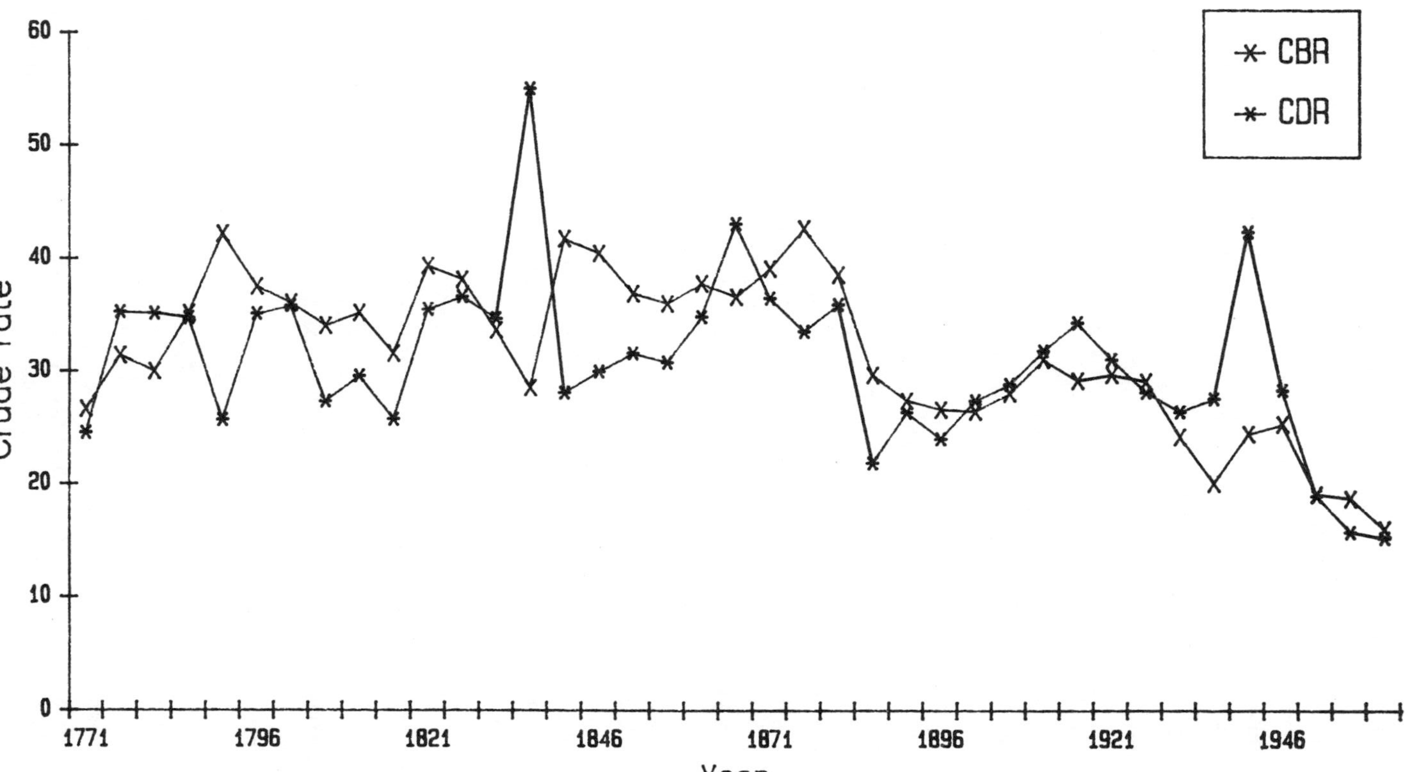

Figure 22-1. Time series of crude birth and death rates in a Japanese village.

Economists calculate that if the value that people place on extended life were imputed and added to national income, measures of economic progress would be increased by one-third or so (Williamson, 1984).

To the people who first recognized these rapid gains, it seemed obvious that medical and public health progress was responsible. The period of most rapid decline since 1850 unquestionably coincided with many medical discoveries and important public health initiatives. However, a distinguished medical historian, Thomas McKeown, has taken issue with the standard interpretation, and much of the subsequent work in the field has been addressed to McKeown's thesis. McKeown and collaborators studied the mortality decline in England and Wales, the first country to have usable national data on causes of death. In the most influential of a series of articles (followed by a book), McKeown and Record (1962) argue that the English mortality decline of the last half of the nineteenth century could not be attributable to medical progress. Among the diseases that declined (especially tuberculosis, which alone accounted for half of the fall) there simply were no advances in medical treatment or prevention. Nor did the organisms themselves appear to mutate, with the possible exception of scarlet fever, a minor contributor to the overall decline. By process of elimination reasoning, they argue that the advances in mortality must have been attributed to improvements in standards of living, and in particular, to nutritional advances. In a later book, McKeown (1976) extends this approach through the twentieth century and reaches similar conclusions, though less quantitatively explicit. For disease after disease, he shows that the bulk of the English decline occurred before any specific medical prophylaxis was developed. He argues that gains in living standards and especially in nutrition must have been responsible for the majority of the declines that occurred.

McKeown was a clear and elegant writer and, until his death this year, was a very influential figure in medical history and British academia. His account has become the standard one, apparently accepted in the main by all but demographers. What problems have demographers raised with his thesis? One objection is that the British mortality decline in the last half of the nineteenth century, where much of the attention has focused, was small in size and atypical in nature. Life expectancy grew by four or five years, less than a fifth of the increase that has occurred since 1850. And tuberculosis, which accounted for half of the decline in this instance, typically accounts for only 10% of the movement from one mortality level to another. This latter evidence is based upon model cause—patterns of mortality based upon data for 43 countries spanning a century (Preston, 1976). Even if McKeown were correct in his analysis of late nineteenth century Britain, the pertinence of the example to world-wide patterns of decline over a broader span of mortality change is questionable.

A second and more important objection is that rising standards of living are unable to account for the bulk of mortality gains in the twentieth century across national populations. If they were responsible, then it should be possible to take the cross-sectional relation between life expectancy and living standards that prevails at a moment in time and attribute changes in life expectancy to movements along the curve, that is, to advances in income. The cross-national relation between the two is in fact very tight, with a correlation coefficient of 0.85 or so, after the variables are transformed. But rather than remaining stationary, the curve itself shifts dramatically in the course of the twentieth century, as demonstrated in Figure 22-2. Countries at a particular income level have a life expectancy at birth that is 9 or 10 years higher in the 1960s than it was in the 1930s. It appears that a shift of equal magnitude may have occurred between the 1900s and the 1930s.

Using this apparatus, one can, for each country or region, attribute the rise in life expectancy to movements along the curve (that is, to income gains) and to shifts in the curve (that is, to exogenous factors). When results are aggregated, it turns out that about 80% of the gain from the 1930s to the 1960s is attributable to the shift in the curve. So income growth seems to have played a minor role in the advance in life expectancy during this period, contrary to McKeown's assertion for England and Wales.

Results are similar when attention is confined to developing countries for the period 1940-75, even when per capita calorie consumption and literacy levels are added to the set of variables describing standard of living. The curve still shifts by about nine years even after these factors are included in it. The shift accounted for about three-quarters of the mortality improvement that occurred during this period in developing countries exclusive of China, which was not included in the analysis (Preston, 1980). Palloni (1981), studying Latin American changes, and Da Vanzo and Habicht (1986), who uses this approach in Malaysia, reach similar conclusions.

If improved standards of living were not primarily responsible for the large declines in mortality during the past century, what produced them? Advances in medicine and public health remain the most obvious alternatives. Within the medical and health arena, it is important to distinguish three major actors: first, the medical establishment, consisting of doctors, nurses, hospitals, and pharmacies; second, public health authorities and activities, which include especially efforts to clean up water, milk, and food supplies, to impose quarantines, to immunize, and to proselytize; and third, the personal health practices of individuals. McKeown put his emphasis on the first, and no one has put forth solid evidence to contradict his conclusion that these were not the prime movers in mortality decline. Most of the effective medicines and drugs were developed too late to be able to explain much of the mortality decline, with the exception of sulpha drugs and antibiotics in the 1940s, which were highly effective in preventing death from pneumonia and several other diseases.

Public health activities and personal health practices are a more likely source of the rapid advances. But in a sense this claim also displaces the focus, because improvements in both were a direct function of the acceptance of the germ theory of disease in the last two decades of the nineteenth century, when the mortality decline in Europe and North America accelerated (Durand, 1980). The germ theory had been an object of discussion for several decades when it received striking empirical support in 1882 through Koch's isolation of the tuberculosis bacillus. The theory led to only one useful drug therapy in the next several decades, diphtheria antitoxin, but it showed how the major diseases were transmitted and put new strategies in the hands of public health authorities. For example, shortly after the theory became accepted, authorities began to inspect milk supplies for bacteria rather than for dilution and odor. Epidemiologic studies confirmed that bacteria counts in milk were closely associated with

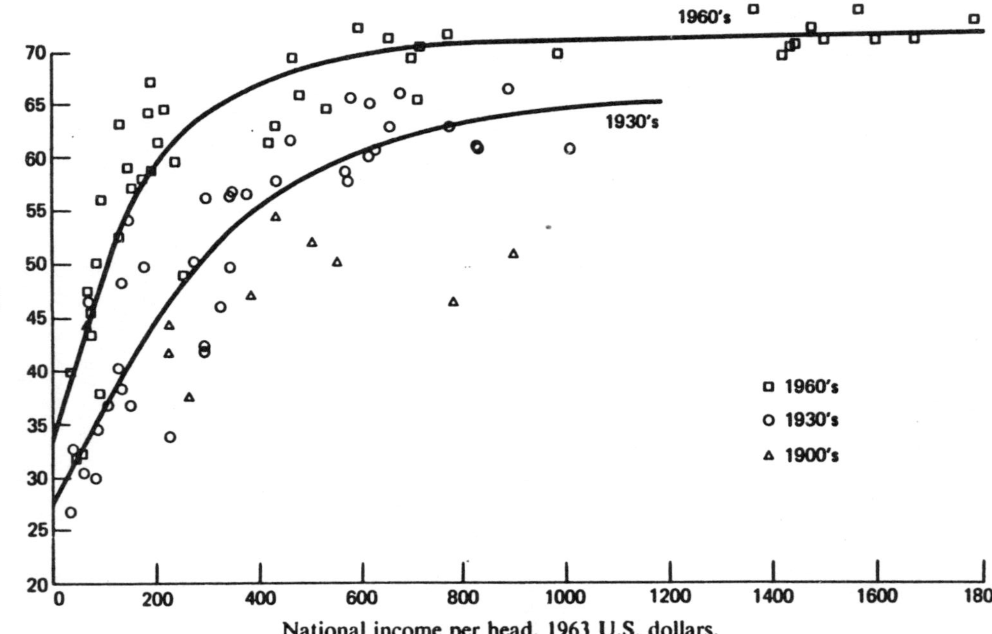

Figure 22-2. Scatter-diagram of relations between life expectancy at birth (e_0^0) and national income per head for nations in the 1900s, 1930s, and 1960s.

rates of sickness and death, and were critically important in the movement for pasteurization (Park and Holt, 1903). Those who were trying to clean up water supplies had a new rationale and criteria for success that they could bring to the job (Chapin, 1901).

It is often hard to realize how primitive the state of medical knowledge was less than 100 years ago. Residues of the theory of the humors could be found in the frequent practice of purging, though bloodletting was subsiding. One of the most popular disease theories of the 1880s in the U.S. was that disease was caused by sewer gas. Great sums were paid to build stench traps so that the offending odors didn't reach the nose (Preston and Haines, 1989). The germ theory allowed much more precise targeting of these hygienic impulses, and provided impetus for the public health movement. By demonstrating the inter-personal nature of infectious disease, the theory also strengthened the grounds for government action in the area of health. And it especially caught the atten-tion of the middle class, who suddenly became more interested in improving disease conditions among the poor. In 1879, no U.S. state legislature had a committee on public health; by 1900, all but five did (Abbott, 1900).

The activities of health boards included a wide range of preventative measures: cleaning water and milk supplies, providing instruction about infant feeding, isolating infectious patients, encouraging vaccination, licensing physicians and midwives, and requiring physical exams in schools (Preston and Haines, 1989). In the United States, the responsibility for these initiatives lay principally with municipalities, and it is at the local level where their efficacy is most readily demonstrated (Condran, Williams, and Cheney, 1984; Sedgwick and MacNutt, 1910). In England, they were more highly centralized, and a recent review of the English evidence finds, that contrary to McKeown, they were the principal factor in the English mortality decline from 1850 to 1914 (Szreter, 1988).

Changes in personal health practices are exceptionally difficult to measure, and perhaps for this reason this factor has been almost entirely neglected in explanations of mortality decline. There were no national surveys of health practices until the 1940s in the U.S. and there have been very few since then that bear upon behaviors relevant to infectious diseases. But by 1910, public health authorities had become convinced that changes in personal health prac-tices were the key to success in the effort to combat ill health, especially among children (Meckel, 1989; Ewbank and Preston, unpublished). Epidemiologic studies had shown, for example, that no matter how clean milk was when it arrived on the doorstep, it was subject to enormous contamination in the home. The major educational efforts made by health authorities over the next three decades emphasized the importance of breastfeeding, heating milk to remove bacteria, keeping flies away from food, isolating sick family members, dispos-ing of waste in a sanitary fashion, washing hands, bathing, and seeking medical attention for the sick (Meckel, 1990). Although empirical research before the germ theory's acceptance had established the efficacy of breastfeeding, most of these personal health practices were founded upon a new understanding of disease processes that the theory supplied. For example, the heating of milk in the home was not initiated until the 1890s and the important role of flies in the transmission of disease to food was not recognized until 1903.

Table 22-1. Comparison of relative mortality in different occupational classes, U.S. 1900 and developing countries in the 1970s. (Ratio, deaths to expected deaths among children ever born; expected deaths are based upon the average child mortality level in a particular country)

		Father's Occupational Category				
	Year	Professional/ Managerial/ Clerical	Sales	Service	Agriculture	Production Workers
United States	1900	.933 (769)	.831 (173)	1.001 (165)	.890 (3375)	1.150 (2967)
Ghana	1971	.621 (2948)	.775 (874)	.665 (1179)	1.128 (30446)	.802 (5081)
Kenya	1978	.652 (425)	887 (388)	.665 (1179)	1.509 (1943)	1.051 (1203)
Lesotho	1977	.743 (66)	1.343 (35)	.894 (85)	1.197 (66)	1.006 (1705)
Southern Nigeria	1972	.726 (536)	.933 (252)	.910 (111)	1.195 (10481)	1.052 (421)
Indonesia	1971	.789 (11721)	1.033 (6850)	1.002 (2402)	1.000 (43467)	1.028 (3965)
Nepal	1976	.784 (187)	.844 (173)	.760 (148)	.995 (3884)	1.253 (423)
S. Korea	1971	.673 (101)	.883 (613)	- -	.956 (1266)	1.088 (491)
Sri Lanka	1975	.556 (225)	.868 (242)	.767 (189)	1.141 (1288)	.938 (771)
Thailand	1975	.606 (132)	.519 (10)	1.061 (72)	1.123 (1378)	.881 (278)
Chile	1970	.629 (455)	1.128 (196)	.866 (135)	1.213 (685)	1.059 (1444)
Peru	1978	.401 (491)	.828 (406)	.732 (209)	1.304 (2197)	.921 (1242)
Mean of 11 Developing Countries	1970s	.652	.913	.841	1.206	1.007

Source: Tabulations of public use data files performed at the University of Pennsylvania.

The educational effort was conducted through an enormous array of educational materials distributed by public health agencies, clinics, physicians, insurance companies, visiting nurses, milk depots, and the Children's Bureau. The Children's Bureau's Infant Care became the largest-selling publication in the history of the Government Printing Office, and by the 1920s the Bureau was receiving over 100,000 inquiries annually from parents seeking health advice. Newspapers and magazines were also full of such advice.

Putting the scattered pieces together to evaluate the contribution of changing personal health practices to mortality decline is probably impossible. Recent indirect evidence from developing countries suggests that variation in personal health practices are quite important. Within countries, women with the same economic and social circumstances will often have child mortality levels that differ by a factor of two depending on what ethnic group they belong to (United Nations, 1985). It is difficult to account for this difference without assuming that health practices differ among the groups. And mother's education has emerged as the strongest predictor by far of child mortality levels in developing countries, despite the fact that father's schooling is more closely associated with the family's economic conditions (United Nations, 1985). It is likely that mother's education is functioning largely as an indicator of child care practices, and the few available inquiries do demonstrate that specific health practices vary systematically with the mother's educational level (Clelland and van Ginneken, 1988).

Some suggestion of the importance of the germ theory of disease for improving those practices is available from comparisons of mortality differentials over time. Table 22-1 compares social class differences in mortality in the United States in the late nineteenth nineteenth century to differentials in contemporary developing countries (Preston, 1985). Identical procedures are applied to microlevel survey data on numbers of children ever born and surviving. Clearly, being in the professional or technical classes in the United States in 1900 bought one much less in the way of mortality reduction. There was still great confusion about what caused disease and how to prevent it. This confusion is perhaps best seen by comparing mortality rates among specific occupations. Table 22-2 shows that offspring of physicians and surgeons actually had higher mortality than children of farmers or coal miners (Preston, 1985). This was an era when public health officials were complaining bitterly that physicians, a much lower status calling then than now, were ignorant of disease causation and obstructing their efforts to advance health. The same result pertains to literacy: children of literate mothers in the United States in 1900 had much smaller advantage than children of literate mothers today in developing countries (Preston, 1985). It is not necessarily the case that literate women in developing countries today know the details of the germ theory, only that the germ theory has led to better recognition of what constitutes hygienic health practices, information to which upper classes have greater access through schools and media and personal networks. As this knowledge spread in the U.S., social class differences in child mortality widened (Ewbank and Preston, unpublished).

Several medical historians such as Charles Rosenberg (1987) and Barbara Rosencrantz (1972) give a less central role to the germ theory in changing medical and public health practice. They find very slow changes occurring in technique for several decades after the germ theory was validated. It is not that the germ theory of disease was accepted and suddenly everything was different, but rather that it accelerated a slow decline that was already underway. It took the better part of a century for the attack on infectious disease mortality to be fully successful. But it seems quite implausible that the attack could have been anywhere near as successful without a clear knowledge of disease causation, and without the social adaptations that this knowledge fostered. Demographic studies provide only circumstantial evidence here: time series of national mortality levels; repeated international cross-sections of income and mortality relations;

Table 22-2. Ratio of child deaths to expected deaths in selected occupations in the U.S., 1900

Father's Occupation	Child Deaths/ Expected Deaths	Number of Expected Deaths
Physicians, Surgeons, Dentists	.927	35.6
Clergymen	.805	32.3
Teachers	.999	26.0
Lawyers	.757	22.4
Pharmacists	.647	13.9
Total, Intelligentsia	.853	130.3
Coal Miners	.903	118.6
Farmers	.863	3058.0

and comparisons of microlevel survey results across time and space. But this adds up to a reasonably convincing case that the great mortality decline of the past century was principally a result of scientific and organizational advance, rather than of economic growth.

As in most studies of demographic change, genetics has not played much of a role in this account. But there is one important loose thread that it might be able to tie together. The decline in tuberculosis in nineteenth-century England, which played such a large role in McKeown's analysis, has not been adequately explained. Perhaps it resulted from improved nutrition, but that case has not been made persuasively. McKeown simply says that, as tuberculosis declined, England was able to feed its growing population; he does not demonstrate that a rise occurred in food consumption per capita. The United States was an exceptionally well fed country in 1900, with a per capita daily consumption of 3,600 calories per adult male equivalent, a substantially higher figure than today. Yet its life expectancy at birth was only 49 years, the same as England's more poorly fed population (Preston and Haines, 1990). But surely McKeown is correct that in the nineteenth century, no medical or public health measures were particularly effective against tuberculosis. In the early part of the twentieth century, the isolation of tubercular patients was begun, which was said to be a major factor in declining tuberculosis death rates in New York City, for example (Preston and Haines, 1990). But little of that was happening in the nineteenth century, and it is doubtful that isolation was a key factor in the twentieth century either. There are occasional suggestions that, through genetic selection, human populations were evolving greater resistance to tuberculosis in the nineteenth and twentieth centuries. But no one has systematically tested this notion. Suggestive evidence is the following. Tuberculosis shows one of the highest familial components of any disease (which may, of course, reflect shared environments as well as shared genes). Because of its high death rate and concentration in the early adult years, prereproductive mortality from tuberculosis provided substantial opportunities for selection. But why should selection have been so intense in the nineteenth century? It has been argued that, because people can remain carriers of it for many decades, tuberculosis exhibited a long epidemic wave associated with urbanization (Grigg, 1958). Urbanization was earliest and fastest in England, which had one of the fastest tuberculosis declines. Rural areas seemed to have lower tuberculosis death rates than urban areas in England and elsewhere. To give credence to this epidemic/selection theory, one would want to demonstrate a period of rising death rates from tuberculosis. English data, which begin in 1841, fail to show this. But evidence of increasing rates in the eighteenth and nineteenth centuries does appear in Scandinavia, Paris, and elsewhere (Grigg, 1958; Dubos and Dubos, 1953; Preston and van de Walle, 1978). So the pieces are present for a genetic investigation of the role of selection in this important unresolved issue in recent population history.

FUTURE RELATIONS BETWEEN DEMOGRAPHY AND GENETICS

At one time, demography and genetics found an important meeting ground in eugenics, the study of population quality or, more generally, of population composition. The eugenics movement was, of course, designed to improve population quality, and paid special attention to reproductive differentials. Geneticists were among the founding fathers of both the Population Association of America and the International Union for the Scientific Study of Population.

After World War II, the dramatic worldwide changes in fertility and mortality redirected demographers' attentions to the causes and consequences of demographic change. The period of rapid growth that resulted from earlier and faster declines in mortality than in fertility caused great concern among policy makers, especially in the United States, and drew demographers into policy studies and program evaluation. The dominant concern was with the volume of vital events rather than with their distribution among social groups. In addition, the low levels of and wide class differentials in fertility during the 1920s and 1930s that generated fears of ''race suicide'' were swept away by the post-war baby boom. The abhorrent eugenics platform of Nazi Germany discredited the entire field of eugenics. And social scientists, both for disciplinary and ideological reasons, chose to ignore any genetic bases of human behavior and to emphasize human plasticity.

But events seem to be reestablishing the type of environment in which the two fields were more closely aligned. Population growth rates have slowed both in developed and developing countries and no longer seem to pose the kind of threat to human welfare that they once did. Studies of identical twins have shown remarkable heritability of character traits that were once considered entirely environmental in origin (e.g., Holden, 1987), and social scientists are beginning to talk about biological bases of behavior and performance (e.g., Rossi, 1984). Finally, reproductive differentials by major social category have reemerged as fertility rates have fallen in developed countries. Table 22-3 shows the current differences in cumulative fertility in the U.S. according to a woman's educational attainment.

It seems likely that demographers will be drawn back into discussions of the intergenerational aspects of population composition, and when they are they will need to assimilate the models from genetics and population biology that have developed in the interim. Vining (1986) has recently discussed the im-

Table 22-3 Mean number of children born to women aged 35-44 in 1987

Educational Attainment	All Women	Black Women
Less than High School	2.92	3.56
Completed High School	2.14	2.38
Some College	1.94	2.12
Completed College	1.76	2.08
Some Graduate School	1.43	1.58

Source: Bureau of the Census, *Fertility of American Women: June, 1987*.   Current Population Reports P-20 No. 427. May, 1988.

plications of current fertility differentials by IQ in the pages of Behavioral and Brain Sciences and his pessimistic conclusions have received highly-publicized endorsement by Richard Herrnstein (1989) in The Atlantic Monthly. But the implications of negative fertility differentials for IQ distributions are less obvious than they may appear to be. For example, any pattern of reproductive differentials by IQ, combined with a stable IQ inheritance matrix and perfect assortative mating, will result in a stable IQ distribution in the population (Preston, unpublished). That is, a negative relation between IQ and fertility at a moment in time does not necessarily imply that the IQ distribution is or will be deteriorating. Genetics has leapt far ahead of demography in developing dynamic models for studying population composition, and the models are applicable whether or not the trait of interest is genetically determined. If demography refocuses some of its attention on intergenerational features of population composition, as seems likely, it will need to rediscover a resource that it has neglected for the past several decades.

Changes in personal health practices are exceptionally difficult to measure, and perhaps for this reason this factor has been almost entirely neglected in explanations of mortality decline. There were no national surveys of health practices until the 1940s in the U.S. and there have been very few since then that bear upon behaviors relevant to infectious diseases. But by 1910, public health authorities had become convinced that changes in personal health practices were the key to success in the effort to combat ill health, especially among children (Meckel, 1989; Ewbank and Preston, unpublished). Epidemiologic studies had shown, for example, that no matter how clean milk was when it arrived on the doorstep, it was subject to enormous contamination in the home. The major educational efforts made by health authorities over the next three decades emphasized the importance of breastfeeding, heating milk to remove bacteria, keeping flies away from food, isolating sick family members, disposing of waste in a sanitary fashion, washing hands, bathing, and seeking medical attention for the sick (Meckel, 1990). Although empirical research before the germ theory's acceptance had established the efficacy of breastfeeding, most of these personal health practices were founded upon a new understanding of disease processes that the theory supplied. For example, the heating of milk in the home was not initiated until the 1890s and the important role of flies in the transmission of disease to food was not recognized until 1903.

REFERENCES

Abbott, S. W., 1900   *The Past and Present Condition of Public Hygiene and State Medicine in the United States.* Wright & Potter Printing Company, Boston.

Bongaarts, J., and R. C. Potter. 1983   *Fertility, Biology, and Behavior.* Academic Press, New York.

Bongaarts, J., 1982   The fertility-inhibiting effects of the intermediate fertility variables. Studies Fam. Plan. 13: 179-189.

Chapin, C. V., 1901   *Municipal Sanitation in the United States.* Snow and Furnhum, Providence, R.I.

Clelland, J.G., and J.K. Van Ginneken, 1988   Maternal education and child survival in developing countries: The search for pathways of influence. Soc. Sci. Med. 27: 1357-68.

Coale, A. J., and P. Demeny, 1966   *Regional Model Life Tables and Stable Populations.* Princeton University Press, Princeton.

Condran, G. A., H. Williams, and R. A. Cheney, 1984   A decline in mortality in Philadelphia from 1870 to 1930: The role of municipal services. The Pennsylvania Magazine of History and Biography 108: 153-177.

Da Vanzo, J., and J. P. Habicht, 1986 Infant mortality decline in Malaysia, 1946 -75: The roles of changes in variables and changes in the structure of relationships. Demography 23: 143-160.

Dubos, R., and J. Dubos, 1953.   *The White Plague.* London.

Durand, J., 1980   Comment, pp. 341-347 in *Population and Economic Change in Developing Countries,* edited by R. A. Easterlin. University of Chicago Press.

Flinn, M. W., 1981 *The European Demographic System, 1500-1820.* Johns Hopkins University Press, Baltimore.

Fries, J. F., 1980   Aging, natural death, and the compression of morbidity. N. Engl. J. Med. 303: 130-35.

Grigg, E. R. N., 1958   The arcana of tuberculosis. Am. Rev. Tuberculosis Pulmonary Dis. 78: 151-172.

Herrnstein, R.J., 1989   IQ and falling birth rates. The Atlantic Monthly, May: 73-79.

Hoffman, F. L., 1896   Race traits and tendencies of the American negro." Publications of the American Economic Association IX(1-3), August: 1-329.

Holden, C., 1987   The genetics of personality. Science 237: 598-601.

McKeown, T., and R. G. Record, 1962.   Reasons for the decline of mortality in England and Wales during the nineteenth century. Pop. Studies 16: 94-122.

McKeown, T., 1976   *The Modern Rise of Population.* Academic Press, New York.

Meckel, R. A., 1990   *Save the Babies: American Public Health Reform and the Prevention of Infant Mortality, 1850-1929.* Forthcoming, Johns Hopkins University Press, Baltimore.

Menken, J., 1987   Proximate determinants of fertility and mortality: a review of recent findings. Sociol. Forum 2: 697-717.

Mosley, W. H., and L. C. Chen, 1984   An analytical framework for the study of child survival in developing countries, pp 25-45 in *Child Survival: Strategies for Research,* edited by W. H. Mosley and L. C. Chen. Supplement to Vol 10, Pop. Dev. Rev.

Nag, M., 1968   *Factors Affecting Fertility in Pre-Industrial Populations.* Yale University Press, New Haven.

Palloni, A., 1981.   Mortality in Latin America: Emerging patterns. Pop. Dev. Rev. 7: 623-49.

Park, W. H., and E. Holt, 1903   Report upon the results with different kinds of pure and impure milk in infant feeding in tenement houses and institutions of New York City: A clinical and bacteriological study. Arch. Pediatr. 20: 881-909.

Preston, S. H., 1976   *Mortality Patterns in National Populations*. Academic Press, New York.

Preston, S. H., and E. Van de Walle, 1978   Urban French mortality in the nineteenth century. Pop. Studies 32: 275-97.

Preston, S. H., 1980   Causes and consequences of mortality declines in less developed countries during the twentieth century. pp 289-360 in *Population and Economic Change in Developing Countries*, edited by R. A. Easterlin. University of Chicago Press.

Preston, S. H., 1985   Resources, knowledge, and child mortality: a comparison of the U.S. in the late nineteenth century and developing countries today. International Population Conference, Florence, 1985. International Union for the Scientific Study of Population 4: 373-88.

Preston, S. H., and M. Haines, 1990   *Fatal Years: Child Mortality in Late Nineteenth Century America*. Forthcoming, Princeton University Press, Princeton, NJ.

Rosenkrantz, B. G., 1972   *Public Health and the State: Changing Views in Massachusetts*. Harvard University Press, Cambridge, MA.

Rosenberg, C. E., 1987   *The Care of Strangers*. Basic Books, New York.

Rossi, A., 1984   Gender and parenthood. Am. Sociol. Rev. 49: 1-19.

Sedgwick, W. T., and J. S. MacNutt, 1910.   On the Mills-Reincke phenomenon and Hazen's theorem concerning the decrease in mortality from disease other than typhoid following the purification of public water supplies. J. Infec. Dis. 7: 489-564.

Sheps, M. C., and J. A. Menken, 1973   *Mathematical Models of Conception and Birth*. University of Chicago Press, Chicago.

Szreter, S., 1988   The importance of social intervention in Britain's mortality decline c. 1850-1914: a reinterpretation of the role of public health. Soc. Hist. Med. 1: 1-37.

United Nations, 1985   *Socioeconomic Differentials in Child Mortality in Developing Countries*. United Nations Population Study No. 97. United Nations, New York.

United Nations, 1986.   *World Population Prospects Estimates and Projections as Assessed in 1984*. United Nations Population Study No. 98. United Nations, New York.

Vining, D., 1986   Social versus reproductive success: the central theoretical problem of human sociobiology. Behav. Brain Sciences 9: 167-216.

Watkins, S. C., and J. A. Menken, 1985   Famines in historical perspective. Pop. Dev. Rev. 11: 647-676.

Williamson, J., 1984   British mortality and the value of life, 1781-1981. Pop. Studies 38: 157-172.

Wrigley, E. A., 1987   *People, Cities, and Wealth: The Transformation of Traditional Society*. Basil Blackwell, New York.

Index